E = EINSTEIN

E = EINSTEIN

HIS LIFE, HIS THOUGHT
AND HIS INFLUENCE ON OUR CULTURE

———○———

Edited by

Donald Goldsmith and Marcia Bartusiak

STERLING

New York / London
www.sterlingpublishing.com

Produced by The Reference Works, Inc.

Harold Rabinowitz, *Director*
Pamela Adler, *Managing Editor*
Patricia Hemminger, *Science Editor*
Elizabeth O'Sullivan, *Photo Editor*

Library of Congress Cataloging-in-Publication Data Available

10 9 8 7 6 5 4 3 2 1

Published by Sterling Publishing Co., Inc.
387 Park Avenue South, New York, NY 10016
© 2006 by The Reference Works, Inc.
Distributed in Canada by Sterling Publishing
c/o Canadian Manda Group, 165 Dufferin Street
Toronto, Ontario, Canada M6K 3H6
Distributed in the United Kingdom by GMC Distribution Services
Castle Place, 166 High Street, Lewes, East Sussex, England BN7 1XU
Distributed in Australia by Capricorn Link (Australia) Pty. Ltd.
P.O. Box 704, Windsor, NSW 2756, Australia

Interior design by Raquel Shapira
Illustrations by Nicholas Meola

Printed in China
All rights reserved

Sterling ISBN 978-1-4027-3787-9 (hardcover)

Sterling ISBN 978-1-4027-6319-9 (paperback)

For information about custom editions, special sales, premium and
corporate purchases, please contact Sterling Special Sales
Department at 800-805-5489 or specialsales@sterlingpublishing.com.

To all those whom Einstein has inspired
—and will continue to inspire

Contents

Einstein riding a bicycle at the home of friends in Santa Barbara, California, 1933. Einstein enjoyed California so much that most observers thought it a foregone conclusion that he would wind up at Caltech.

TIMELESS EINSTEIN

Albert Einstein, recognized worldwide as the greatest physicist of the twentieth century, died more than 50 years ago. This interval might seem sufficiently long for public awareness of Einstein to have faded gently into nearly forgotten history. Instead, Einstein's image and reputation have only gained in fame with the passage of time. Today everyone "knows" two things about Einstein: He was a shaggy-haired genius, and he produced the world's most famous equation—indeed, the world's only widely-known equation: $E = mc^2$.

These perceptions may be limited, but at least they are correct. Einstein was a scientific genius, and in his later years he wore his hair long, at least by the standards of the day. He did write $E = mc^2$ (although he used different symbols that mean the same thing). This straightforward equation deserves its fame for the tremendous insight into nature that it provides. $E = mc^2$ describes the amount of energy (E), the energy of mass, that resides in every object that possesses an amount of mass, m. Because the speed of light, denoted by c, is an extremely large number, its square is even larger, which implies that even a small object possesses an enormous energy of mass. A single

coin, for example, contains within itself an amount of energy sufficient to supply all of the energy requirements of the United States for several minutes. If converting this type of energy into more useful forms of energy were a simple process, we would have no need to burn coal or oil. On the other hand, if this were so, nature might well have long ago converted all the matter residing in our neighborhood into kinetic energy, destroying by this process all objects with mass. Had Albert Einstein done no more than discover that $E = mc^2$, his fame would be assured, and he would rank among the scientific greats. Every molecule in our bodies contains nuclei forged inside stars as they turned some of their mass into the energy that makes them shine. $E = mc^2$ unlocks not only the secrets of starlight but also the cosmic code of element production, which explains how dying and exploding stars produced the raw material that made our planet and the creatures who live upon it.

Einstein's energy equation appeared in 1905 almost as a sidelight of his much greater triumph, the special theory of relativity. Called "special" because it deals only with objects moving at a constant velocity, this theory analyzes the nature of space and time, which seem eternal and obvious to us because we rely—inevitably but wrongly—on our own perceptions to draw conclusions about the entire cosmos. Einstein demonstrated that different observers will measure space and time differently, and that they have a perfect right to do so, despite the contradictions that this conclusion seems to provoke. When tested, however, special-relativity theory has proven correct. In the same year 1905, Einstein also published a new method for determining the dimensions of molecules, and showed that light and other forms of electromagnetic radiation must behave like tiny bullets of energy, called photons, and not only as waves. This demonstration laid the foundation of quantum mechanics, the physics that explains how the smallest particles move and interact.

Had Einstein disappeared after 1905, he would still rank as the greatest of modern physicists. To be sure, everything he found in that "wonder year" would in all likelihood have been discovered a few years later by other gifted researchers. But Einstein's next great contribution to modern physics might have taken decades for others to achieve: the general theory of relativity, which deals with objects whose velocities change under the influence of gravity. In this theory, generated through years of heroic individual effort, with only some mathematical assistance from his friends and colleagues, Einstein produced a new way to understand the effects of gravity: Gravity bends space, and bent space tells objects how to move. Because general relativity describes all of space, it offers new insights into the universe as a whole. For example, general relativity predicts that the universe must always exist either in a state of expansion or in one of contraction. In 1929, more than a decade after Einstein published his work, Edwin Hubble verified this result by discovering the expansion of the universe.

Nor was this all. In 1917, before his doubts about the quantum theory turned him into a dissenter within the community of science, Einstein made another great contribution to quantum mechanics by providing the basic theory of how atoms interact with beams of light. Today we enjoy direct application of Einstein's thoughts in the lasers of CD and DVD players that afford musical or visual delight.

This summary of the highlights of Einstein's work helps to explain why scientists continue to revere him. But if Einstein had only set the world of science on its ear, or had merely revolutionized our technology, he would be as little recognized by today's public as Thomas Edison or Marie Curie. Einstein, however, still speaks to millions who know almost nothing about his contributions to science. This arises in part from Einstein's role as an earlier version of Stephen Hawking—a totemic figure, the smartest person in the world, even if no one can tell you what he has

Einstein makes the front page by talking to reporters about atomic energy, December 29, 1934.

done. Beyond this, the public recognizes, dimly but enduringly, that Einstein played an important role in the development of twentieth-century consciousness, as an outstanding example of liberal intelligence, as an opponent of (and refugee from) political repression in Nazi Germany, as the man who could explain the power of the atom bomb, as an early and enthusiastic Zionist, and as a firm believer in humanity's continuing ability to improve itself through international cooperation.

How and where did Einstein acquire the personality and capability to make himself a man of such qualities, as well as a pre-eminent scientist? His parents, only two generations removed from Germany's ghettos, helped him to gain a sound education, but Einstein's suspicion of, and resistance to, all forms of authority seem to have arisen on their own from an early age, nurtured by a nearly automatic reaction against German insistence on correct behavior and national superiority, and sealed, one may argue, by the eight months he spent as a high-school dropout in northern Italy, wandering for some weeks through the glorious countryside. Einstein later recalled that he found himself "surprised" by ordinary Italians' "high level of thought and cultural content, so different from the ordinary German," and described these as "happy months... my most beautiful memories."

Among all philosophers, Einstein identified most closely with Spinoza, likewise a devoted opponent of attempts to enforce conventional thinking and behavior. The resistance of Einstein's family to his proposed marriage to a fellow physics student only increased his desire to wed her (never mind that his parents were essentially right in anticipating no great success for this match); the resistance of established physicists toward advancing his career, motivated in part by his Jewish origins, only induced him to work harder on his own, so that he managed to produce the great discoveries of 1905 with a hard-won day job in the Swiss patent office and a wife and young son at home. Even as his discoveries made him famous among scientists, Einstein maintained a calm aloofness as he climbed the job ladder, and in 1914, as he reached the top in Berlin, he did not hesitate to alienate most of his colleagues by endorsing an anti-war petition to counter the war fever then rampant.

Soon after World War I, when the confirmation of his theory of general relativity catapulted him to fame, Einstein came up with one more "invention"—himself. He invented the archetype of the world's greatest scientist, which he played thereafter for newspapermen and photographers with easy composure, instinctively recognizing that the public cared less about science than about their vision of scientists. When the rise of Nazism forced him out of Europe, Einstein went to England—where he had previously charmed even those scientists who mistrusted him as a foreigner and a Jew—before establishing a new life at Princeton, where he gave direct support to dozens of fellow émigrés.

Happily left to himself in a gentler age of publicity, Einstein continued his attempts to find a theory that would explain all the forces of nature—a quest that remains incomplete, but which Einstein saw was right for him to attempt, not least because his career was already secure. When the newborn state of Israel offered to make Einstein its second president, he wisely declined, aware of his limitations and his desires. And when people from all segments of the public wrote him letters, he answered them by the hundreds—a sweet touch from a long-vanished epoch.

Einstein was a self-contained, self-confident, charming and complex man who brought us tremendous insights into the nature of physical reality. His persona has survived, amplified and distorted by the forces of publicity, for more than half a century, so that in many ways, Einstein now seems a more vibrant and relevant historical personality than he did thirty years ago. We hope that this careful look at him will increase admiration for this man, shaped by history but supremely individualistic, living within his mind without losing sight of the world outside it, using that mind to see the cosmos in ways never before imagined, and dreaming of the day when nature would yield her ultimate secrets to those who would earn this victory by their talented and unceasing devotion to scientific inquiry.

PART ONE

EINSTEIN'S LIFE

Einstein's modest home at 112 Mercer Street. in Princeton. A quiet, unassuming house in which a pillar of modern scientific thought grappled with the mysteries of the universe. (Preceding page: Einstein as a teenager.)

Albert Einstein:
A Laboratory in the Mind

―――――○―――――

by Owen Gingerich

The long conical shadow of the moon was scheduled to sweep from Brazil to Africa on 29 May 1919. Near each end of the path of the eclipse, astronomers waited in tense expectancy to test a bold new theory of gravitation that challenged the time-honored laws of the English mathematician Sir Isaac Newton.

This new theory had been evolving for nearly a decade in the mind of a young German-born physicist, Albert Einstein. When it was finally published in 1916 as the "general theory of relativity"—to distinguish it from his earlier "special theory of relativity"—Einstein was only 37, eight years younger than Newton was when he wrote his *Principia* with its famous law of universal gravitation. Einstein described gravitation as a natural property of curved space. Just as a marble placed on a stretched rubber membrane would deform the rubber surface, so would a massive object bend or warp the space around it. Thus, Einstein predicted the sun bends the space around it, and the distortion could be detected by measuring the deflection of starlight passing near it. Of course, the blinding glare of the sun normally masks the distant stars behind it. But during an eclipse, stars whose line of sight lies near the sun can be photographed.

Waiting to test Einstein's theory on the tiny Portuguese island of Príncipe were Arthur S. Eddington, a great enthusiast of the theory of relativity, and his young assistant, E. T. Cottingham. Before embarking on the expedition, Cottingham had asked Sir Frank W. Dyson, director of Great Britain's Royal Greenwich Observatory and official organizer of the expedition, "What will it mean if we get double the [predicted] Einstein deflection?" Dyson replied, "Then Eddington will go mad and you will have to come home alone." But when Eddington had completed the measurements

Reprinted from *The Great Copernicus Chase and Other Adventures in Astronomical History* by Owen Gingerich, with permission of the author. Originally published in *Science Year* 1979.

of his first photograph plate on Príncipe, he turned to his companion and said, "Cottingham, you won't have to go home alone."

Eddington's formal announcement of the expedition's results the following September provided dramatic verification of Einstein's esoteric theory. Overnight, "Einstein" became a household word, and the man was plunged into a fame from which he never escaped.

Einstein knew that his theory of gravitation would have several consequences that could be verified by experiments. One was that the planets' elliptical orbits should slowly change their orientations beyond what Newtonian theories accounted for. Most conspicuously, the long axis of Mercury's orbit should rotate around the sun by 43 arc seconds (43/1296000 of a circle) more than the 500 arc seconds per century predicted by Newton's laws. Indeed, astronomers had measured such a rotation and had been puzzled by it.

Another consequence was that in the warped space-time near a massive object, clocks would run more slowly. One form of clock is a vibrating atom, and the color of any spectral line radiated by the atom indicates the clock's rate. Since the atoms on the surface of the sun or a star act like tiny clocks in a strong gravitational field, they should run slightly slower than their counterparts on earth. As a result, their spectral lines should be slightly redder. This effect is slight on the sun and on most stars, and measurements were not sensitive enough to confirm it when the general relativity theory was published. The slowing of time has since been confirmed. Extremely accurate maser clocks carried in high-flying aircraft (where the gravity is less than at the Earth's surface) run faster than identical clocks in the laboratory. Another experiment testing this theory is so sensitive that it has detected the difference in the gravity between the first and the fourth floors of a Harvard University physics laboratory.

Who was this remarkable physicist so abruptly cast into the limelight by the scientific events of 1919? Although Einstein had received a prestigious research professorship at the University of Berlin at the comparatively early age of 34, as a youth he did not appear to be precocious. In fact, his parents had worried that he might be retarded because he did not begin talking until he was three years old. Nevertheless, young Albert was fascinated by physical and mechanical phenomena. Later, when he was in his mid-20s and working in the Swiss Patent Office in Bern, he could understand the intricacies of the submitted inventions so quickly that he finished a normal day's workload in a few hours. This left him plenty of time to ask himself "simple" questions about the nature of space, time, and matter.

Many years later, when Einstein set down some autobiographical notes, he did not bother with such details as, "I was born on 14 March 1879 in the German town of Ulm." Nor did he document his moves to Munich; Milan, Italy; or Switzerland. Instead he recorded his sense of wonder about things when, for example, as a child of 4 or 5, his father showed him a magnetic compass.

Einstein also described how he developed a deep religiosity as a boy, although his Jewish parents did not practice any religion. He abruptly lost this in "an orgy of freethinking" at age 12. Much later, as a result of his experiences in Germany between World Wars I and II, he identified with his Jewish heritage and became an active Zionist. But it was the contemplation of the universe that induced a deep reverence in Einstein. "The road to this paradise was not as comfortable and alluring as the road to the religious paradise," he wrote, "but it has shown itself reliable, and I have never regretted having chosen it."

In the fall of 1895, the teen-aged Albert took the exams at the renowned Swiss Polytechnic Institute in Zurich—and flunked them. A year later, however, he was admitted and in 1900 was awarded a degree. But his independent spirit had so alienated his teachers that the door to a university career seemed closed. Thus, Einstein took the job in the Swiss Patent Office in 1902. During the seven years he worked there, Einstein developed the concept most irrevocably linked with his name—the special theory of relativity. This theory rests on two postulates, or assumptions: first, the laws of light and optics are the same for any reference frame in which Newton's laws also hold; second, the velocity of light in empty space is always the same, regardless of the motion of its source. Einstein published these postulates in 1905 in a paper entitled "On the Electrodynamics of Moving Bodies."

At first glance, the two postulates seem innocent enough. Yet they contradicted the two simplest models for the behavior of light. First, if light waves were like sound waves, there must be a medium through which they travel. The theory of electromagnetic waves, worked out in the 1860s by British mathematician James Clerk Maxwell, assumed the existence of an ethereal medium in which light waves traveled and which provided a preferred, fixed reference frame for them. But this concept of light traveling in an invisible ether is denied by the first postulate of relativity, which declares that all observers must use exactly the same law of light and optics. Even if two observers are moving with respect to each other, the laws apply for each one as if each were at rest. According to relativity and in contradiction to Maxwell's theory, there can be no preferred reference frame—no ether—for light that is "really" at rest.

Many experiments in the late 1800s had been designed to detect an ether in which light might move. The most notable one was carried out by American physicist Albert A. Michelson, who won the 1907 Nobel prize for physics, and his colleague Edward W. Morley. Without exception, these experiments failed to detect any ether.

A second simple model pictures light as particles shot out from its source. In this case, we might expect the particle of light to travel with the combined speed of the source of light itself. An analogy would be a baseball thrown forward on a moving train; another, a spacecraft that races toward Mars with the combined velocity of the rocket's thrust and the earth's orbital motion. But for light, such a combined velocity is denied by the second postulate of relativity. The truth of this postulate is verified by observations of binary stars—pairs of stars that orbit around each other. As a companion star orbits its primary, it alternately approaches and recedes from the earth. According to the model, light should be alternately speeded up and slowed down so that its rays in traveling the vast distance to the earth would arrive out of their original sequence. But they arrive in sequence. Analysis of X-ray binary stars in 1978 has confirmed this second postulate of relativity to an accuracy of 1 part in 10 billion.

In exploring the consequences of the two postulates of special relativity, Einstein realized that our notion of time was intimately bound up with our concept of space, and that he had to be exceedingly careful in judging two events to be simultaneous. One can imagine the impatience of the German physics professors in their stiff starched collars when Einstein elaborated these details in his paper: "If, for instance, I say, 'That train arrives here at 7 o'clock,' I mean something like this: 'The pointing of the small hand of my watch to 7 and the arrival of the train are simultaneous events.'"

By pressing his careful reasoning to its logical conclusion, Einstein found a startling result: in trying to synchronize two clocks that are far apart from each other, we get different readings when we do it by carrying a third clock between the two than we do if we try to synchronize them with light signals. Moreover, time appears to run more slowly for a moving clock than for a clock at rest. Einstein's theory yielded additional astonishing results. Moving rods appear to contract, compared with identical rods at rest, and moving objects become more massive than similar objects at rest. But all of these concepts are relative, because it is impossible to find absolute rest or absolute motion. Whether a clock or rod is at rest or in motion depends on the point of view of the observer.

Einstein's conclusions appear to fly in the face of common sense—at the very least, they are contrary to ordinary experience. Conspicuous changes in clocks, rods, or masses occur only at velocities near the speed of light, something rarely encountered in everyday life. After scientists began working with objects at very high speeds, it became apparent that nature really does work according to Einstein's theory of relativity.

For example, as atomic particles are accelerated in synchrotrons and linear accelerators, they become more massive and, beyond a certain point, additional energy can produce almost no further increases in speed. The atomic particles resist being accelerated past the velocity of light.

Another example is provided by the quasars, whose spectral lines are spectacularly shifted toward the red. According to the ordinary Doppler effect, which explains why a radiating object moving away from an observer appears to be radiating at a lower frequency, these far-distant objects must be rushing away from us at two or three times the speed of light. A relativistic Doppler effect, however, shows that while they are moving rapidly, they do not exceed the speed of light.

In 1922, only a few years after relativity had entered the public vocabulary, Einstein won the 1921 Nobel prize for physics. Ironically, the citation did not specifically mention relativity. Instead, it honored another discovery he made in the same wonder year of 1905. The prize was awarded "for his services to theoretical physics and, in particular, for his discovery of the law of the photoelectric effect."

In the photoelectric effect, light shining on a metal releases electrons from the surface. Einstein proposed that the phenomenon could be explained quantitatively if light was treated as a series of discrete, or separate, packets of energy, or quanta. Because well-known optical phenomena, such as diffraction and polarization, could only be described by picturing light as a wave, Einstein's quanta at first seemed contrary to established physical principles.

Indeed, American physicist Robert A. Millikan found Einstein's photoelectric hypothesis so unbelievable that he conducted detailed experiments beginning in 1913 to settle the matter. But instead of disproving the theory, Millikan's work confirmed it in every detail. Millikan himself won the Nobel prize for physics in 1923. So when Einstein won his prize in 1922, the photoelectric effect had been verified experimentally, while much of his relativity theory had not.

Although Einstein placed relativity foremost among his achievements and chose to speak on this topic when he accepted his Nobel prize, he made numerous other advances in the quantum theory of atomic structure. Quantum theory had its foundation in the work of German physicist Max Planck, who, in 1900, discovered the law that described how radiant energy is distributed at different temperatures—the so-called black body radiation. This work earned him the Nobel prize for physics in

1918. Planck had to assume that radiating atoms could not vibrate with any arbitrary energy, but only with specific discrete amounts, or quanta. In explaining the photoelectric effect, Einstein noticed that he could apply these same quantized energy levels to light. Two years later, by a further brilliant recourse to quantum theory, he explained unresolved discrepancies in experiments on the heat capacity of solids.

The quantum view of nature received another boost in 1913 with Danish physicist Niels Bohr's new interpretation of the atom. Bohr, who won the 1922 Nobel prize for physics for his work, theorized that electrons traveling about an atomic nucleus would move only in certain definite, or quantized, orbits. In Bohr's picture, a specific quantum, or photon, of light could be emitted or absorbed when an electron changed from one allowed orbit to another. Einstein realized that Planck's radiation law would not work on the basis of only absorption or spontaneous emission. A third process called stimulated emission was required.

Stimulated emission provides the basis for two important modern inventions, the microwave maser and its optical counterpart, the laser. These devices use photons, created when excited atoms in the material drop back to their unexcited state, to stimulate other excited atoms to release identical photons. This rapidly multiplies the number of photons, increasing the intensity of microwave or light signals.

Laser beams have many uses, from drilling tiny holes in sapphires for watch bearings to transmitting telephone and television signals on glass fibers. Masers, operating by the same principle in the microwave region of the electromagnetic spectrum, are used as exceedingly accurate atomic clocks.

In the early 1920s, Einstein supported the crazy ideas of the French physicist Louis de Broglie, winner of the Nobel prize for physics in 1929. De Broglie had proposed that the specific atomic orbits suggested by Bohr could be pictured as standing waves. A permitted orbit could contain only complete multiples of its waveform. Thus, electron particles were intimately bound up with waves, just as light itself exhibited these dual particle and wave aspects that seemed to be contradictory.

Einstein's recommendations had a catalytic effect in the physics community. Almost overnight, the field of wave mechanics was born. Shortly thereafter, the waves were reinterpreted as probability waves, and German physicist Werner Heisenberg (who won the 1932 Nobel prize in physics) introduced his famous "uncertainty principle." According to this view, complete information concerning the position and motion of atoms was forever unobtainable. Einstein was repelled and distressed at this loss of determinism. At two major meetings of physicists in 1927 and 1930, he argued vehemently with Bohr and Heisenberg against the emerging interpretation of atomic physics.

Failing to convince them, he retired from the field he had done so much to create, remonstrating, "God does not play dice."

In the early 1930s, political events in Germany took an ominous turn. Einstein was in Pasadena, California, when Adolf Hitler came to power, and he realized at once that he could not go back to Germany. He resigned from the Prussian Academy of Sciences in March, 1933, while his writings were being burned by the Nazis in their anti-Semitic frenzy. Einstein went to Belgium for a short period, and then briefly visited Oxford University in England. In October, he accepted an appointment at the newly founded Institute for Advanced Study in Princeton, New Jersey, where he stayed until his death in 1955.

While he was at Princeton, the wide-ranging implications of yet another of his 1905 papers became clear. In a sequel to his special relativity article, Einstein had shown that a further consequence was the equivalence of mass and energy. Ironically, his first derivation of this relationship contained a flaw in logic and, in effect, he assumed the answer he intuitively sought. In 1906, he gave a correct derivation. The following year, he wrote what has become the most famous equation of physics: $E=mc^2$—the energy of a body is equal to its mass times the square of the velocity of light.

How the mass of atomic nuclei supplied the prodigious energy that powers the stars became known through the work of Eddington in the 1920s. In 1919, Eddington wrote: "If, indeed, the subatomic energy in the stars is being freely used to maintain their great furnaces, it seems to bring a little nearer to fulfillment our dream of controlling this latent power for the well-being of the human race—or for its suicide."

By 1939, physicists in the United States and Germany had discovered the specific nuclear reactions that power the stars. Almost simultaneously, other physicists realized that a quite different set of reactions, involving the *fission* (splitting) of uranium atoms, could also release vast amounts of energy. Uranium held not only the promise of a vast new energy source, but also the grim specter of an atomic bomb. Einstein's colleagues convinced him to write to President Franklin D. Roosevelt about building a nuclear weapon.

Many of the physicists who worked on the bomb really hoped that the task would prove impossible, but such was not the case. On 6 August 1945 a nuclear bomb was dropped on Hiroshima, Japan. One of Einstein's biographers, Banesh Hoffmann, writes in *Albert Einstein, Creator and Rebel*: "Einstein's secretary heard the news on the radio. When Einstein came down from his bedroom for afternoon tea, she told him. And he said, 'Oh weh!' which is a cry of despair whose depth is not conveyed by the translation 'Alas.'"

Einstein and Lasers—Einstein's Afterthoughts

"A splendid light has dawned on me about the absorption and emission of radiation…" wrote Albert Einstein to Michele Besso in 1916. Einstein had recognized the possibility of a phenomenon called stimulated emission. In 1964 American physicist Charles H. Townes, along with Soviet physicists, Aleksandr M. Prokhorov and Nikolay G. Basov, received the Nobel Prize for applying the idea in the laser, shorthand for "light amplification by stimulated emission of radiation."

Einstein's idea was this: incoming photons could stimulate electrons to release identical photons as they dropped to lower-energy orbits. The key was to use photons the energy of which matched the quantum drop between orbits. The catch was that photon emission would not be amplified unless most atoms were already in an excited state—a situation known as population inversion—otherwise photons would be absorbed to excite electrons. Population inversion does not occur spontaneously on Earth, but naturally occurring lasers have been observed in distant galaxies in the infrared, ultraviolet, and microwave regions.

The key to the design of lasers is a source of excitation energy and mirrors. A population inversion is produced by shining light on the lasing medium or passing an electric

Right: An incoming photon stimulates an electron to drop to a lower energy level, releasing another photon with the same frequency. That these frequencies can only assume certain values was Planck's original postulate. **Below:** Inside a laser, reflected photons initiate a chain reaction to produce more and more photons that exit the laser in perfect phase.

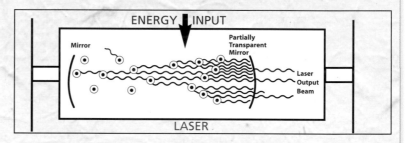

discharge through it. Mirrors at each end of a laser tube reflect photons released when excited electrons spontaneously drop to a lower energy level. The reflected photons become incoming photons for other excited electrons and trigger a chain reaction that amplifies those photons with a frequency fixed by the particular quantum drop.

The ensuing laser beams are monochromatic waves that move in the same direction and in phase. This means that the crest of the wave affiliated with one photon occurs at the same time as the crests of the waves of all other photons. This results in a

powerful beam of electromagnetic radiation. Different lasing materials can produce beams of X-rays, ultraviolet and infrared radiation, as well as visible light. Laser applications include surgery, precision cutting tools, CD and DVD readers, supermarket checkout scanners, and holograms—which are now being used as security devices on credit cards and currency.

The maser, the microwave equivalent of the laser, was developed in 1953 before the laser. Masers are especially important in atomic clocks that have been used to verify some predictions of Einstein's theory of general relativity. — PH

Einstein's E = mc², like most scientific knowledge, is morally neutral. It has indeed opened a Pandora's box before which we stand with much trepidation. The dreadful, haunting fear of nuclear warfare casts a sobering pall over all international relations.

Similarly, the fear of a catastrophic nuclear accident poisons the potential for abundant power for a civilization rapidly depleting its conventional energy resources. But $E=mc^2$ also forms the basis of a beautiful understanding of the evolution of stars, laying open the past and future of our sun, and revealing the physical reasons for the wide variety of stars spanning the night skies. It is we, and not Einstein, who must be held responsible to use $E = mc^2$ wisely.

During the 1970s, scientists began to search for yet another phenomenon predicted by Einstein's general theory of relativity—gravitational waves. If a massive object that bends the space in its vicinity also changes its arrangement—for example, in the collapse of a supernova, or exploding star—surrounding space will be deformed by the propagation of gravitational ripples that travel with the speed of light. Sensitive machines are being built in several laboratories throughout the world to detect gravitational waves. Meanwhile, analysis of the acceleration of pulses from a binary pulsar by radio astronomers at the University of Massachusetts in Amherst in 1978 provided indirect evidence of gravity waves.

Astronomers are currently in hot pursuit of yet another phenomenon described by general relativity—the black hole. In the 1700s, the British astronomer John Michell and French astronomer Pierre Simon Laplace independently considered the possibility of an object so dense and massive that the escape velocity from its surface would equal or exceed the velocity of light. Light itself could leave from such an object, but gravity would eventually pull it back again.

Relativistic calculations show, however, that light could not even leave from such an object—hence the name "black hole." Although black holes can never be observed directly, their gravitation can betray their presence by the intense warp they create in space. Likely candidates for black holes include the collapsed companion of an X-ray binary, Cygnus X-l; the cores of several compact globular star clusters; and a massive condensed object within our own galaxy.

Just as objects bend the space in their vicinity, the cumulative effects of the many masses embedded in space can curve it on a large scale. In his early work on general relativity, Einstein realized that over cosmic distances space might not conform to the principles of Euclidean geometry—that is, the sum of the angles of a very large triangle might not equal exactly 180°.

When Einstein announced his general theory of relativity in 1916, astronomers had little conception of distances beyond a few thousand light years. By the 1920s, they began to realize that what appeared as the spiral nebulae consisted of remote stellar systems millions of light years away. In the 1930s, they realized that these far-flung galaxies were rushing away from one another at immense velocities and that the universe was rapidly expanding.

Will the universe expand forever, or will it eventually coast to a stop and then begin to contract? If the universe is dense enough, gravitation can overcome the expansion and pull it back until it ultimately collapses. Such a universe is said to be "closed," and the curvature of its space produces triangles containing more than 180°. If the density is too low, however, gravity will be insufficient to halt its outward rush. Such a universe is said to be "open," and its space is curved in such a way that triangles contain less than 180°.

So far, the observations point toward an open universe, providing a real challenge for a small but dedicated group of scientists who believe on aesthetic grounds that the universe should be closed. [Editor's Note: See "Beyond the Big Bang" in this volume for current thinking on this subject.] In 1978, many of them hoped that the discovery of a faint background of X-ray radiation would furnish evidence for previously overlooked matter—mass that would raise the measured density past the minimum needed for closure. In 1979, these hopes were dashed by results from a new orbiting X-ray telescope appropriately named the Einstein satellite. The new data show that the background X-rays come from distant quasars that were already counted in the mass totals and not from previously overlooked hot gas.

Part of Einstein's genius was his incredibly well-tuned intuition and his remarkable sense of the aesthetic. He believed his general theory of relativity had to be right because it was so beautiful mathematically. It is fascinating to note that many competing gravitational theories have been postulated in the past 20 years, and that virtually all of them have been eliminated by more refined experiments. At the same time, Einstein's relativistic theory of gravitation has come through every test unscathed.

One of Einstein's University of Berlin students, Ilse Rosenthal-Schneider, reported that one day in 1919 when she was studying with him, "He suddenly interrupted the discussion of the book, reached for a telegram that was lying on the window sill, and handed it to me with the words, 'Here, this will perhaps interest you.' It was Eddington's cable with the results of the eclipse measurements. When I was giving expression to my joy that the results coincided with his calculations, he said, quite unmoved, 'But I knew that the theory is correct'; and when I asked, what if there had been no confirmation of his prediction, he countered: 'Then I would have been sorry for the dear Lord—the theory is correct.'"

Einstein's aesthetic sense drove him to search even further for a harmonic unity between the gravitation of space and the electrical structure of matter. Although he failed in this quest for a unified field theory, several leading theoretical physicists in several parts of the world are continuing to search for it.

Einstein is a towering figure in science because he found and attacked a broad range of crucial problems and discovered deep underlying harmonies that other physicists missed. But even more important, while many of the other giants of his age won their laurels with a dazzling display of new discoveries in the laboratory, Einstein's workshop was in the mind. His experiments were almost always thought experiments. Like Copernicus before him, who had begun a revolution with theories "pleasing to the mind," Einstein astonished his contemporaries with entirely new ways of looking at phenomena that had been known for decades, or even centuries.

Einstein's work, like a shaft of sunlight in the great cathedral of science, suddenly illuminates the structure and lets us realize that this beautiful edifice is man-made. Scientific theory is not an absolute, but a human construction, molded according to our taste and perspectives. Einstein himself clearly summarized this understanding of science when he wrote, "The sense experiences are the given subject matter. But the theory that shall interpret them is man-made. It is the result of an extremely laborious process of adaptation: hypothetical, never completely final, always subject to question and doubt."

Relativity and quantum mechanics have radically altered our picture of nature. But Einstein's insight into these concepts reminds us of the remarkable impact that a single gifted individual can make in our scientific view of the world.

Notes and references

Gerald Holton's several essays and lectures on Einstein have guided my interest; two from which I have profited especially are "Einstein, Michelson, and the 'Crucial' Experiment" and "On Trying to Understand Scientific Genius," both reprinted in *Thematic Origins of Scientific Thought: Kepler to Einstein* (Cambridge, Massachusetts, revised edition, 1988). Holton gives the Ilse Rosenthal-Schneider quotation about the eclipse telegram on pp. 254-5; I once had occasion to question Einstein's long-time secretary, Helen Dukas, about it, and she replied that it seemed quite authentic, as Einstein had more than once expressed himself similarly. Two excellent biographies of Einstein are Abraham Pais, *Subtle is the Lord: the Science and Life of Albert Einstein* (Oxford, 1982) and Banesh Hoffmann, *Creator-Rebel: Albert Einstein* (New York, 1972). The Eddington incident at Príncipe in 1919 is described in A. Vibert Douglas' biography, *Arthur Stanley Eddington* (London, 1957), p.40.

Einstein playing his beloved violin, accompanied by his friend Ehrenfest at the piano.
From an original watercolor by Maryke Kammerlingh-Onnes.

Albert Einstein in Leiden

―――――――○―――――――

By Dirk van Delft

Albert Einstein liked coming to Leiden, the Dutch city particularly known for its venerable university. Vienna-born theoretical physicist Paul Ehrenfest, professor at Leiden University since 1912, was one of his closest friends.[1] In July 2005, Rowdy Boeyink, a history-of-science student, stumbled across a long-lost, handwritten Einstein manuscript in the Ehrenfest Library of Leiden University's Lorentz Institute for Theoretical Physics. The manuscript was part two of a paper entitled "Quantum Theory of Monatomic Ideal Gases," which Einstein presented at a 1925 meeting of the Prussian Academy of Sciences.[2]

That paper has a special significance. It contained Einstein's last great discovery—Bose–Einstein condensation, as the effect came to be called. Unearthed amid the celebrations of the World Year of Physics, the centenary of Einstein's 1905 *annus mirabilis*, the 16 handwritten pages attracted international media attention.[3] That same month, Boeyink also found among Ehrenfest's papers typescripts of two more Einstein papers—one from 1914, the other from 1920. Both manuscripts contain interesting differences from the version that eventually appeared in print.

Last October I came across reprints of 22 articles by Einstein from the period 1902–15 in the archive of the Huygens Laboratory, home to experimental physics at the university. In some cases, including the famous 1905 article on the special theory of relativity, the reprints feature handwritten "improvements" by Einstein himself.

Einstein and Ehrenfest

How did the Einstein manuscripts and the hand-annotated reprints end up in Ehrenfest's library? They reflect the intimate bond between Einstein and Ehrenfest. How did that bond develop? What was Einstein's connection with Leiden and with two other leading scientists at the university, astronomer Willem

A relaxed and confident Albert Einstein as painted in 1920 by Harm Kamerlingh Onnes, nephew of the discoverer of superconductivity.

de Sitter and Heike Kamerlingh Onnes, the discoverer of superconductivity?

In 1912 Ehrenfest succeeded Hendrik Antoon Lorentz (1853–1928) as professor of theoretical physics at Leiden.[4] Lorentz would have preferred to get Einstein, but Einstein opted for Zürich. Ehrenfest, who got his Ph.D. in Vienna under Ludwig Boltzmann, had been working at the University of St. Petersburg in Russia since 1907. In Leiden, the Ehrenfests moved into a Russian-style villa designed by Ehrenfest's Russian wife Tatiana Afanashewa, a mathematician. They brought with them from St. Petersburg the tradition of a weekly colloquium, with frank and open discussion of the latest developments in physics. At first the colloquia met at the Ehrenfest home, but later they were held at the institute for theoretical physics.

The formal Lorentz would keep silent until he had his thoughts in order, but the high-spirited Ehrenfest viewed boisterous debate as absolutely essential. He was informal with students, his lectures and presentations were lively and lucid, and his German was peppered with colorful expressions like *Das ist wo der Frosch ins Wasser springt* (that's where the frog jumps into the water).

According to his student Hendrik Casimir, Ehrenfest was "a passionate admirer of the beautiful and the profound."[5] Yet he was increasingly weighed down over the years by doubts about his ability to keep pace with developments in the new physics, particularly quantum physics. Those doubts contributed to the deep depression that tragically led him to take his own life in 1933, at age 53. Boltzmann, his great teacher, had set him a sad example by committing suicide in 1906.

Nonetheless, Einstein and Ehrenfest were soul mates. They first met in February 1912 in Prague, where Einstein was a professor until he moved to Zürich later that year. The friendship clicked from the very beginning. They interrupted their intense discussions on subjects like the ergodic principle or gravitation to play Brahms sonatas, with Einstein on the violin and Ehrenfest at the piano.

After spending a week at Einstein's home, Ehrenfest confided to his diary that he "was terribly happy… Yes, we will be friends." The affection was mutual. In a 1934 eulogy Einstein wrote: "Within a few hours we were true friends—as though our dreams and aspirations were meant for each other."

Einstein first visited Leiden in February 1911. He had accepted a student invitation to give a lecture there because he was keen to meet Lorentz, a father figure whose work he valued highly. He also wanted to meet Kamerlingh Onnes and see his celebrated cryogenics laboratory. Interestingly, Kamerlingh Onnes had failed to respond to a 1901 letter from Einstein applying for a graduate assistantship. Einstein, who had just graduated from the Zürich Polytechnic, enclosed a stamped, self-addressed postcard and his recently published paper on intermolecular forces. The postcard disappeared into a folder in Onnes's home.

By 1910, Einstein was no longer an obscure supplicant. The research reports on critical opalescence that Kamerlingh Onnes sent to Prague that year with an eye to Einstein's arrival were of common interest to both men. They elicited from Einstein a consignment of reprints of his own work.

Einstein and his wife Mileva stayed with Lorentz during that week in February 1911. Back home again in Prague, Einstein wrote to say how much he had enjoyed the hospitality and the scientific discussions. He expressed great interest in Onnes's research into the temperature dependence of the electrical resistance of metals. At the time, Kamerlingh Onnes was on the brink of observing superconductivity in mercury, a discovery he would present at the first Solvay conference in Brussels in the autumn of 1911. Einstein was a participant in that historic meeting.

During the next 20 years Einstein visited Leiden often. He enjoyed the scientific give and take. In Berlin, where he was based from 1914 to 1932, the theoretical physicists were generally unreceptive to wide-ranging discussions about fundamentals. But in Leiden he could talk in a relaxed fashion with Ehrenfest about his work on general relativity. In March 1914, shortly before his move to Berlin,

Einstein spent a week at Ehrenfest's home. The two friends spoke at length about problems with quantum theory and statistical mechanics. Afterwards, Ehrenfest received a warm letter from Berlin in which, as a sign of close friendship, Einstein no longer addressed him with the formal *Sie* but rather with *du*.

A neutral respite

After World War I began with the German invasion of Luxembourg and Belgium in August 1914, Einstein gratefully seized upon an invitation from Ehrenfest to visit the neutral Netherlands. Einstein abhorred the wave of nationalism that was sweeping the German academic community. As one of the few prominent academics who refused to sign the "Manifesto to the Civilized World," a pompous justification of the invasions that invoked Beethoven and Goethe, Einstein was desperate for a respite from all the patriotic hysteria. "Every fiber of my being itches to get away from here," he wrote his friend.

But traveling in wartime was not easy; it was many months before Einstein could arrange the necessary documents. He finally arrived in Leiden in September 1916 for a two-week visit. In Haarlem, he and Ehrenfest looked up the semi-retired Lorentz, with whom he had been corresponding intensively about general relativity, which had been published in its completed form the previous November. Back again in Berlin, Einstein wrote to his old friend Michele Besso that he had "spent unforgettable hours with Ehrenfest and especially with Lorentz, not only stimulating but also refreshing. I feel in general that I am incomparably closer to these people [than I was before]." During his stay at Ehrenfest's home, Einstein had opened his host's eyes to the music of Bach. For several months thereafter, Ehrenfest seemed to be more taken up with choral music than with physics.

Because there was no free exchange of scientific ideas in wartime Europe, it was the Dutchman de Sitter who passed on the new general theory of relativity to England. At Arthur Eddington's request, de Sitter wrote a three-part paper entitled "On Einstein's Theory of Gravitation, and Its Astronomical Consequences" for the *Monthly Notices of the Royal Astronomical Society*.

Einstein first met de Sitter during his 1916 Leiden visit. The two spoke at length about general relativity and the residual elements of absolute space and time that it still preserved.[6] There ensued a detailed correspondence about cosmological solutions of the theory's field equations. That was the beginning of relativistic cosmology. Responding to Einstein's closed, static universe and its cosmological constant, de Sitter in March 1917 introduced his eponymous empty-universe solution.

The Lost Manuscript

On his visits to Leiden, Albert Einstein left three documents with Paul Ehrenfest. They eventually disappeared between the pages of journals in Ehrenfest's study. When Ehrenfest's daughter Tanja died in 1984, his books and papers were left to Leiden University's Lorentz Institute. There the piles of paper stayed undisturbed for decades until Rowdy Boeyink, working on his master's thesis, began leafing through them one by one.[14]

On his first visit to Leiden in 1914, Einstein left behind a copy of his last Zurich paper. Entitled "Covariance Properties of the Gravitation Theory Based on the Generalized Theory of Relativity." It was to appear in May in the *Zeitschrift für Mathematik und Physik.* Co-authored with mathematician Marcel Grossmann, it was a follow-up to a paper they had written together the previous year.

The 13-page typescript, with

Long-lost manuscript of Einstein's 1925 paper "Quantum Theory of Monatomic Ideal Gases" which predicts Bose-Einstein condensation, left in Leiden with Ehrenfest.

neat formulae handwritten by Grossmann, contains corrections made by Einstein and a supplement all of which were incorporated into the published version. The typescript features comments and drawings in Ehrenfest's hand that bear

witness to the intensive discussions during which the two friends grappled with ideas that would result, in November 1915, in the completed theory.

When Einstein visited Leiden in May 1920, he once again left behind a typescript. This time it was the page proofs of the paper "Propagation of Sound in Partly Disassociated Gases," which the Prussian Academy had published in April. Thanks to his recently acquired stardom, Einstein received many requests and suggestions for what he termed "scientific diversions." One of those diversions involved researching the question of whether the speed of sound is frequency-dependent. The resulting article on gases did not attract much attention—the paper's conclusion was that there was no such dependence.

On his penultimate visit to Leiden in February 1925, Einstein left behind his handwritten manuscript predicting Bose-Einstein condensation. It differs only in minor respects from the landmark version of the theory ultimately published.

Eddington, who was "immensely interested" in the new ideas, led the expedition that observed the deflection of starlight passing by the Sun during the solar eclipse of May 1919, thus confirming an important prediction of general relativity. Einstein learned of the happy news from Lorentz in September. With the official announcement of the Eddington expedition's result on 6 November 1919, Einstein become a superstar.

At the time, Ehrenfest, Lorentz, and Kamerlingh Onnes were busy trying to bring Einstein to Leiden as a guest professor.[7] They also tried to lure him to accept a full professorship. "The matter is very simple," Ehrenfest wrote to him on 2 September after consulting with his colleagues. "If you just say yes, we can… arrange everything extremely quickly in accordance with your wishes." Ehrenfest held out the prospect of an ideal existence in Leiden: Einstein could determine his own salary ("Our *maximum* of 7500 guilders [per annum] is your *minimum*"); he did not have to deliver lectures; and he could have unlimited travel. Ehrenfest stressed that Kamerlingh Onnes warmly supported the offer. "Please remember that you would be surrounded here by people who are fond of you personally, and not just of the stream of ideas that flow from you." Everything was possible, so long as Ehrenfest and company could say that Einstein was now at Leiden.

But Einstein declined the offer. It was certainly attractive, and he loved Leiden, but he couldn't simply follow his heart. Max Planck had pleaded with him to stay in Berlin. Overcome by the wretched state of the postwar capital city, Einstein opted for loyalty. He did, however, yearn to look up his Leiden friends once more. In the second half of October, he stayed with Ehrenfest. Back in Berlin, three days after the front page of the London *Times* reported the verification of general relativity, Einstein wrote him a moving letter. "It feels as though you are a part of me, and that I belong to you. From now on we will stay in close personal contact with each other. That will do us both good—each of us feels less out of place in this world because of the other."

When it was clear that Einstein would not accept a full professorship, Leiden switched to plan B—the guest professorship. On 9 November 1919, Kamerlingh Onnes, a shrewd organizer who was much better than Ehrenfest at getting his way with the authorities, approached the foundation that funded special chairs at the university. Who, he asked rhetorically, could lay greater claim to a special chair than Einstein, a "star of the first magnitude," a man who could be "compared to Newton"? Onnes's timing was perfect. Just days earlier Einstein had rocketed to international fame. "A few weeks ago we had the great Swiss Einstein in our midst," Onnes began. (A year after the armistice of 11 November 1918, the word "Swiss" was no idle addition, even in neutral Holland.) Onnes shrewdly added that "Einstein was led to his discoveries by building on Lorentz's work in Leiden." He suggested appointing Einstein for three years at a visitor's salary of 2000 guilders per annum, and the foundation accepted.

Ehrenfest informed Einstein of the offer on 24 November. It would involve being in Leiden for three or four weeks a year. That appealed to Einstein; he described it as a "comet-like existence." To the funding foundation, Onnes had further accentuated Einstein's importance for Leiden by mentioning his own low-temperature work. The theory of relativity was of course of the greatest importance, he said, but "in other areas too, Einstein has produced work of great significance." He pointed to Einstein's contributions to quantum theory. "It was investigations at low temperatures that first shed light on many of the phenomena to which [quantum] theory relates." For that reason, Onnes wrote, "Prof. Einstein attaches great importance to the work of the cryogenics laboratory," and he emphasized his own eagerness to benefit from Einstein's "teaching."

In December 1919, Lorentz made the official offer to Einstein, emphasizing once again that Kamerlingh Onnes would be keen to discuss with Einstein matters arising from the cryogenics laboratory. Einstein accepted the offer and announced that his inaugural lecture—at Lorentz's request—would be about "ether and the theory of relativity." By "ether" Einstein meant, in this case, the gravitational field.

"The Red Countess"

There was, however, a snag. Lorentz's hope that the Dutch government would quickly confirm the appointment was disappointed. Ehrenfest wrote Einstein in March that the royal decree was still "lying around in government departments," but that an inaugural lecture on 5 May was surely possible. He took the trouble to acquaint Einstein with some of Leiden's academic customs. Wearing the obligatory academic gown during the inaugural lecture made "speaking with one's hands" impossible. And after the talk, before the host's speech of thanks, Ehrenfest suggested, Einstein would do well to take his time drinking a glass of water so that the older gentlemen in the room could be shaken from their slumbers.

In fact, Einstein did not present his inaugural lecture until 27 October 1920. Why the long delay? It seems to have been a case of mistaken identity. The authorities in The Hague thought they were dealing with Carl Einstein, the German writer, art historian, and leftist revolutionary. That other Einstein, according to the committee that was to advise the minister of internal affairs on the appointment, was in Brussels, allegedly living in sin with the "Red Countess" Ada von Hagen, a well-known propagandist of the time. The shocked minister demanded a full explanation from the foundation.

"Prof. Einstein does not go in for countesses," replied Professor Cornelis van Vollenhoven to the minister in a letter recently discovered by science historian Jeroen van Dongen.[8] "He has never lived at the address you mention. He is married to a Jewish woman whose maiden name is also Einstein. He did not live or stay in Brussels during the war."

Although van Vollenhoven's letter seemed to convince the minister, the royal decree was not enacted until 21 September. Meanwhile, at a meeting of the Society

Einstein's Friends: Besso, Grossmann and the Queen Of Belgium

Though Einstein's relationships with family members could often be strained, he maintained strong personal friendships for many years and was a conscientious correspondent, spending several hours a day answering letters. His closest lifelong friend was probably Michele Angelo Besso, whom Einstein met in 1900 when the two men roomed in the home of Jost Winteler, headmaster of the school Einstein attended as he prepared for his university entrance exams. Six years Einstein's senior, Besso was trained as an electrical engineer and the two men worked in the Swiss Patent Office in Bern. Besso served as a valuable sounding board for Einstein during the months he was preparing his first papers in relativity. In the ensuing years, Besso served as an intermediary during Einstein's difficult divorce from Mileva—often chiding Einstein for being cruel and inattentive—and he helped Einstein's son, Eduard, when he suffered a mental collapse in 1932. It was about Besso that Einstein wrote (in a letter to Besso's wife after Besso had died), "He has preceded me a little by parting from this strange world. This means nothing. To us believing physicists, the distinction between past, present and future has only the significance of a stubborn illusion."

Another lifelong friend was his college classmate, Marcel Grossmann, a very proper professor at ETH who helped Einstein develop his

At far right, Einstein with the Queen of Belgium; left, Michele Besso and his wife, Anna Winteler; and below, Paul Ehrenfest at the piano, Einstein and Ehrenfest's son Paul Jr. on Einstein's lap

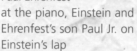

mathematical skills and co-authored several important papers in general relativity. Grossmann's life was cut short when he was stricken with multiple sclerosis and died in 1936. In addition to being a helpful collaborator for Einstein, Grossmann also helped Eduard during his illness. As was the case with Ehrenfest, Einstein remained grateful to Grossmann—he claimed it was the need to express his gratitude that prompted Einstein to write his brief "Autobiographical Notes" in 1955—and cherished their friendship his entire life.

Two women stand out as lifelong friends of Einstein: Elisabeth, Queen of Belgium, with whom he carried out a rich correspondence his entire life; and Helen Dukas, his nurse and later his personal secretary, and finally keeper of the voluminous archives that Einstein left behind. Visitors to the Einstein home (which included a long list of notables, including his Princeton neighbor, the controversial author and fellow supporter of Hebrew University, Immanual Velikovsky) often thought Einstein treated Dukas harshly and dismissively. But in the end, Dukas cared for Einstein's fragile health and safeguarded his privacy and reputation, which placed restrictions on Einstein's life that, from all accounts, irked him no end.

Einstein was also friendly throughout his life with several scientists, most notably Niels Bohr, with whom he carried on a lifelong debate about quantum theory of nearly Talmudic proportions; and Kurt Gödel, the great logician, who was also at the Institute for Advanced Study and accomapanied Einstein every day for years, discussing... well, see Jim Holt's essay on page 247.　　— HR of German Natural Scientists in Berlin's Philharmonic Hall, Einstein was branded a "publicity-seeking dog," a "plagiarist," a "charlatan," and a "scientific Dadaist." Einstein, never shy, attended the meeting—reportedly with occasional amusement—and hit back hard with a response in the *Berliner Tageblatt*.[9] As soon as Ehrenfest heard of the commotion in Berlin, he once again offered Einstein—on his own authority—a full professorship, but Einstein stuck to the guest arrangement.

Kamerlingh Onnes seized the opportunity of Einstein's debut as guest professor to organize a mini-conference on magnetism for October 1920. The only experimental work of Einstein's scientific career, conducted in 1915 with Lorentz's son-in-law Wander De Haas, had been on magnetism. In addition to Onnes and Einstein, the other conference participants were Ehrenfest, Lorentz, Johannes Kuenen, Paul Langevin, and Pierre Weiss. The experimenter Weiss, who had just transferred from Zürich to what was now, once again, French Strasbourg, had in the past brought his heavy magnet with him from Zürich to conduct low-temperature magnetic investigations with Onnes's group. Since 1914, however, magnetic research at Leiden had fallen into something of a decline, and Onnes was seeking to revive it.

In 1920 Ehrenfest, who had been working primarily on econometrics during the war, began to devote himself to paramagnetism. So he was looking forward to the mini-conference. "I'm dying to discuss these matters," he wrote to Einstein, who in turn regarded paramagnetism as "fully ripe for theoretical attention." After the mini-conference, Onnes wrote a report on it for the 1921 Solvay conference. Ehrenfest summarized his thoughts in a paper on the paramagnetism of solid substances.[10] Einstein had his hands full with general relativity.

Superconductivity

Einstein did, however, lecture on superconductivity at Leiden in November 1921. This time he was invited to stay at Kamerlingh Onnes's home. The host was curious about Einstein's views on quantum effects in superconductors. Einstein returned to that topic in 1924, giving Onnes several formulae based on speculative arguments of uncertain significance that one would have to test in the laboratory. Onnes replied that although he still had some reservations, Einstein's suggestions had brought him "immense joy" and that the idea that quantum rules help determine the time that molecules spend in close proximity had been a "lightning flash" for him. But those notions remained vague. He was sure, he told Einstein, that they would lead to "something beautiful." But at age 71, he was too old to do anything with them.

Kamerlingh Onnes was no Ehrenfest. Many years later Einstein recalled that scientific discussions with him were always rather awkward. There was nothing wrong with his intuition, said Einstein, but he could not find the right words to express himself and he was not very receptive to others' lines of thinking.

In November 1922, Einstein set out his ideas on superconductivity in an article for the festschrift celebrating the 40th anniversary of Onnes's professorship.[11] Following discussions with Ehrenfest, Einstein had arrived at a model of "chains of atomic electrons running almost in single file," as he explained it in a postcard to his friend. In the superconducting state, he went on, these chains would be "stable and undisturbed." Einstein suggested testing his theory by measuring the self-induction of a non-superconducting coil placed beneath a short-circuited superconducting coil. His festschrift article does not contain this somewhat vague suggestion, but he did stick to his electron-chain conjecture. However, after Kamerlingh Onnes found superconductivity across a lead–tin interface, Einstein did have to retract his hypothesis that the electron chains could not consist of different types of atoms.

Surprisingly, Einstein's festschrift paper did not cite a contribution by Onnes to the 1921 Solvay conference.[12] In it, Onnes had also come up with the idea—in much greater detail than Einstein—of electrons moving via low "threads" from atom to atom. But Einstein had not attended the 1921 Solvay conference in Brussels, so he may not have known about Onnes's contribution.

Over the years, Leiden benefited seven times from Einstein's appearances as guest professor. In addition to typescripts and manuscripts, in 1921 he left his fountain pen behind as a gift at Ehrenfest's home. Ehrenfest treasured it, noting that Einstein had used it to write all his articles and calculations on general relativity.

Einstein's signature can be found in the visitors' book at the Spinoza house in Rijnsburg, a village just outside Leiden. He made the visit in 1920 with Onnes's nephew Harm Kamerlingh Onnes, a well-known painter. The 17th-century philosopher Baruch (Benedict) de Spinoza, whom Einstein greatly admired, had lived in that peasant's cottage. There, when not writing, Spinoza ground lenses for a living. In a poem that Einstein, a prolific versifier, wrote that year, he said of Spinoza:[13]

Wie lieb ich diesen edlen Mann,
Mehr als ich in Worten sagen kann
(How I love this noble man/
More than I can say in words.)

Einstein visited Leiden for the last time in April 1930. Shortly after he left Germany in December 1932 for a visit to the U.S., the Nazis came to power and

Einstein never returned. The newly established Institute for Advanced Study in Princeton, New Jersey, was to be his scientific home for the remainder of his life.

On the afternoon of 25 September 1933, Einstein's dear friend Ehrenfest picked up his younger son Vassily, who had Down syndrome, from the institution in Amsterdam that housed him. In a park nearby, the tormented man drew a revolver, shot the boy, and then killed himself.

References

1. See http://www.lorentz.leidenuniv.nl/history/einstein/einstein.html
2. A. Einstein, *Sitzungsberichte Preuss. Akad. d. Wissenschaften* **1**, 3 (1925).
3. D. van Delft, "De laatste Einstein," *NRC Handelsblad*, 20 August 2005, p. 1.
4. See Martin Klein, *Paul Ehrenfest. Volume 1: The Making of a Theoretical Physicist*, North Holland, Amsterdam (1970).
5. H. Casimir, *Haphazard Reality: Half a Century of Science*, Harper & Row, New York (1893).
6. A. Einstein, *The Collected Papers of Albert Einstein*, vol. 8A, R. Schulmann, A. Kox, M. Janssen, J. Illy, eds., Princeton U. Press, Princeton, NJ (1998), ed. note, p. 351.
7. D. van Delft, *Heike Kamerlingh Onnes: Een Biografie*, Bakker, Amsterdam (2005). English translation forthcoming.
8. C. van Vollenhoven, letter to the Minister of Internal Affairs, 27 March 1920, National Archive, The Hague. See also *De Volkskrant*, 14 May 2005, p. K5.
9. A. Einstein, *The Collected Papers of Albert Einstein*, vol. 7, M. Janssen et al., eds., Princeton U. Press, Princeton, NJ (2002), p. 344.
10. P. Ehrenfest, *Proc. R. Netherlands Acad. Arts Sci.* **23**, 989 (1921).
11. A. Einstein, in *Natuurkundig Laboratorium der Rijksuniversiteit te Leiden in de Jaren 1904–1922*, IJdo, Leiden, the Netherlands (1922) p. 429. An English translation by B. Schmekel is available at http://arxiv.org/abs/physics/0510251].
12. H. Kamerlingh Onnes, *Rapport IIIième Conseil Solvay*, International Institute of Refrigeration, Paris (1923), p. 165.
13. See http://www.Albert Einstein.info/db/ViewImage.do?DocumentID=17814&Page=1
14. See http://www.phys.uu.nl/~Boeyink/

Portrait of a family in distress. Eduard Einstein, Mileva Maric Einstein and Hans Albert Einstein in 1914.

The Forgotten Wife

—◦—

by Andrea Gabor

What a strange thing must be a girl's soul: Do you really believe that you could find permanent happiness through others, even if this be the one and only beloved man?

—Albert Einstein to Marie Winteler, his first love

During, the fall of 1891, a quiet, dark-haired girl entered the Royal Classical High School in Zagreb. She was painfully shy and not particularly pretty, and she walked with a noticeable limp. But no one who knew her could help but be impressed by sixteen-year-old Mileva Maric. For she possessed an unmistakable spark of genius, a dedication to mathematics and science that was unusual for anyone at that age, and in that time and place, extraordinary for a girl. As a result she had received a special dispensation to attend the all-boys' school and thus had become one of the first girls in the Austro-Hungarian Empire to sit in a classroom with boys.

Mileva Maric took her place in the Zagreb high school just as Maria Sklodowska (later Marie Curie) was getting her degree in physics from the Sorbonne. Indeed, until she fell in love with Albert Einstein a few years later, the brilliant Serbian student would follow a path remarkably similar to that of her Polish contemporary, leaving behind home and family to become one of the first women ever to study physics at a university.

Mileva Maric was born in a rural Serbian outpost of the Habsburg monarchy in Vojvodina, a province of what was then Southern Hungary, and an ethnic melting pot of Croats, Slovaks, Hungarians, Romanians, and Gypsies. Mileva's father, Miles Maric, was a successful and well-to-do civil servant. He and his wife,

Marija, would eventually have three children. Mileva, or "Mitza," as she was affectionately called, was the oldest. Born in 1875, she was eight years older than her sister, Zorka, and ten years older than her brother, Milos junior. And she was said always to have been her father's favorite.

When Mileva was a child, it seemed as though there was nothing she put her mind to that she couldn't do. She showed an uncanny knack for mastering everything from mathematics and languages to handicrafts and painting. By the age of eleven, during her first year at a secondary school for girls, Mileva had become the top student in her class.

Before long, the Marics had extracted a long list of special exemptions and dispensations from school authorities and government bureaucrats, which allowed Mileva to break through the rigid gender barrier in Austro-Hungarian education. First, in 1890, when Mileva turned fifteen, her parents sent her to a school in the neighboring village of Sabac, across the border in Serbia, where there were no rules explicitly excluding girls and where the curriculum was probably stronger in the sciences than it would have been at an all-girls' school. Mileva, who already had shown herself to be something of a prodigy when it came to mathematics, also began to show an aptitude for languages and art. She produced sophisticated, highly detailed sketches of local village scenes. And at the secondary school in Sabac, where typically only one foreign language was taught, she got permission to study French as well as German, and soon became fluent in both languages. When she began attending school in Zagreb, where Greek was also a requirement, she mastered that language too.

It was in Zagreb that Mileva Maric probably first developed a taste for physics and first demonstrated the grit to pursue an unladylike vocation in the sciences. Mileva began her studies in Zagreb as a "private student," which probably meant that she received tutoring but did not, at first, attend classes at the gymnasium, the college-preparatory school, which was reserved for male students. Her relative isolation would not have been entirely unwelcome: Since childhood, Mileva had suffered from a congenital hip deformity that caused her to limp. She had been forced to suffer the taunts of classmates, who ridiculed her as much for her extraordinary intelligence as for her physical handicap. She grew up painfully shy and sensitive, yet also exceptionally determined. At a time when the few women who dared to break through rigid barriers to pursue a higher education were often harassed by male students and faculty, Mileva applied for, and won, permission to attend physics lectures at the gymnasium. Recalling his own days at a Viennese gymnasium in the 1930s, Gerald Holton, an Einstein scholar, said of Maric's feat: "The idea of a girl sitting in on a class [was] so, oh, bizarre... the poor girl would have been the subject of attacks in the early years and seduction in the later years." Yet although she was the only woman in her class, Mileva was awarded the highest grades for her work in both math and physics.

If she was to pursue a higher education a woman of Mileva's generation needed brains, bravery, and an ability to envision a life that ran counter to the norms of her day. However, even for such a woman the chances of succeeding, as Mileva was to discover, were slim indeed.

For it was one of the ironies of nineteenth-century Europe that the very philosophical changes that celebrated individualism and the freedom of human will, that extolled the cult of the hero and established the rights of man, specifically excluded women from the new spirit of liberty. From Rousseau and Hegel to Darwin and Hume, the great thinkers of the eighteenth and nineteenth centuries supported the view of women as inferior beings. Even John Locke, who championed everyman's equal right to his "natural freedom, without being subjected to the will or authority of any man," put women in the same category as animals and upheld "the Subjection that is due from a Wife to her Husband."

Such thinking came to be institutionalized in law—and hence, also, in society. When Maric embarked on her career, the increasingly bureaucratized world of the nineteenth century had imposed more, rather than less, explicit restrictions on women. The Code Napoleon, for example, classified married women—along with children and the insane—as legal incompetents; it also vastly expanded a husband's power over his wife, eliminating her right to control her own bank account or even to keep her correspondence private. And universities that had not explicitly barred women from attending in the eighteenth century did so in the nineteenth. Thus when she set off for Zurich at the turn of the century, Maric did so not only to complete the final step of her education but also to study in one of the only cities in Europe where women were permitted to attend and graduate from the university.

It is against this backdrop that Mileva Maric and Albert Einstein met in Zurich in 1896 and embarked on a bohemian, and decidedly modern love affair. They were fellow classmates at the prestigious Swiss Federal Polytechnic, which is better known as the ETH, its German acronym. Mileva was the only woman in her class and the fifth woman ever to attend the school. And although she was shy and studious, her first year in Switzerland was academically successful. Though most of what is known about Maric's life during this period comes from recently published

Einstein, Mileva and Hans in 1904

correspondence between her and Einstein, it is clear that Maric took to her first year of studies with quiet self-assurance and a belief that a world of opportunity was open to her. What lends exceptional poignancy to the couple's courtship and marriage is that from the beginning, Mileva, who thought at first that she would never marry, seemed to sense the danger to her professional dreams that was posed by her passionate feelings for her brilliant and effusively affectionate "Johnnie." (The nickname Johnnie is a diminutive for Johann and probably a play on the fact that, in German, Johann is a common name for a servant and for everyman.)

Initially, Mileva had not planned to study physics. After completing her secondary education, she entered the University of Zurich, in the summer of 1896, as a medical student—the Zurich university system having become the first in Europe to grant admission to women, in 1867. An interest in medicine ran in the Maric family. Her brother, Milos, would become a well-known expert in histology. And both Mileva and her younger son, Eduard, would develop a keen—though ultimately ill-fated—interest in psychiatry. But for reasons that have never been fully explained, Mileva switched her studies from medicine to math and physics at the ETH, a fact that probably hastened the demise of her career.

It is tempting to surmise that Mileva abandoned her medical studies in response to some trauma she experienced at the hands of hostile classmates or teachers during a time when she was still very much alone in a strange city. Yet she showed few signs of vulnerability during her first years as a university student; on the contrary Mileva was almost cocky in her zest for independence and learning. Nothing seemed to scare her as she blithely marched from one male-dominated discipline to the next.

What is most notable about Mileva's youth, in fact is not her dedication to one specific field of study but rather her wide range of interests and her boundless zest for new academic experiences. Long after she had committed herself to studying physics, for instance, Mileva maintained an abiding interest in psychiatry, regaling Einstein with reports of the latest studies in the field, such as the new experiments in hypnotism, just as he showered her with news of the latest discoveries in physics. For if Einstein was consumed by physics, Maric was drawn to math, physics, medicine, music, and, eventually, the raising of both children and exotic cacti.

Legend has it that the couple met when Einstein asked Maric how she had arrived at the solution to a particular problem—probably in mathematics—for which he himself had not found the answer. In the intimacy of a small class of five students, Maric and Einstein, who took several courses together, couldn't help but get to know each other through their shared interest in physics. Einstein, who at seventeen was still virtually a boy, must have been somewhat in awe of his unusual female classmate, who was three and a half years his senior. They discovered a shared love of music and the outdoors. During their first year together, Albert often joined Mileva and her friends for musical evenings. Albert played the violin, while Mileva sang or accompanied him on the piano. Although in later life he developed an aversion to mountains, as a young student in love, he spent some of his happiest weekends hiking with Mileva through the Alps.

Maric told Einstein early in their relationship that she doubted she would ever marry, because although she insisted that "a woman can make a career just like a man," she apparently also believed the two enterprises to be mutually exclusive. What she didn't say, but probably assumed, was that the combination of her physical deformity and what was at the time considered to be her decidedly "unwomanly" intellectual interests made her a less than desirable prospect for marriage. Indeed one of their classmates once remarked to Einstein that he would never consider marrying a woman who wasn't completely healthy. (To which Einstein is said to have replied laconically: "But she has such a sweet voice.")

It is worth noting that from the beginning, Mileva and Albert were drawn together not only by love but also by the disapproval of others. They saw themselves united in adversity, besieged on all sides—by their parents, who opposed their relationship for a long list of reasons headed by their religious differences; by some of their professors who didn't know what to make of the eccentric young couple; and finally even by some of their friends who resented the attention the lovers lavished on each other.

Yet even though Maric was very attracted to Einstein and aware that she had few romantic prospects, she wasn't ready to settle down, either to her studies at the ETH or to her relationship with Einstein. By fall of the following year, she had transferred to the University of Heidelberg, one of the oldest and most venerable academic institutions in Europe, where she audited courses for one semester. (Heidelberg was among the many European universities in which women were still not permitted to matriculate, although since 1891 the institution had permitted women to sit in as auditors.) According to Desanka Trbuhovic-Gjuric, a Serbian historian who wrote one of the only accounts of Maric's life, the young woman left Zurich because she feared the consequences of her feelings toward Einstein.

Certainly, the evidence suggests that Maric had good reason to fear her relationship with Einstein. Within a year of her return from her semester in Heidelberg, Mileva would undergo a shocking metamorphosis from a seemingly independent, ambitious, and consummately self-assured young woman to one racked by doubts, disappointment and resignation.

But at the writing of Maric's first letter to Einstein from Heidelberg, her decision to leave the ETH almost certainly stemmed more from an irrepressible urge for academic adventure than from an instinct for self preservation. In fact, Mileva was having such a good time that it took her weeks even to get around to writing her boyfriend. And when she did, her correspondence brimmed not only with self-confidence and a hunger for discovery but also with a joie de vivre that was decidedly lacking in her later years: "I don't think the structure of the human skull is to be blamed for man's inability to understand the concept of infinity. He would certainly be able to understand it if, when young, and while developing his sense of perception, he were allowed to venture out into the universe rather than being cooped up on earth or, worse yet, confined within four walls in a provincial backwater. If someone can conceive of infinite happiness, he should be able to comprehend the infinity of space—I should think it much easier. And human beings are so clever and have accomplished so much, as I have observed once again here in the case of the Heidelberg professors."

In that first letter, Mileva also recounted her enthusiasm for a four-hour lecture she had just attended on the kinetic theory of gases, given by Philipp Lenard. The renowned theoretical physicist, who would become an enthusiastic supporter of the Nazis and a nemesis to Einstein, was nevertheless to be an important influence in Einstein's early work. For example, it was Lenard who discovered that a light beam trained on a piece of metal can dislodge electrons from the surface, an insight that paved the way for Einstein's explanation of the "photoelectric effect," an early milestone of quantum theory for which Einstein would win the Nobel Prize. Lenard also would become known for his explanation of Brownian motion, a theory that explained the unceasing and irregular motion of minute bodies suspended in liquid and that in turn would help lay the foundation for Einstein's later work on the electron theory of metals.

Maric's return the following February was due, at least in part, to Einstein's entreaties to his "little runaway." Indeed, there seemed to be no stopping the momentum of their relationship. Although it wasn't until a year later that Mileva began referring to Albert in the familiar *du* form—a usage that connoted far more intimacy at the time than it does today—the couple probably already had become lovers. By September of that year, 1899, when they were both on vacation, Albert wrote to her from Milan, where he stayed with his family, that he wished he could get back sooner to "our place" in Zurich, "the nicest and coziest place I can think of," probably a reference to the Bachtold rooming house where Mileva lived at the time. Upon his return to Zurich Einstein planned to look for a new room nearby, but to avert "start[ing] any rumors" he would not move into the Bachtold residence itself.

Another profound change was beginning to take place, though, one that would foreshadow the growing difficulties Mileva would have in maintaining an ongoing commitment to her work. That summer, she stayed with her parents in the family home outside Novi Sad, in Vojvodina. It was hot and buggy. And because of an outbreak of scarlet fever and diphtheria, she never ventured into town. Instead, Mileva spent much of her vacation cramming for the first of two major examinations in October, which she would have to pass in order to get her diploma. Partly because of the ETH lectures she had missed during her semester in Heidelberg, she felt insecure about her command of the material in at least two classes. But her letters also hint at a more general sense of foreboding, which seems to go well beyond the natural anxiety of a student who is trying to make up coursework. Toward the end of the summer, she wrote to Einstein: "I'll probably be back in Zurich on the 25th [of September], but rather than looking forward to it, I'm returning with mixed feelings." Her apprehensions seem especially puzzling given the numerous educational hurdles she already had vaulted over.

Einstein, for his part, tried to allay her fears. "I can't think of any other assurances to offer you, other than to say that you shouldn't let this little exam bother you too much," he wrote to her. "That should be easy for you—especially with such harmless competition."

It turned out that, at least for the moment, Mileva's fears were ill founded. Despite missing one semester, she passed her first year's examinations, her highest grade 5.5 out of a possible 6. By contrast, Einstein's final grade in physics was 5.25, and he received his only 6 in electrical engineering.

Yet sometime between that first jaunty letter from Heidelberg and the summer of 1899, something had shaken Mileva's confidence. Had she returned from her leave of absence to a less than friendly reception from her classmates and teachers? Had she compared herself to Einstein's Promethean genius and found herself wanting? Or,

Women in Early Twentieth Century Science: Lise Meitner

Very few women, with the exception of Nobel laureates Marie Curie (1867–1934) and Maria Goeppert-Mayer (1906–1972), were able to succeed in the early twentieth century male-dominated world of science. Although almost 20,000 women were enrolled in American universities by the turn of the twentieth century, Zurich was the only European city where universities were officially open to women. Prejudice against women scientists was such that the Royal Society in London and the Academie des Sciences of Paris, both founded in the mid-seventeenth century, had no female representation through the ninteenth century. These institutions were not to admit women until 1945 and 1962, respectively.

A notable exception was Lise Meitner (1878-1968), born in Vienna one year before Einstein and the second woman to graduate with a doctorate from the University of Vienna in 1906. Berlin University, where she continued her studies with Max Planck was similarly biased—women could only attend courses if the lecturer agreed. Meitner was obliged to do her laboratory work in a convert-

Lise Meitner, about 1928—with cigarette in hand, every bit the cosmopolitan woman.

ed carpentry studio in the basement because women were not formally accepted at the university.

Meitner's early achievements included the isolation of the isotope protactinium-231 in 1917, followed by elucidation of nuclear isomerism and beta decay. In 1926, because of her achievements, Meitner finally became the first female full professor of physics at the University of Berlin.

Her 30-year collaboration with experimental chemist Otto Hahn led to the discovery of nuclear fission. Meitner introduced the term in her explanation of the process published in 1939 in the German scientific magazine *Naturwissenschaften*. It was this paper that prompted American scientists to urge Einstein to write to President Roosevelt warning of the possibility of a German atom bomb.

For reasons that have never been fully justified, it was Hahn alone who received the Nobel Prize in 1944 for the discovery of nuclear fission. (Meitner is therefore often called, "the greatest physicist never to win the Nobel Prize.") Finally in 1966 Meitner's contribution was officially recognized when she, Hahn and Fritz Strassmann were given the Enrico Fermi Award.

Although there are now more female scientists, bias against women still remains. As recently as 1997 one study (*Nature* 387, 341–343), reported that women applying to the Swedish Medical Research Council for post-doctoral fellowships needed, on objective criteria, to be 2.2 times better than men for acceptance
—PH

having fallen in love with Einstein, and believing in the incompatibility of marriage and a career, had she begun to doubt the strength of her professional convictions?

All three factors probably helped fuel Mileva's growing sense of insecurity. In addition, she was particularly plagued by the thundering disapproval of Einstein's parents. As far as Pauline and Hermann Einstein were concerned, Mileva had no redeeming qualities. They objected to her because she was not Jewish, because she was older than he, and because she was both physically lame and an intellectual. "Like you, she's a book," Pauline Einstein warned her son. "And you should have a wife." While Einstein actually seemed to enjoy goading his parents and flaunting his relationship, the elder Einsteins' disapproval weighed heavily on Maric.

Without ever having met the Einsteins, Mileva surmised Pauline's power. A woman who had held her family together over the years in the face of considerable obstacles, Pauline herself had made a match that was destined to disappoint her. Hermann was easygoing and passive. He had little business sense and was to run numerous commercial ventures into the ground. Pauline, who had come from a wealthy family, found it difficult to tolerate her husband's successive failures; not only did she have to suffer the social stigma of reduced circumstances but both she and members of her family invested, and lost, large sums in several of Hermann's business schemes. After each failure, Hermann uprooted his family; he moved them around Germany, Switzerland, and Italy in search of new opportunities. The relocations were often traumatic. Before the family moved to Italy, for example, in 1894, Albert and his sister, Maya, were forced to watch the contractor who had bought the comfortable villa where they had grown up, in Munich, demolish it, then replace it with an unsightly apartment complex. To make matters worse, Albert, who was only fifteen, was to be left behind to finish his gymnasium education. The plan, which reflected Pauline's driving and unsentimental ambitions for her son, ultimately backfired; for Albert dropped out of school and crossed the Alps to rejoin his family. Where Albert was concerned, Pauline seemed dedicated to nothing so much as ensuring that he would develop the backbone that she felt his father lacked.

Maric lived in fear of Pauline Einstein, whose tirades against her Einstein recounted in almost sadistic detail. On one occasion he described a "scene" in which Pauline "threw herself onto her bed, buried her head in the pillow, and wept like a child." After regaining her composure, Einstein said, his mother asserted: "You are ruining your future. No decent family will have her." Not surprisingly, despite Einstein's assurances that his mother would not confiscate his mail, sometimes Maric wrote to him via registered post to make sure her letters would not be intercepted.

The more insecure Maric became in her relationship with Einstein, the more she came to identify her interests with his, ultimately putting Einstein's welfare ahead of her own. Whether out of love or insight, she had become convinced of Einstein's potential long before his professors recognized it, and served as an indefatigable booster and helpmate to her beloved Johnnie. Einstein was, in fact, a mediocre student and one of the few ETH graduates who did not receive a position as an assistant upon graduation, which made it difficult for him to get a job. Maric's determination to see Einstein succeed was, at least in part, a result of her own scientific zeal. In championing Einstein's unrecognized genius, Maric also undoubtedly was influenced by the prevailing cultural norms, which put the needs of men ahead of those of women, and perhaps even by a desire to emulate Pauline Einstein's leonine protectiveness.

Since the beginning of their relationship, Maric and Einstein had studied together over vast quantities of freshly brewed coffee. Gradually, and long before they ever married, Maric also came to assume a more traditional, wifely role. Already during the 1899-1900 school year, "Johnnie" would "pound the books as usual while poor Dollie [would] cook." Einstein, who had seemed detached from the world even as a child and, in later years, would become legendary for his rumpled appearance and for eschewing socks and slippers as an unnecessary distraction, craved Maric's maternal attentions. Even at that early stage, he was already so engrossed in his work that he rarely ate unless Mileva fed him, a fact that may have accounted for his chronic stomach problems. Maric saw less of her friends, who came to feel that Einstein was taking advantage of her, so much so that in July of 1900, Mileva's friend Milana Bota, who had initially liked Einstein, wrote a letter to her own mother complaining that she "rarely sees [Mileva], because of the German [Einstein], whom she [Milana] has come to hate."

Eventually, Maric's protectiveness toward Einstein, who had difficulties with some of his professors, spilled over into their schoolwork and may have damaged her own academic relationships. For example, Einstein eventually developed serious problems with Heinrich Weber, the professor of physics at the ETH; by the time he graduated, there was so much bad blood, in fact, that Einstein was convinced that the reason he had trouble finding a job after graduation was that Weber was sabotaging his efforts. Maric also had her conflicts with Weber, who served as a thesis adviser to both young people. But her relationship with the professor was much better than Einstein's, who had been the only graduate from his class to he turned down for a post as an assistant. By contrast, Maric held a position in Weber's laboratory in 1901 and received an excellent evaluation for her work, according to Desanka Trbuhovic-Gjuric, Maric's biographer. Although the ETH has no record of an official appointment with a salary, it is possible that Maric worked in an unofficial capacity without pay. This scenario would not have been unusual given the fact that most women of her generation worked for little or no pay, if they were allowed into a laboratory at all.

What is clear is that Maric tried several times to intercede with Weber on Einstein's behalf—an effort in which she was unsuccessful and one that may have eroded her relationship with the professor. In the summer of 1901, Mileva wrote to her friend Helene Kaufler Savic: "I've already quarreled with Professor Weber two or three times, but now I am already used to such things. Because of him I have suffered a lot… We still do not know what destiny has determined for us [Albert and Mileva]."

Maric gradually assumed that her destiny was tied to Einstein's. As if in confirmation, Maric suffered her first academic setback in the summer of 1900, when she failed her final exams. Although the final grades for both Maric and Einstein fell below the 5 point average that was necessary to pass, Einstein's 4.9 got rounded up to 5 so that he just barely squeaked by. Maric's 4, on the other hand, meant that she failed outright; once again, she had gotten high marks in physics, but it was a miserable 2.5 average in the Theory of Functions that dragged down her final grade.

When it came to judging her academic performance, probably no one was harder on Maric than she was herself. In the spring of 1901, after Einstein had left Zurich to look for a job in Milan, Maric stayed behind at the ETH, feverishly studying to retake her exams. Just about the only respite she allowed herself was the weekends she spent with Einstein. During one exceptionally romantic trip to Lake Como, where the couple took a boat ride and visited the Villa Carlotta, with its magnificent gardens bursting with azaleas, camellias, and rhododendrons, Maric seemed to almost forget her academic travails. But in May she received another blow, in the form of a letter from home that was probably written by her father. Although the letter itself is lost, Maric's reaction to it suggests that it must have taken her to task both for her recent academic failure and for her relationship with Einstein, which her parents disapproved of almost as much as Einstein's did. It filled her with the sort of despair that would become all too characteristic of her state of mind in the coming years. "I received a letter from home today that has made me lose all desire, not only for having fun, but for life itself," Mileva wrote her beloved Johnnie. "I'm going to lock myself up and work hard, because it seems I can have nothing without being punished."

After years of academic triumph, Maric had lost her momentum. Mysteriously, in the summer of 1901, Maric failed the second round of examinations, and at about the same time, she also gave up the work on her dissertation.

Her friends knew that there was something terribly wrong, but Mileva could not bring herself to confide in anyone. For months, she stopped writing to her friend Helene Savic. "I just wasn't able to bring myself to do it in those awful times," she told Einstein. "I wrote a long letter once and poured my heart out to her, but then I tore it up." Indeed, almost no one but Einstein and the couple's parents knew that sometime in the early summer, just a few weeks before she was to sit for the second round of examinations, Mileva discovered that she was pregnant. The timing of the pregnancy could only underscore her foreboding that she was destined to be "punished."

If her own growing sense of self-doubt had begun to slow her progress then surely an unwanted pregnancy must have led her to doubt seriously the feasibility of her professional dreams. Although scientists and historians have pointed to Maric's failing her exams as proof of her intellectual inferiority, this seems hardly fair or logical, especially in light of the fact that Einstein's own performance at the ETH was relatively poor! Certainly, it is hard to imagine that the girl who had repeatedly distinguished herself as a top student in her native Serbia, who had passed the difficult ETH entry examination (which Einstein had failed the first time he took it), as well as the first round of university examinations, and who won at least some kudos from the hard-to-please Weber was simply not gifted enough to pass her final test. Robert Schulmann, an Einstein scholar, suggests that because the finals included an oral component, she might have been subject to the prejudice of her examiners. It is even more likely that Mileva's poor performance was due to anxiety brought on by both the discovery of her pregnancy and the actual physical discomforts of her condition, which in Mileva's case continued well past her first trimester. For Mileva, illness was often accompanied by severe emotional stress, a fact that later in her life would lead to a series of nervous breakdowns that eventually culminated in her death in 1948.

The pregnancy and her failure to complete her diploma ensured that Maric was now more dependent on Einstein than ever. In the intensely conservative world of the Swiss and Austro-Hungarian middle class, which was Maric's world, there could be no greater stigma than being an unwed mother. Even a stronger person— one who was not already overshadowed by the egotistical Einstein—would have had difficulty getting through this problematical period with her ambitions intact. But as things stood, Maric now came to rely on Einstein not only to save her good name by marrying her but also to provide, through his work, the only outlet left to her in science.

For Maric, the first years of the new century couldn't have seemed more bleak, and she clung to Einstein as to life itself. She suffered through the July exams alone while Einstein spent the month on holiday with his mother and sister. Afterward, she was forced to skulk home alone to her parents; for by then Einstein had taken up a temporary, poorly paid job as a substitute teacher in Schaffhausen, twenty miles north of Zurich. Mileva pleaded with him to help cushion her arrival at home: "Write my old man soon, sweetheart; because I'd like to leave on Saturday already, and they should have a letter before I arrive home.... Just write a short letter to my Papa... Will you send me the letter so I can see what you've written?"

While she had once frowned upon marriage as unsuitable for a woman of her ambitions, it now remained Mileva's last hope. Yet the prospect of being reunited with Einstein still seemed remote. He had not yet received a position that would allow him to support a wife and child. At the same time, Mileva knew that Pauline Einstein was campaigning against her as hard as ever, an obstacle she tried desperately to overcome. "You should also remember that your parents have a false impression of me and that its in my power to place myself in a more favorable light," she wrote to Einstein, outlining a strategy. "I think it will take much time and goodwill for reconciliation, but I'm sure it will happen... I've even come up with some techniques to set the thing in motion. For example, if I could ingratiate myself with an acquaintance of theirs whom they look up to a little, then they would already be defeated for the most part (or so I think). I also have a few other ideas on this."

Her best intentions came to nothing, however. Sometime during the fall, when Mileva was in her fifth or sixth month of pregnancy, Pauline Einstein, who would never reconcile herself to the relationship between Mileva and her son, sent a scathing letter to the Maric family, which impugned Mileva's character and heightened her fear that Einstein's family would somehow find a way to stand between them. Maric, whose only contact with Einstein since learning of her pregnancy had been through the mails now took a train to Switzerland; anxious that his parents not learn of her trip, she booked herself into a small hotel in Stein am Rhein, a few miles outside Schaffhausen.

That November, as she waited for Einstein to visit her in Stein am Rhein, Maric's position must have seemed nothing short of desperate to her. She was by now seven months pregnant, the child visibly rounding out her belly. Yet Einstein left her waiting several days before he scraped together the train fare to come and see her, a fact that should have given her an inkling of his limited capacity for commitment. Although he claimed to be short of money, it is much more likely that

Mileva and Relativity

In a poll taken in 2004, over 70 percent of those asked thought that Einstein's wife had contributed to his research. The poll followed a 2004 U.S. documentary on Einstein's first wife, Mileva Maric, that was perhaps prejudiced toward this view. The bias against women scientists in the early twentieth century would tend to support such a claim, but what is the evidence?

One dispute centers upon the signature on three of the 1905 papers. D. Trbuhovic, author of the 1969 Serbo-Croatian biography of Maric, "In the Shadow of Albert Einstein," writes in support of the view that Maric was equally responsible for some of Einstein's ideas: "[Abraham] Joffe notes that the three epochal articles by Einstein from 1905 were signed 'Einstein-Maric'." Indeed, Russian physicist Abraham Joffe did attest to this in a 1955 memorial for Einstein, but rather than assuming Maric a co-author put the joint signature down to Swiss custom: "In 1905, there appeared three articles in the *Annalen der Physik*…. the author of these articles—unknown at the time—was the clerk at the Patent Bureau Einstein-Marity (Marity—the maiden name of his wife, which by Swiss custom is added to the husband's family name)."

The question of Mileva's role was reopened in 1986, when letters between Mileva and Albert surfaced as part of the estate of Einstein's elder son Hans Albert. In several letters to Mileva, Einstein definitely refers to "our work." In particular, in March 1901 he wrote, "How happy and proud I shall be when together we shall conclude victoriously our work on relative motion." But opponents to the idea of Maric's intellectual collaboration with Einstein point out that she never claimed any credit for the work herself. In letters to an intimate friend in 1906 she complains, "the papers he has written are already piling quite high."

The truth about Mileva's contribution may never be known but comparisons are often made between the Einsteins, the Curies and Paul and Tatiana Ehrenfest. All three wives were

Young Mileva Maric in 1899—a brilliant and promising student, in a field not very hospitable to women in the early 1900s.

Slavic with university training, but Marie Curie and Tatiana Ehrenfest were given support and credit for their intellectual achievements by their husbands. Einstein, by contrast, did not acknowledge any assistance Mileva may have given him, and at the same time thanked only his friend Michele Angelo Besso in his 1905 paper on special relativity.   — PH

he was less than eager to see her, for Einstein's salary, which amounted to 150 francs plus room and board should have been able to cover transportation.

And so Mileva waited, anxious over Albert's repeated delays, admonishing him to keep her visit a secret even from his sister, Maya—the only member of the Einstein family who supported their relationship—for fear that Albert's parents would learn of it. "I'm afraid that something could happen again, as it always has in the past.… Just don't give her my address sweetheart, I'm terribly worried about it. . . Don't write your parents anything about me. No more fights; I dread the mere thought of it."

Throughout what was becoming known as the "Dollie affair," Maya, who was two years younger than her brother, acted as the family mediator. During Albert and Mileva's courtship, it was Maya who tried—usually unsuccessfully—to ease the tensions between her mother and her brother over the impending marriage. And when Pauline attacked Mileva, Maya was quick to come to her sister-in-law's defense. However, in the end, Einstein's ties to his sister turned out to be far closer than his relationships with either of his wives. By December, the young couple's situation had improved somewhat. Although Maric, who had by then returned home to Serbia, was by this time bedridden with complications of her pregnancy, and Einstein, as usual, had forgotten to remember her birthday, there was good news on the horizon. Einstein had learned that he would get a job in the Swiss Patent Office, a position that would give him a steady income and finally allow him to marry. And sometime at the end of January, following a difficult labor, Mileva gave birth to a baby girl, whom she named Lieserl (Little Lisa). Although Einstein was still in Switzerland, it now seemed to be only a matter of time before the couple could be married. "In two months' time we could find our lives brilliantly changed for the better, and the struggle would be over," wrote Einstein to Maric a month before she was to give birth. "I'm dizzy with joy when I think about it… Together we'd surely be the happiest people on earth."

Yet it took over a year before the promised marriage came off. And when the couple were finally married, on January 6, 1903, the wedding took place under a cloud. In June of 1902, Einstein finally went to work in the Swiss Patent Office. In the fall, Hermann Einstein, who suffered from heart disease, became gravely ill and died, but not before finally granting permission for his son to marry Mileva. In the weeks before the wedding, Einstein busied himself with sorting out the estate of his father, whose business affairs were, as always, in a state of disarray. Now he also had to worry about supporting his mother, who had never made peace with her son's marriage and who had been left with nothing but debts.

In the meantime, one serious "problem" that Einstein couldn't bring himself to focus on was what to do about his daughter. Since the couple was unmarried, and both Pauline Einstein and the Marics continued to oppose their relationship, there had been talk about putting the child up for adoption, a solution that neither Mileva nor Albert seemed to welcome. Yet it is unlikely that he ever even saw his daughter—there is no record of Einstein having visited Serbia in the months after

Lieserl's birth or of Maric traveling with her outside of Serbia. And Einstein appears to have left the fate of his daughter up to Maric and her parents. "Ask your Papa," he suggested in a letter written in December of 1901, shortly before Lieserl's birth. "He's an experienced man, and knows the world better than your overworked, impractical Johnnie."

By the time Maric and Einstein were married, shortly after the New Year, the Lieserl problem had still not been resolved. Far from enjoying the anticipation of her long-awaited wedding, Maric seemed to her friends deeply preoccupied in the months before her marriage. If anyone asked her what was wrong, all Mileva would say was that the problem was too "intensely personal." What she didn't need to say, her friends knew, was that Einstein was at the center of whatever it was that bothered her; some even surmised that the romance might be over. Few, if any, of their friends knew the extent of her troubles. Isolated in Serbia, Mileva appears to have kept the birth of her child a secret. Nor would any of her friends have known that the child was stricken by scarlet fever soon after the wedding and might have suffered lasting side effects.

What finally became of Lieserl may never he known. No record of her birth or death survives. Lieserl may have died as a consequence of her illness. But if she was put up for adoption, it was probably due in large part to Einstein's reluctance to compromise his work in the interests of caring for a small and possibly sickly child. Or Einstein may have feared that the stigma of an illegitimate child, especially one born to a Slav, would hurt his career in the decidedly conservative world of the Swiss civil service. In the late summer of 1903, a bittersweet exchange between Mileva and her new husband marks the last known written reference to Lieserl and hints at Mileva's anguish.

Maric was in Budapest, probably at the beginning of her second pregnancy, suffering the first bouts of morning sickness, when she wrote a short, plaintive note to Einstein: "Dear Johnnie... It's going quickly, but badly. I'm not feeling well at all... Write me soon, okay? Your poor Dollie." Einstein replied: "I'm not the least bit angry that poor Dollie is hatching a new chick. In fact, I'm happy about it and had already given some thought to whether I shouldn't see to it that you get a new Lieserl.... I'm very sorry about what has befallen Lieserl. It's so easy to suffer lasting effects from scarlet fever. If only this will pass. As what is the child registered? We must take precautions that problems don't arise for her later."

To her academic failure, the trials of her courtship with Einstein, and the shame of an illegitimate pregnancy, Maric could now add the loss of her first child. If

Maric became "gloomy" and "distrustful," as Einstein's biographers report, who could have blamed her? Yet, at least during the next few years, Maric seems to have handled her grief primarily by pouring all her dreams, love, and ambition into her husband. As long as she could maintain a scientific dialogue with Einstein and feel that she was helping to develop a new understanding of the universe and the scientific interests they had shared since the earliest days of their relationship, then perhaps all her suffering would have been worth it.

Einstein continued to promise Maric a life of shared work. "We'll be students (horrible dictu) as long as we live and won't give a damn about the world.... When you're my dear little wife we'll diligently work on science together so we don't become old philistines, right?" he had assured her just a month before Lieserl's birth.

For a time Einstein kept his promise. During the early years of their marriage, which was also the most productive period of Einstein's career, Einstein credited Maric with "solv[ing] all of his mathematical problems," a fact that is confirmed by their son Hans Albert, as well as by at least one student who lived for a time with the Einsteins.

Exactly how much she was to contribute to Einstein's work has become the subject of considerable controversy. Much of the debate revolves around fragmentary evidence suggesting that the original version of Einstein's three most famous articles, on the photoelectric effect, on Brownian motion, and on the theory of relativity, were signed Einstein-Marity, the latter name being a Hungarianized version of Maric. Although the original manuscripts have been lost, Abraham F. Joffe, a member of the Soviet Academy of Sciences, claimed that he saw the original papers when he was an assistant to Wilhelm Röntgen, who belonged to the editorial board of *Annalen der Physik*, which published the articles. (An article in a 1955 Soviet journal of physics quotes Joffe, now deceased, as ascribing the 1905 papers to "Einstein-Marity.")[Editor's note: See sidebar on page 52]

Although there is no evidence to suggest that Maric came up with any of the original insights for the three most famous papers attributed to Einstein, she probably proofread the articles and performed the mathematical calculations for some of them. Svetozar Varicak, a student who lived with the Einsteins for several months in about 1910, remembered how Maric, after a day of cleaning, cooking, and caring for the children, would then busy herself with Einstein's mathematical calculations, often working late into the night. Varicak said he remembered feeling "so sorry for Mileva" that he sometimes helped her with the housework. At around this time, Maric also joined Einstein and Friedrich Adler for discussions in the quiet attic of the Zurich apartment building where the Einsteins and the Adlers both lived; on those afternoons, Katja Adler, Frederick's wife, watched the children.

During the early years of her marriage, Maric also spoke frequently to her family and friends about collaborating with her husband. She told Milana Bota, for example, about the work she was doing with Einstein. And in 1905, just after the completion of "On the Electrodynamics of Moving Bodies," the initial paper on special relativity, while the Einsteins were on vacation in Serbia, Maric boasted to her father and Desana Tapavica Bala, who was married to the mayor of Novi Sad: "Just before we left for Novi Sad, we finished some important work that will make my husband world famous."

To whatever extent Maric managed to pursue science, at least during the early years, she had to all intents and purposes completely subsumed her ambitions and interests to those of her husband. Einstein had begun teaching at the University of Zurich, and she regularly attended his lectures. And even though she would soon be pregnant with her second son, Eduard, and had no household help, Maric took in boarders as a way to supplement the family's income and to free Einstein from financial worries.

Yet her efforts were not to be appreciated. Einstein's reputation had skyrocketed since 1905. And the more successful he became, the more he neglected his marriage. "Albert has devoted himself completely to physics and it seems to me that he has little time if any for the family," Mileva confided to her friend Helene Savic. Einstein spent more and more time away from home, in the company of scientists—gatherings to which Maric was not invited. Always absentminded, Einstein also began to show signs of the callousness and insensitivity that would become much more pronounced as he grew older. In the fall of 1909, Einstein took up a correspondence with Anna Meyer-Schmid, an old girlfriend; Maric was so incensed when she intercepted one of Meyer-Schmid's letters that she wrote a letter to the woman's husband, complaining of the "inappropriate" correspondence.

Relations between husband and wife deteriorated after the family moved to Prague, where Einstein had been offered a position at the university, in about 1911. Neither Maric nor Einstein liked Austria-Hungary's Bohemian capital, a city marked by stark disparities between great wealth and grinding poverty, as well as by growing nationalist hostilities. But the move was hardest on Maric, who was particularly sensitive to the tensions between the Czechs, with whom she identified, and the city's nationalistic German elite. Prague was so polarized, in fact, that there was one university for Germans, where Einstein taught, and one for Czechs (and incidentally, there was anti-Semitism everywhere).

What's more, with the birth of her children, and the burdens of running a household, Maric had become the "philistine" Hausfrau that Einstein had always shunned and that she herself had vowed never to become. She spent most of her time at home with her two small children, especially Eduard, who was an exceptionally sickly infant. Trapped in a city she disliked and a life she scorned, Maric became lonely and depressed.

As for Einstein, he had little sympathy for the difficulties his wife was experiencing. He complained about Maric's moodiness. And he spent more time away from home than ever. While they were living in Prague, Einstein traveled almost constantly. "It must have been very interesting in Karlsruhe," Maric wrote her husband in October. "I would like to have been there and listened a little, and seen all these fine people… It is such an eternity since we saw each other, I wonder if you will recognize me." Yet, come November, Einstein took off once again, this time to the Solvay Conference in Brussels, a convocation of the world's leading physicists.

The turning point in the Einsteins' marriage came in 1912. At first things seemed to be going better after Einstein accepted a new job at the ETH, which took the family back to Zurich. Day-to-day life in their beloved city, where the Einsteins had many friends, became more pleasant as they resumed their musical evenings with friends like Adolf Hurwitz and their excursions into the Alps.

But the fragile rapprochement would not last long. In the spring of 1912, during a trip to Berlin, Einstein had become reacquainted with his cousin Elsa Lowenthal and declared his love for her—for "I must love somebody," he wrote her in April, in a letter that began a secret correspondence that was to last nearly two years. As if this betrayal were not enough, Einstein maligned his wife in the letters to his lover, calling Maric "the sourest sourpot that there has ever been" and describing his relationship with her as that of a man and his "employee."

Thus, for Maric, it must have been particularly poignant when, in 1913, she met Marie Curie during a trip to Paris, one of the last journeys she would make with her husband. Curie and the Einsteins hit it off so well that they organized a family vacation together (the widowed Curie had two daughters) later that year in the Alps. Unlike Einstein's male colleagues, Marie Curie is likely to have treated Maric with the sort of respect that the young Serbian housewife rarely received anymore. With her two Nobel Prizes, Curie must have been an inspiration, but also a painful reminder of Maric's own failed dreams and strained marriage.

Yet Maric wasn't the only woman in her circle whose ambitions had been thwarted by marriage and whose marriage, ultimately, would end in tragedy. In

Prague, one of the only people with whom Maric became friendly was a Russian physicist, Tatjana Afansijeva, who collaborated with her husband Paul Ehrenfest for years, until the couple's estrangement eventually led to Ehrenfest's suicide in 1933. Maric would also have felt a kinship for Clara Immerwahr, the first woman to receive a doctorate in chemistry from the University of Breslau, whose research on behalf of her husband, Fritz Haber, an expert on gases, went largely unrewarded. Although no correspondence survives between the two wives, Maric and Immerwahr knew each other well. When the Habers' marriage finally collapsed, it was Immerwahr who committed suicide. Haber, a friend of Einstein's who, as director of the Kaiser Wilhelm Institute of Physical Chemistry and Electrochemistry, was instrumental in bringing Einstein to Berlin, always blamed himself for his wife's death. Maric also had much in common with Katja Adler, who had given up her studies in physics to marry Friedrich.

Though Maric loved her husband and was not yet prepared to give up on their marriage, she could not live with him on his terms any longer. Just eighteen months after settling back in Zurich, Einstein had accepted a position at the Prussian Academy of Science in Berlin, over Maric's objections. Both of them shared an abiding dislike of Germany, Einstein so much so that he had taken the extraordinary step of renouncing his German citizenship when he was a teenager, an act that left him stateless for several years, until he became a Swiss citizen in 1901. Still, Einstein was drawn to the august Prussian academy, and, initially, Maric agreed to accompany him in April of 1914.

But Maric detested Berlin; she liked neither the city nor its inhabitants, and being there was made all the worse by the fact that Einstein's relatives, including his fearsome mother, lived there. Even at their lowest points, Einstein blamed the bad relationship between his wife and Pauline Einstein on his mother's "perfidious... hatred" of Maric. When Maric traveled to Berlin to look for an apartment a few months before they were to move, Einstein suggested that she avoid his family altogether. Instead, Maric stayed with Immerwahr and Haber. By the time Paul Ehrenfest visited the Einsteins in Berlin, a few months later, he found that both Maric and Hans Albert were miserable in their new home. Maric remained as isolated as ever, while her son chafed at the rote learning and discipline he encountered in school. In July, Maric decided to return to Zurich with her sons.

Despite their problems, Maric assumed that the separation was temporary. Bidding farewell to his family at the train station, Einstein had wept bitterly. Maric was convinced that Einstein would eventually leave the hated Berlin and rejoin them. For months, Maric and her sons lived in a rooming house and waited anxiously for his return; it was a period that Hans Albert remembered as "the worst time." When even the outbreak of World War I failed to bring Einstein home, Maric was forced to face the hard truth. "I don't expect to ask you for a divorce," Einstein wrote to his wife in July 1914. "All I ask you is that you send me news of my beloved boys every two weeks." Although Maric sought to foster a long-distance relationship between father and sons, the boys, especially Hans Albert, resented their father's absence. And Einstein soon came to blame Maric for the rift that developed with his sons; he accused her of intercepting his messages and threatened that he would sever all contacts with the family if Maric so much as commented to them on one of his letters.

The break with Einstein represented the end of Maric's hopes and dreams—not because she had chosen to fulfill herself, as have so many bourgeois women over the years, through the career of her husband, but because she had left herself few alternatives. By marrying Einstein, Maric had attached herself not only to one of the greatest and most forceful minds of the twentieth century, but also to a potent combination of gender, motherhood, and isolation. While a stronger personality might have overcome even these obstacles, Maric had a pessimistic nature that, ultimately, tipped the balance and ensured her defeat. Her ambition—which once had so resembled Einstein's—could have allowed her to find some satisfaction in his success, if only he had continued to allow her to play even a small, unobtrusive role in it.

As Einstein's own success and intellectual thirst grew, his ability to embrace the needs of even those closest to him virtually evaporated. Einstein himself acknowledged that he had become obsessively absorbed with his ideas, to the exclusion of almost everything else. This man who had no interest in food or clothes, who had to be prodded to brush his unruly shock of hair, who eschewed socks as an unnecessary hindrance, stripped away emotional ties just as ruthlessly. "I'm not a family man," Einstein confessed when he was already in Berlin, separated from his first wife and two sons and remarried to a woman who placed no intellectual demands on him. "I want to know how God created this world... I want to know his thoughts, the rest are details." Max Born, a friend and rival, both as a physicist and as a humanist, would say of Einstein: "For all his kindness and sociability, and love of humanity, he was nevertheless totally detached from his environment and the human beings included in it."

Unlike Pierre Curie, who amplified his wife's already strong sense of self-confidence

Albert and Elsa aboard the SS Rotterdam en route to the United States.

and professional drive by ceaselessly reinforcing her ambitions and who saw his own happiness in their shared success, Einstein lacked the generosity of spirit—and the vision to keep even a small space cleared for Maric on the vast canvas of his scientific universe. The final severing of intellectual ties between husband and wife probably occurred around 1913, when Einstein began collaborating with Marcel Grossmann on the general theory of relativity, the collaboration is particularly noteworthy since, according to Einstein's biographer Peter Michelmore, Maric was "as good at mathematics as Marcel [Grossmann]."

Although well educated, Maric had few prospects of employment since she had never finished her degree or held a full-time job. Like many single mothers over the ages, she also found herself in deep financial trouble. With the start of World War I and the disruption in the mails, Maric received Einstein's checks only sporadically and often was unable to pay her bills. As the war progressed, the devaluation of the German mark only worsened her situation. Maric began giving music lessons and tutoring in mathematics. And she was often reduced to borrowing money from friends.

The progress of the war brought only more grief. Maric's brother, Milos, was taken prisoner on the Russian front and disappeared in the Soviet Union; although he became a respected professor in Moscow, the family neither saw him again nor, for years, was aware of his fate. In 1915, Maric would have learned that Einstein had rented an apartment around the corner from his cousin Elsa. The arrangement must have been a painful reminder of her own cozy living arrangements with "Johnnie" during their student days in Zurich. Worse still, the relationship with Elsa inevitably led Einstein to demand a divorce and to wage an escalating battle for the custody of their children.

It was the fear of losing her children, more than anything else, that probably led

Maric to a physical and mental breakdown in the summer of 1916. Although Einstein later acknowledged that he had the greatest confidence in Mileva's role as a mother and eventually would thank her for trying to bolster the relationship between father and sons, in the years leading up to their divorce he harassed her relentlessly for turning the boys against him. And although he rarely showed any interest in Eduard, whose emotional fragility he considered a repulsive sign of weakness, Einstein schemed repeatedly to move his older son, Hans Albert, to Berlin. When the pressure finally led to Maric's collapse, Einstein who happened to be in Zurich at the time, refused to visit his wife, telling his friend Michelangelo Besso that he was convinced she was merely feigning illness as a way to stall their divorce. Einstein finally relented when she was hospitalized.

What exactly was wrong with Maric has never been determined. Some friends assumed she had suffered a series of strokes. Einstein himself believed—or hoped—that she was suffering from tuberculous meningitis, a potentially fatal infection of the brain membranes. Yet over the next several months, Maric gradually got better. She would never fully recover, however, and was to be hospitalized repeatedly during subsequent years.

Einstein enthusiasts might be tempted to attribute the role Einstein played in the dissolution of his first marriage to his "monomaniacal" devotion to his work. But there is considerable evidence that Einstein's growing antipathy for Maric was grounded in malice and misogyny as much as tunnel vision. Certainly, Einstein displayed a particular dislike for intelligent women. In Berlin, he told Esther Salman, a female physics student, that "very few women are creative." Salman objected, noting that Marie Curie was surely an exception. Einstein, who perhaps failed to recognize his own brand of single-focused genius when encountering it in a woman, dismissed the female Nobel laureate as having "the soul of a herring." By the end of his life, his sister, Maya, is said to have been one of the only women Einstein treated with consideration and respect.

Yet it was a tribute to Maric's enduring love and commitment, not only to her sons but also to Einstein, that the couple eventually achieved a reconciliation of sorts. By 1918, a year before their divorce, Einstein would stay with Maric and the children whenever he visited Zurich, even though he was already living with Elsa. And although Einstein's children, particularly Hans Albert, never overcame the resentment they felt for him, it was not for lack of effort on the part of Maric, who always encouraged the boys to be proud of their father.

Most biographers have treated Maric, at best, as a footnote barely worth mentioning in the great man's life or, at worst, as a "gloomy" and "laconic" shadow that clouded his existence. This was due, first, to Einstein's unwillingness to discuss his personal life and, later, to a conscious effort, both by Einstein's executors and by scientists, to protect Einstein's monolithic reputation from even a hint of human frailty, let alone personal scandal. Indeed, in 1958, the executors, Otto Nathan and Helen Dukas, who had been Einstein's secretary, and his stepdaughter Margot blocked the publication of a book by Hans Albert and Frieda Einstein, which was based on Einstein's lifelong correspondence with Maric and his sons; the book, which undoubtedly would have revealed the mortal man behind the great genius, has still not been published. Yet during Maric's lifetime, long after the couple's divorce, Maric continued to command the respect and sympathy of a surprisingly wide array of mutual friends. Scientists like Michelangelo Besso and Fritz Haber, whose sympathies might otherwise have been expected to go to their much-respected colleague Einstein and who served as go-betweens for the couple during their divorce often defended Maric's motives, her lifestyle and her role as a mother. Their neighbors the Hurwitzs also remained close to Maric until the time of her death.

Though Maric achieved some stability for herself and her children after her divorce, she would soon face another domestic tragedy. In 1920, in the dead of winter, Maric was called home to Novi Sad by a family emergency. Her sister, Zorka, who had stayed with her in Zurich for a few years during Maric's separation, had begun to go insane. Maric arrived home to find her sister paranoid and hostile; the only creatures to whom Zorka showed any affection were cats.

Eduard was only ten years old at the time. But within a few years Maric would come to recognize in him the signs of mental instability that she had seen in her sister that winter. As a small child, Tete, as he was called by his parents, had begun to suffer severe earaches that made his whole head throb. Gradually, the pain took on a more sinister form, and it seemed as though the sensitive, high-strung Tete was being tormented by ear-splitting noises inside his head. As a teenager, the thin, handsome boy grew fat and deeply troubled. He suffered mood swings that swept him from lethargy to nearly suicidal hysteria, as he ranted and raved in an effort to still the voices in his head. At one time, Tete, who read voraciously and developed an uncanny ear for music was widely believed to have inherited his father's genius; he had even enrolled at the University of Zurich intending to study psychiatry. Yet before long, his intellectual aspirations devolved into a passion for pornography and the biographies of great men. The youth who had once adored the great man who was his father and whose visits to Berlin as a teenager were often followed by desperate, passionate letters pleading for fatherly love and reassurance, now turned on Einstein, berating him in a deluge of hate mail.

Einstein, for his part, came less and less frequently to Zurich. He vacillated between feeling guilty for having abandoned his children, Tete in particular, and blaming Maric and her family's history of mental illness for his younger son's fate. As a man of old-world upbringing, he probably couldn't help but see his son's vulnerability as a sign of weakness, an almost inexcusable character flaw. "Who knows if it would not have been better if he had left the world before he had really known this life," observed Tete's father even before the appearance of any overt signs of mental illness. When Einstein emigrated to the United States in 1933, fleeing the Nazi tide that swept across Germany, the goodbye he said to Tete and Maric was final; he would never see either of them again.

Separated now by an ocean and eventually, by another world war, Maric was left to cope with her son's mental collapse. Maric's mother died in 1935, Zorka three years later. Under their divorce settlement, Einstein had given Maric the thirty thousand kronor that he had received for the Nobel Prize. But the costs of Tete's medical care remained an enormous burden to her during these Depression years. His fits of insanity forced her periodically, to send Tete to the Burgholzli, a mental institution in Zurich. At other times, she took him on trips into the mountains, in the hope of calming his nerves. Eventually, she was forced to sell two of the three real estate properties she had purchased with the proceeds of the Nobel Prize. And in order not to lose the building on Huttenstrasse where she and Tete lived, Maric finally agreed to transfer ownership of the property to Einstein, although she was allowed to keep the rental income from apartments in the building. But in 1947, Einstein sold the house without consulting Maric. Under the terms of the sale, Maric was permitted to remain in her apartment for the rest of her life. Indeed, because she retained Einstein's power of attorney in Switzerland, she was able to keep the proceeds of the sale. But the episode represented a final devastating act of betrayal by the one man Maric had ever loved, the man to whom she had entrusted, so completely her life and happiness. To make matters worse, in 1938, Hans Albert, his wife, Frieda, and their son, Klaus, had emigrated to the United States. Shortly after their arrival, Klaus, Maric's first grandchild, who was only about six years old, suddenly became ill and died.

In May of 1948, Tete was home with Maric when, toward nightfall, the demons suddenly returned. As he searched frantically for some imaginary missing object, he began destroying everything within reach. He hurtled books dishes, mementos, at the walls and finally collapsed on the floor, weeping inconsolably. In the midst of Tete's fit, Maric suffered an emotional breakdown and was rushed to the hospital.

The few friends who came to visit her in the hospital were struck by the severity of Maric's latest breakdown. Utterly confused, she began to complain about Einstein, the hospital, the nurses. She rang her call button so relentlessly the staff was forced to disconnect it. Over and over she begged to be sent to the Burgholzli, perhaps because she recognized the extent of her mental collapse, perhaps because Tete was there now. Finally, on August 4, 1948, Mileva Maric died alone in her hospital room, physically, mentally, and financially broken.

In his later years, Einstein liked to boast that he had survived not only Nazism but also two wives. For in the end, Elsa's lot had not been much happier than Mileva's. Due in large part to Einstein's interest in other women, his second marriage showed signs of strain even before the couple emigrated to the United States, according to Konrad Wachsman, the architect who designed their summer house in Caputh, Germany. "Women were drawn to the world-famous professor like iron filings to a magnet," write Roger Highfield and Paul Carter, two Einstein biographers. According to the accounts of numerous acquaintances, Einstein did not discourage women's advances. He often escorted female companions to the opera and took them on sailing trips near his summer house. These excursions frequently resulted in jealous outbursts from Elsa, who, however, ultimately accepted her husband's marital detours.

Nor was Einstein particularly sensitive to his wife's most basic emotional needs. In 1934, soon after the Einsteins moved to Princeton, Elsa traveled to Paris and found that Ilse, her eldest daughter from a previous marriage, was dying. She returned home with Ilse's ashes and was soon herself bedridden with severe heart and kidney problems. Throughout the long months of Elsa's illness, which culminated in her death in 1936, Einstein stuck "frightfully to his [scientific] problems." In a letter to a friend, Elsa confided: "I have never before seen him so engrossed in his work." After her death, Einstein corresponded with Max Born about his work at Princeton and, with shocking indifference, about the loss of his wife: "I have settled down splendidly here [at Princeton]," Einstein wrote. "I hibernate like a bear in its cave, and really feel more at home than ever before in all my varied existence. This bearishness has been accentuated further by the death of my mate, who was more attached to human beings than I."

Nor was Einstein moved by the plight of his son Tete, who remained at the Burgholzli for seventeen years after Maric's death. In his lucid moments, Tete spoke of his brother, Hans Albert, who visited Tete occasionally during his trips to Europe after the war, and of his father, who never came to see him and rarely wrote. But after Maric's death, Tete could never bring himself to speak of his mother again.

When Tete died finally, in October of 1965, his death certificate identified him as the son of the late Albert Einstein, who hadn't seen his son in over three decades. Maric, who lived for years under the shadow of Eduard's insanity and died, as she had lived, struggling to ease his pain, received no mention at all.

The only one whose memory of Mileva Maric remained fresh and loving and lucid was her son Hans Albert. Hans Albert Einstein's own life, in marked contrast to that of his father was a testimony to his regard for strong and intelligent women. Hans Albert blamed Einstein for his parents' divorce and retained a certain bitterness toward his father throughout his life. The eminent scientist objected when his son became a hydraulic engineer rather than a theoretician. And the tensions between father and son were exacerbated when Einstein opposed his son's marriage. Ironically, Einstein's "explosions" over Frieda, the young woman Hans Albert begin to court while he was a student at the ETH, mirrored Pauline Einstein's attacks on Mileva's age and health. In fact, while Frieda was nine years older than Hans Albert, short and plain in appearance, she was also said to be highly intelligent. Soon after Frieda's death, in 1958, Hans Albert married Elizabeth Roboz, a neurochemist who remained active in her profession.

Mileva Maric belonged to the first generation of women who tried to make a place for themselves in the scientific community. Neither society nor the professions made it likely that even the most obsessive female scientist would succeed at her chosen vocation. The combination of Maric's unusual marriage, her peripatetic interests, and her tendency toward defeatism put success even further beyond her reach. It wasn't until the 1970s, when it became more common for women to seek a higher education in the sciences and when nepotism rules began to crumble in the United States under the weight of equal-opportunity lawsuits, that women got their first real chance to work in science. It was a world that the young Mileva Maric might have fantasized about during her student days in Serbia and Zurich. It was a dream, however, that all but vanished by the time she reached womanhood in the shadow of Albert Einstein.

Conrad Habicht, Maurice Solovine and Albert Einstein—the self-styled Olympia Academy—in about 1903. Solovine had answered an advertisement Einstein had placed in the newspaper seeking students to tutor in mathematics and physics—and entered the pages of history as a result.

Albert Einstein as a Philosopher of Science

by Don A. Howard

Nowadays, explicit engagement with the philosophy of science plays almost no role in the training of physicists or in physics research. What little the student learns about philosophical issues is typically learned casually, by a kind of intellectual osmosis. One picks up ideas and opinions in the lecture hall, in the laboratory, and in collaboration with one's supervisor. Careful reflection on philosophical ideas is rare. Even rarer is systematic instruction. Worse still, publicly indulging an interest in philosophy of science is often treated as a social blunder. To be fair, more than a few physicists do think philosophically. Still, explicitly philosophical approaches to physics are the exception. Things were not always so.

"Independence of judgment"

In December 1944 Robert A. Thornton had a new job teaching physics at the University of Puerto Rico. He was fresh from the University of Minnesota, where he had written his PhD thesis on "Measurement, Concept Formation, and Principles of Simplicity: A Study in the Logic and Methodology of Physics" under Herbert Feigl, a noted philosopher of science. Wanting to incorporate the philosophy of science into his teaching of introductory physics, Thornton wrote to Albert Einstein for help in persuading his colleagues to accept that innovation. Einstein replied:

> I fully agree with you about the significance and educational value of methodology as well as history and philosophy of science. So many people today—and even professional scientists—seem to me like someone who has seen thousands of trees but has never seen a forest. A knowledge of the historic and philosophical background gives that kind of independence from prejudices

Reprinted with permission from *Physics Today*, December 2005 pp. 34-40, Copyright © 2005, American Institute of Physics of his generation from which most scientists are suffering. This independence created by philosophical insight is—in my opinion—the mark of distinction between a mere artisan or specialist and a real seeker after truth.[1]

Einstein was not just being polite; he really meant this. He had been saying the same thing for nearly 30 years. He knew from his experience at the forefront of the revolutions in early 20th-century physics that having cultivated a philosophical habit of mind had made him a better physicist.

A few years after his letter to Thornton, Einstein wrote in a contribution to *Albert Einstein: Philosopher-Scientist*, "The reciprocal relationship of epistemology and science is of noteworthy kind. They are dependent upon each other. Epistemology without contact with science becomes an empty scheme. Science without epistemology is—insofar as it is thinkable at all—primitive and muddled." [2]

In a 1936 article entitled "Physics and Reality," he explained why the physicist cannot simply defer to the philosopher but must be a philosopher himself:

> It has often been said, and certainly not without justification, that the man of science is a poor philosopher. Why then should it not be the right thing for the physicist to let the philosopher do the philosophizing? Such might indeed be the right thing to do at a time when the physicist believes he has at his disposal a rigid system of fundamental concepts and fundamental laws which are so well established that waves of doubt can't reach them; but it cannot be right at a time when the very foundations of physics itself have become problematic as they are now. At a time like the present, when experience forces us to seek a newer and more solid foundation, the physicist cannot simply surrender to the philosopher the critical contemplation of theoretical foundations; for he himself knows best and feels more surely where the shoe pinches. In looking for a new foundation, he must try to make clear in his own mind just how far the concepts which he uses are justified, and are necessities.[3]

Already in 1916, just after completing his general theory of relativity, Einstein had discussed philosophy's relation to physics in an obituary for the physicist and philosopher Ernst Mach:

> How does it happen that a properly endowed natural scientist comes to concern himself with epistemology? Is there not some more valuable work to be done in his specialty? That's what I hear many of my colleagues ask, and I sense it from many more. But I cannot share this sentiment. When I think

about the ablest students whom I have encountered in my teaching—that is, those who distinguish themselves by their independence of judgment and not just their quick-wittedness—I can affirm that they had a vigorous interest in epistemology. They happily began discussions about the goals and methods of science, and they showed unequivocally, through tenacious defense of their views, that the subject seemed important to them.[4]

Notice that philosophy's benefit to physics is not some specific bit of philosophical doctrine such as the antimetaphysical empiricism championed by Mach. It is, instead, "independence of judgment." The philosophical habit of mind, Einstein argued, encourages a critical attitude toward received ideas:

> Concepts that have proven useful in ordering things easily achieve such authority over us that we forget their earthly origins and accept them as unalterable givens. Thus they come to be stamped as "necessities of thought," "a priori givens," etc. The path of scientific progress is often made impassable for a long time by such errors. Therefore it is by no means an idle game if we become practiced in analyzing long-held commonplace concepts and showing the circumstances on which their justification and usefulness depend, and how they have grown up, individually, out of the givens of experience. Thus their excessive authority will be broken. They will be removed if they cannot be properly legitimated, corrected if their correlation with given things be far too superfluous, or replaced if a new system can be established that we prefer for whatever reason.[4]

Here Einstein is describing the kind of historical–critical conceptual analysis for which Mach was famous. This mode of analysis is at the heart of the arguments for the special and general theories of relativity, and of many of Einstein's other revolutionary works.[5] How did he become this kind of philosophical physicist? Reading Mach was one way, but not the only way.

Early acquaintance with philosophy

Einstein was typical of his generation of physicists in the seriousness and extent of his early and lasting engagement with philosophy. By the age of 16, he had already read all three of Immanuel Kant's major works, the *Critique of Pure Reason*, the *Critique of Practical Reason*, and the *Critique of Judgment*.[6] Einstein was to read Kant again while studying at the Swiss Federal Polytechnic Institute in Zürich,

The Vienna Circle

The Vienna Circle, founded in 1920 by Moritz Schlick, included philosophers, scientists, and mathematicians who met regularly in Vienna for almost 20 years. Its members included Rudolf Carnap, Philipp Frank, Kurt Gödel, Otto Neurath, and Hans Hahn. A sister group, the Society for Empirical Philosophy, met in Berlin with Carl Hempel and Hans Reichenbach. They formulated the "verifiability principle" which asserted that only statements supported by experience and observation had any meaning. Much of the group's interest lay in clarifying the language of science by reducing the content of scientific theories to truths of logic and mathematics, coupled with statements based on sense experience. The Vienna Circle also advocated a doctrine of "unified science," by which they meant that physical, biological and social sciences share the same language, laws and methods.

At different times, the philosophy of the Vienna Circle was called logical positivism, logical empiricism, scientific empiricism and neopositivism. Positivism, associated with French philosopher Auguste Comte (1798–1857),

regards only information gathered by experience meaningful, and gives no credence to a priori or metaphysical ideas. Logical Positivism is a movement associated with German philo-

Moritz Schlick and his children, about 1926.

sopher Rudolf Carnap (1891–1970). While the philosophy holds to the idea that only the "positive" data of experience can make a statement true, it also acknowledges the truth

of logic and mathematics.

The influence of the Vienna Circle spread to other countries through its publications and by contact with its proponents. In 1929, the group published their intentions in a manifesto: *Scientific Conception of the World: The Vienna Circle.* That same year, Schlick traveled briefly to the United States as visiting professor at Stanford University. Other publications included the review *Knowledge* and a series of monographs published under the title, *Encyclopedia of Unified Science.*

In 1936, Mortiz Schlick was shot by a deranged student—purportedly the boyfriend of another student—and by 1938, the group had disbanded under political pressure. Some of its members, such as Gödel, emigrated to the United States; others fled to Great Britain.

The philosophy of the Vienna Circle exerted great influence (and continues to exert an influence) on modern science and on the analytical school of philosophy, which became a dominant approach in philosophy as practiced in Great Britain and the United States.   — PH

where he attended August Stadler's lectures on Kant in the summer semester of 1897. Stadler belonged to the Marburg neo-Kantian movement, which was distinguished by its efforts to make sense of foundational and methodological aspects of current science within the Kantian framework.[7]

It was also at university that Einstein first read Mach's *Mechanics* (1883) and his *Principles of the Theory of Heat* (1896), along with Arthur Schopenhauer's *Parerga and Paralipomena* (1851). It was probably also there that he first read

Friedrich Albert Lange's *History of Materialism* (1873), Eugen Dühring's *Critical History of the Principles of Mechanics* (1887), and Ferdinand Rosenberger's *Isaac Newton and His Physical Principles* (1895). All those books were, at the end of the century, well known to intellectually ambitious young physics students.

A telling fact about Einstein's acquaintance with philosophy at university was his enrollment in Stadler's course on the "Theory of Scientific Thought" in the winter semester of 1897. The course was in fact required for all students in Einstein's division at the Polytechnic. Think about that: Every physics student at the Polytechnic, one of the leading technical universities in Europe, was required to take a course in the philosophy of science. Such an explicit requirement was not found at every good university, although in 1896 Mach was named to the newly created chair for the "Philosophy of the Inductive Sciences" at the University of Vienna, and students learning physics under Hermann von Helmholtz in Berlin got a heavy dose of philosophy as well. Even if not every university had a specific requirement in the philosophy of science, the Zürich curriculum tells us that good young physicists were expected to know more than just a smattering of philosophy.

Einstein's interest in philosophy continued after graduation. At about the time he started his job in the patent office in Bern in 1902, Einstein and some newfound friends, Maurice Solovine and Conrad Habicht, formed an informal weekly discussion group to which they gave the grandiloquent name "Olympia Academy." Thanks to Solovine, we know what they read.[8] Here is a partial list:

Richard Avenarius, *Critique of Pure Experience* (1888).
Richard Dedekind, *What Are and What Should Be the Numbers?* (2nd ed., 1893).
David Hume, *A Treatise of Human Nature* (1739; German translation 1895).
Ernst Mach, *The Analysis of Sensations and the Relation of the Physical to the Psychical* (2nd ed., 1900).
John Stuart Mill, *A System of Logic* (1872; German translation 1887).
Karl Pearson, *The Grammar of Science* (1900).
Henri Poincaré, *Science and Hypothesis* (1902; German translation 1904).

Those are titles one would have found on the bookshelf of many bright young physicists at the time. That Einstein and friends read them for pleasure or self-improvement shows how common it was in the scientific culture of the day to know such books and the ideas they held.

The philosophical seeds sown at the Polytechnic and the Olympia Academy were soon to bear fruit in Einstein's 1905 paper on the special theory of relativity and in many other places in his scientific work. But they would bear additional fruit in Einstein, himself, becoming an important philosopher of science.

Relations with philosophers

Einstein's philosophical education made a profound difference in the way he did physics. But his interest in the philosophy of science went further. By the 1930s he had become an active participant in the development of the freestanding discipline of the philosophy of science. His role evolved largely through his personal and professional relations with many of the era's most important philosophers, mainly the founders of the tradition known as logical empiricism.

Einstein's personal acquaintance with prominent philosophers of science began early and somewhat by accident. Friedrich Adler was also a physics student in Zürich in the late 1890s.[9] Although Adler studied at the University of Zürich, not the Polytechnic, he and Einstein became friends. The friendship was renewed in 1909 when Einstein moved back from Bern to Zürich to take up his first academic appointment, at the University of Zürich, a position for which Adler had been the other finalist.

By then, Adler had become a well-known defender of Mach's empiricism, especially after the searing criticism that Max Planck leveled at Mach in a 1908 lecture on "The Unity of the Physical World Picture." The close relationship with Mach led Adler to publish, in 1908, a German translation of Pierre Duhem's influential 1906 book, *Aim and Structure of Physical Theory*.

From Duhem Einstein learned a version of what is known as conventionalism. Henri Poincaré, another well-known conventionalist, famously argued that the geometer's conventional definition of "straight line segment" as "the path of a light ray" made Euclidean geometry safe from straightforward empirical refutation, say by line-of-sight triangulation of three mountain peaks, because anyone impressed by the simplicity of Euclidean geometry could save it by simply changing the definition of straight line.

Duhem's conventionalism differed somewhat from Poincaré's. He argued that what was conventional was not the choice of individual definitions, but rather one's choice of a whole theory. According to Duhem, it is always whole theories and never individual scientific claims that one tests. Duhem's "holistic" conventionalism was to become deeply woven into Einstein's mature picture of the structure of theories and the way they are tested.

It was also in 1909 that Einstein's fame made possible his first meeting with Mach.

There was mutual respect on both sides. When Einstein left the German University of Prague in 1912, he nominated Philipp Frank as his successor. Frank was a Mach disciple who was to become an important member of the so-called Vienna Circle of logical empiricists. Frank's 1947 Einstein biography is well known.[10]

Einstein's move to Berlin in 1914 further expanded his circle of philosophical colleagues. It included a few neo-Kantians like Ernst Cassirer, whose 1921 book, *Einstein's Theory of Relativity*, was a technically sophisticated and philosophically subtle attempt to fit relativity within the Kantian framework. General relativity presented an obvious challenge to Kant's famous assertion that Euclidean geometry was true a priori, the necessary form under which we organize our experience of external objects.

Hans Reichenbach, a student socialist leader in Berlin at the end of World War I, went on to anchor the Vienna Circle's Berlin outpost and become logical empiricism's most important interpreter of the philosophical foundations of relativity with books like his 1928 *Philosophy of Space and Time*. He had been Einstein's student in Berlin, and Einstein was so impressed by his abilities as a philosopher of physics that when the conservative Berlin philosophy department refused Reichenbach a faculty post in the mid-1920s, Einstein contrived to have a chair in the philosophy of science created for him in the university's more liberal physics department.

Without question, the most important new philosophical friend Einstein made during his Berlin years was Moritz Schlick. He was originally a physicist who did his PhD under Planck in 1904. Schlick's move to Vienna in 1922 to take up the chair in philosophy of science earlier occupied by Mach and Ludwig Boltzmann marks the birth of the Vienna Circle and the emergence of logical empiricism as an important philosophical movement. Prior to the work of Reichenbach, Schlick's 1917 monograph *Space and Time in Contemporary Physics* was the most widely read philosophical introduction to relativity, and Schlick's 1918 *General Theory of Knowledge* had a comparable influence on the broader field of the philosophy of science.[11]

Einstein and Schlick first got to know one another by correspondence in 1915, after Schlick published an astute essay on the philosophical significance of relativity. For the first six years of their acquaintance, Einstein showed high regard for Schlick's work, but by 1922 the relationship had started to cool. Einstein was dismayed by the Vienna Circle's ever more stridently antimetaphysical doctrine. The group dismissed as metaphysical any element of theory whose connection to experience could not be demonstrated clearly enough. But Einstein's disagreement with the Vienna Circle went deeper. It involved fundamental questions about the empirical interpretation and testing of theories.

Schlick, Reichenbach, and Einstein agreed that the challenge facing empiricist philosophers of physics was to formulate a new empiricism capable of defending the integrity of general relativity against attacks from the neo-Kantians. General relativity's introduction of a hybrid spacetime with varying curvature was a major challenge to Kantianism. Some of Kant's defenders argued that general relativity, being non-Euclidean, was false a priori. More subtle and sophisticated thinkers like Cassirer argued that Kant was wrong to claim a priori status for Euclidean geometry but right to maintain that there is some mathematically weaker a priori spatial form, perhaps just a topological form.

Mach's philosophy was not up to the task. It could not acknowledge an independent cognitive role for the knower. Schlick, Reichenbach, and Einstein, on the other hand, agreed that the Kantians were right to insist that the mind is not a blank slate upon which experience writes; that cognition involves some structuring provided by the knower. But how could they assert such an active role for the knower without conceding too much to Kant? They were, after all, empiricists, believing that the reasons for upholding general relativity were ultimately empirical. But in what sense is our reasoning empirical if our knowing has an a priori structure?

Schlick and Reichenbach's eventual answer was based mainly on Poincaré's version of conventionalism. They argued that what the knower contributes are the definitions linking fundamental theoretical terms like "straight line segment" with empirical or physical notions like "path of a ray of light." But, they contended, once such definitions are stipulated by convention, the empirical truth or falsity of all other assertions is uniquely fixed by experience. Moreover, since we freely choose only definitions, the differences resulting from those choices can be no more significant than expressing measurement results in English or metric units.

Einstein also sought an empiricist response to the Kantians, but he deeply disagreed with Schlick and Reichenbach. For one thing, he, like Duhem, thought it impossible to distinguish different kinds of scientific propositions just on principle. Some propositions function like definitions, but there was no clear philosophical reason why any one such proposition *had* to be so regarded. One theorist's definition could be another's synthetic, empirical claim.

As used by philosophers, "synthetic," as distinguished from analytic, means an assertion that goes beyond what is already implied by the meanings of the terms being used. An analytic assertion, by contrast, is a claim whose truth depends solely on meaning or definitions. A central empiricist tenet is that there are no synthetic a priori truths.

A deeper reason for Einstein's dissent from Schlick and Reichenbach was his worry that the new logical empiricist philosophy made science too much like engineering. Missing from the empiricists' picture was what Einstein thought most important in creative theoretical physics, namely, "free inventions" by the human intellect. Not that the theorist was free to make up any picture whatsoever. Theorizing was constrained by the requirement of fit with experience. But Einstein's own experience had taught him that creative theorizing could not be replaced by an algorithm for building and testing theories.

How would Einstein reply to Kant? He deployed Duhem's holism in a novel way. When a theory is tested, something must be held fixed so that we can say clearly what the theory tells us about the world. But Einstein argued that precisely because theories are tested as wholes, not piecemeal, what we choose to hold fixed is arbitrary. One might think, like Kant, that one fixes Euclidean geometry and then tests a physics thus structured. But we really test physics and geometry together. Therefore, one could just as well hold the physics fixed and test the geometry. Better just to say that we are testing both and that we choose among the possible ways of interpreting the results by asking which interpretation yields the simplest theory. Einstein chose general relativity over rivals equally consistent with the evidence because its physics plus non-Euclidean spacetime geometry was, as a whole, simpler than the alternatives.

Such questions might seem overly subtle and arcane philosophical issues better left alone. But they cut to the heart of what it means to respect evidence in the doing of science, and they are questions about which we still argue. As theoretical physics moves ever deeper into realms less firmly anchored to empirical test, as experimental physics becomes ever more difficult and abstruse, the same questions over which Schlick, Reichenbach, and Einstein argued become more and more acute.

When theory confronts experience, how do we apportion credit or blame for success or failure? Can philosophical analysis supply reasons for focusing a test on an individual postulate, or should judgment and taste decide what nature is telling us? The logical empiricists were seeking an algorithm for choosing the right theory. But Einstein likened crucial aspects of the choosing to the "weighing of incommensurable qualities."[12] In one sense, Einstein lost the argument with Schlick and Reichenbach. By mid-century, their logical empiricism had become orthodoxy. But Einstein's dissent did not go unnoticed, and today it lives again as a challenge to another Kant revival.[13]

Philosophy in Einstein's physics

How did Einstein's philosophical habit of mind lead to his doing physics differently? Did it, as he believed, make him a better physicist?

Most readers of Einstein's 1905 special-relativity paper note its strikingly philosophical tone. The paper begins with a philosophical question about an asymmetry in the conventional explanation of electromagnetic induction: A fixed magnet produces a current in the moving coil by an induced electromotive force in the coil. A moving magnet, on the other hand, is said to produce a current in a fixed coil through the electromagnetic field created by the magnet's motion. But if motion is relative, why should there be any difference? The paper goes on to fault the idea of objective determination of simultaneity between distant events for similarly philosophical reasons; nothing other than the simultaneity of immediately adjacent events is directly observable. One must therefore stipulate which distant events are deemed simultaneous for a given observer. But that stipulation must rest on a conventional assumption about, say, the equal speeds of outbound and inbound light signals.

There is dispute among historians and philosophers of physics about exactly what philosophical perspective is involved here. Some explicitly conventionalist language in the paper suggests Poincaré as a source. Einstein himself credited principally Hume and secondarily Mach. In any case, the strikingly philosophical tenor of the 1905 relativity paper is unmistakable.

Einstein's philosophical sources are less obscure with regard to his lifelong commitment to the principle of spatial separability in the face of quantum mechanical nonlocality. We know that Einstein read Schopenhauer while a student at the Zürich Polytechnic and regularly thereafter. He knew well one of Schopenhauer's central doctrines, a modification of Kant's doctrine of space and time as necessary a priori forms of intuition. Schopenhauer stressed the essential structuring role of space and time in individuating physical systems and their evolving states. Space and time, for him, constituted the *principium individuationis,* the ground of individuation. In more explicitly physical language, this view implies that difference of location suffices to make two systems different in the sense that each has its own real physical state, independent of the state of the other. For Schopenhauer, the mutual independence of spatially separated systems was a necessary a priori truth.

Did that way of thinking make a difference in Einstein's physics?[14] Consider another famous paper from his annus mirabilis, the 1905 paper on the photon hypothesis, which explained the photoelectric effect by quantizing the way electromagnetic energy lives in free space. A photoelectron is emitted when one quantum of electromagnetic energy is absorbed at an illuminated metal surface, the electron's energy gain being proportional to the frequency of the incident radiation. What most struck Einstein about the behavior of these energy quanta was that in the so-called Wien regime near the high-energy end of the black body spectrum, they behave like mutually independent corpuscles by virtue of their occupying different parts of space.

Einstein argued that assuming the validity of Boltzmann's entropy principle ($S = k \log^W$) for radiation fields in the Wien regime implies a granular structure to such radiation. Thanks to the Boltzmann principle's logarithmic form, the additivity of the entropy S is equivalent to the factorizability of the joint probability W for two spatially separated constituents of the radiation field to occupy given cells of phase space. The factorizability of a joint probability is one classical expression for the mutual independence of events.

But there was a problem: The same reasoning that implied a quantal structure for radiation in the Wien regime also implied that, outside that regime, the assumed mutual independence of photons must fail. The assumption of mutually independent photons does not yield a derivation of the full Planck formula for the energy density of black body radiation. Einstein realized that fact, and for nearly 20 years he sought to understand how it could be.

As early as 1909, Einstein toyed with assigning a wave field to each particlelike photon to account for interference, an obvious failure of mutual independence. That's where the idea of wave–particle duality began. Only late in 1924, when Einstein first read Satyendra Bose's new derivation of the Planck radiation formula, did he grasp that what was implied was a new quantum statistics, in which particles fail to be independent not because of some exotic interaction but because their identity makes them indistinguishable.[15]

Thanks to Bose, Einstein realized that failure of the mutual independence of spatially separated light quanta would be an enduring feature of the emerging quantum theory. But from Schopenhauer he had learned to regard the independence of spatially separated systems as, virtually, a necessary a priori assumption. As the new quantum formalism appeared in the mid-1920s, Einstein sought either to interpret it in a manner compatible with spatial separability or to show that if quantum mechanics could not be so interpreted, it was fatally flawed. In 1927, Einstein produced a hidden-variables interpretation of Erwin Schrödinger's wave mechanics. But he abandoned the effort prior to publication when he found that even his own hidden-variables interpretation involved the kind of failure of spatial separability that Schrödinger later dubbed "entanglement."

Einstein's most famous assault on the quantum theory was his 1935 "EPR" paper with Boris Podolsky and Nathan Rosen, which sought to demonstrate that quantum mechanics was an incomplete theory. Many readers find the EPR argument convoluted. Few are aware that Einstein repudiated the paper soon after publication, writing to Schrödinger in June of 1935 that the paper was actually written by Podolsky "for reasons of language," and that he was unhappy with the result because "the main point was buried by excessive formalism."

The argument Einstein intended starts from an assumption that he called "the separation principle." Spatially separated systems have independent realities, and relativistic locality precludes superluminal influences between spacelike separated measurement events. Therefore quantum mechanics must be incomplete, because it assigns different wavefunctions, hence different states, to one of two previously interacting systems, depending on what parameter one chooses to measure on the other system. Surely a theory cannot assign two or more different states to one and the same physical reality unless those theoretical states are incomplete descriptions of that reality.[16]

The important point here is that Einstein regarded his separation principle, descended from Schopenhauer's *principium individuationis*, as virtually an axiom for any future fundamental physics. In later writings he explained that field theory, as he understood it—after the model of general relativity, *not* quantum field theory—was the most radical possible expression of separability. In effect, such classical field theories treat all point events in the spacetime manifold as mutually independent, separable systems endowed with their own separate, real physical states.

Einstein's deep philosophical commitment to separability and his consequent lifelong disquiet about quantum mechanics is nowhere more clearly expressed than in a long note he wrote to Max Born in 1949. Einstein asks, "What must be an essential feature of any future fundamental physics?" His answer surprises many who expect him to say "causality."

> I just want to explain what I mean when I say that we should try to hold on to physical reality.

> We are... all aware of the situation regarding what will turn out to be the basic foundational concepts in physics: the point-mass or the particle is surely

not among them; the field, in the Faraday–Maxwell sense, might be, but not with certainty. But that which we conceive as existing ("real") should somehow be localized in time and space. That is, the real in one part of space, A, should (in theory) somehow "exist" independently of that which is thought of as real in another part of space, B. If a physical system stretches over A and B, then what is present in B should somehow have an existence independent of what is present in A. What is actually present in B should thus not depend upon the type of measurement carried out in the part of space A; it should also be independent of whether or not a measurement is made in A.

If one adheres to this program, then one can hardly view the quantum-theoretical description as a complete representation of the physically real. If one attempts, nevertheless, so to view it, then one must assume that the physically real in B undergoes a sudden change because of a measurement in A. My physical instincts bristle at that suggestion.

However, if one renounces the assumption that what is present in different parts of space has an independent, real existence, then I don't see at all what physics is supposed to be describing. For what is thought to be a "system" is, after all, just conventional, and I do not see how one is supposed to divide up the world objectively so that one can make statements about the parts.[17]

That is how a philosopher-physicist thinks and writes.

Too much philosophizing?

One might respond to Einstein's argument by saying that it proves what's wrong with importing too much philosophy into physics. Einstein was probably wrong to doubt the completeness of quantum mechanics. The entanglement that so bothered him has emerged in recent decades as the chief novelty of the quantum realm.

But such a reaction would reflect a serious misunderstanding of the history. Einstein was wrong, but not because he was a philosophical dogmatist. His reasons were scientific as well as philosophical, the empirical success of general relativity being one among those scientific reasons. What the philosophical habit of mind made possible was Einstein's seeing more deeply into the foundations of quantum mechanics than many of its most ardent defenders. And the kind of philosophically motivated critical questions he asked but could not yet answer were to bear fruit barely 10 years after his death when they were taken up again by another great philosopher–physicist, John Bell.

References

1. A. Einstein to R. A. Thornton, unpublished letter dated 7 December 1944 (EA 6-574), Einstein Archive, Hebrew University, Jerusalem, quoted with permission.

2. P. A. Schilpp, ed., *Albert Einstein: Philosopher-Scientist*, The Library of Living Philosophers, Evanston, IL (1949), p. 684.

3. A. Einstein, *J. Franklin Inst.* **221**, 349 (1936).

4. A. Einstein, *Phys. Zeitschr.* **17**, 101 (1916).

5. A. Pais, *'Subtle is the Lord: The Science and the Life of Albert Einstein*, Oxford U. Press, New York (1982), is still the best intellectual biography of Einstein.

6. For details on Einstein's early philosophical reading, see D. Howard, "Einstein's Philosophy of Science," in *The Stanford Encyclopedia of Philosophy*, E. N. Zalta, ed.,

7. M. Beller, in *Einstein: The Formative Years, 1879–1909*, D. Howard, J. Stachel, eds., Birkhäuser, Boston (2000), p. 83; D. Howard, in *Language, Logic, and the Structure of Scientific Theories*, W. Salmon, G. Wolters, eds., U. of Pittsburgh Press, Pittsburgh, PA (1994), p. 45.

8. M. Solovine, ed., *Albert Einstein. Lettres à Maurice Solovine*, Gauthier-Villars, Paris (1956).

9. D. Howard, *Synthese* **83**, 363 (1990).

10. P. Frank, *Einstein: His Life and Times*, Knopf, New York (1947).

11. See D. Howard, *Philosophia Naturalis* **21**, 616 (1984).

12. A. Einstein, *Autobiographical Notes: A Centennial Edition*, P. A. Schilpp, trans. and ed., Open Court, La Salle, IL (1979), p. 21.

13. A widely debated recent work of Kantian revival is M. Friedman, *Dynamics of Reason*, CSLI Publications, Stanford, CA (2001).

14. D. Howard, in *The Cosmos of Science*, J. Earman, J. D. Norton, eds., U. of Pittsburgh Press, Pittsburgh, PA (1997), p. 87.

15. D. Howard, in *Sixty-Two Years of Uncertainty*, A. Miller, ed., Plenum, New York (1990), p. 61.

16. A. Einstein to E. Schrödinger, unpublished letter dated 19 June 1935 (EA 22-047), Einstein Archive, Hebrew University, Jerusalem, quoted with permission; D. Howard, *Stud. Hist. Phil. Sci.* **16**, 171 (1985); D. Howard, in *Philosophical Consequences of Quantum Theory: Reflections on Bell's Theorem*, J. T. Cushing, E. McMullin, eds., U. of Notre Dame Press, Notre Dame, IN (1989), p. 224.

17. M. Born, ed., *Albert Einstein–Hedwig und Max Born. Briefwechsel 1916–55*, Nymphenburger, Munich (1969), p. 223.

The Crimean Expedition

---○---

by Amir Aczel

"I am glad that our colleagues are busying themselves with my theory—even if it's with the hope of killing it."

—Albert Einstein, in a letter to Erwin Finlay Freundlich, August 7, 1914.

The Crimea, August 1, 1914

As Germany declared war on Russia, a German scientist was caught by the Russians on the Black Sea and was transferred to Odessa. Erwin Finlay Freundlich was suspected of being a German spy. He was traveling with strange looking equipment—a telescope. His equipment was confiscated and he was held until the end of August, when he and his team were exchanged for high-ranking Russian officers held by the Germans. Throughout his imprisonment, Freundlich maintained that he was a scientist and that he had come to observe an eclipse. Back in Berlin, Freundlich called on Albert Einstein. Why did Freundlich risk his life to travel to a nation at war with his own? What did he intend to do there? And what was his relationship with Einstein, a German scientist who had renounced his German citizenship only to reclaim it later and move to Berlin?

Shortly after his meeting Pollak in Berlin, Erwin Freundlich began his collaboration with Einstein, who was still in Prague. The two met in Berlin in April 1911 when Einstein worked out the gravitational lensing problem.[1] A year later, during his honeymoon, Freundlich and his bride met Einstein while on their visit to Zürich. As the newlyweds' train arrived at the Zürich station in early September, 1913, they saw Fritz Haber on the platform waiting to meet them. Haber was then the director of the Kaiser Wilhelm Institute, and with him, in untidy sporty clothes and a straw hat, was Albert Einstein.

Einstein invited Freundlich and his bride to accompany him to Frauenfeld, where he was to give a talk about relativity. From there, they traveled to the shore of Lake Constance, and later back to Zürich. Einstein was earnestly discussing the problems of the theory and ways to verify results with Freundlich throughout the entire time. On November 8, Einstein received a letter from a Professor Campbell of the Lick Observatory in California, to whom he had written requesting that photographs of the stars near the sun be taken by the observatory during an eclipse and sent to Freundlich for analysis. The analysis did not bear fruit.

Einstein's relationship with Freundlich is mostly known from a surviving collection of 25 letters that Einstein wrote to the younger astronomer over a period of 20 years, from 1911 to 1931.[2] The letters tell a fascinating story, whose full details have until now not been told. It is a story about the vagaries of fate. It is a story of how the world's greatest theoretical physicist desperately wished for experimental confirmation of his hypothesis and hoped to obtain it through the work of an eager young astronomer. It is a story about the evil of war and the evil of politics and how both stood in the way of humanity's quest for knowledge. And yet it is also a story about luck, faith, confidence, and the mercurial nature of human relationships.

As soon as he received word from Pollak about the interest in his work by the young astronomer, Einstein wrote back to Freundlich. His letter was very polite, almost ingratiating, in its language and especially its address. In his first letter, and in many to follow, Einstein—who by then was already a well-known physicist, if not the world figure he would become within less than a decade—addresses the neophyte astronomer as "Highly Esteemed Mr. Colleague." He then continues to thank Freundlich profusely for his great interest in such an important problem (the theory of general relativity). He encourages him to make every effort to find observational evidence for the predictions of the theory, saying that astronomers can provide a great service to science by finding such evidence. There is a hint of desperation in Einstein's tone, and, reading his letters, one clearly senses that he would do anything to prove by physical means that his theory was indeed right.

Space is curved around massive objects, and a light ray passing by such an object will be bent. In addition, a ray of light climbing up a gravitational field will lose energy, as evidenced by a shift in frequency toward the red end of the spectrum (a gravitational redshift)—just like a person gets tired clambering up a winding stairway. Einstein concentrated his attentions on the light-bending phenomenon he was sure existed in nature. He asked the young astronomer if there was a way of detecting such an event.

In September, Einstein responded stoically to what seems to have been his first disappointment with Freundlich's efforts on his behalf. He wrote: "If only nature had given us a planet bigger than Jupiter!—but nature did not give us the possibility of help in these discoveries." In looking for confirmation of the theorized phenomenon of bending of light rays, it seems that Freundlich chose to look at light passing by the planet Jupiter on its way to Earth. He did not find such an effect. In the hindsight of close to a century, it is easy to see why Freundlich failed. The bending effect is relatively small, and Jupiter—while much larger than Earth—is only one thousandth as massive as the Sun. The planet's mass is not enough to allow measurement of the bending of light rays around it.

On September 21, Einstein had new ideas. He realized that the smallest massive object around which the bending of light had a chance of being detected would have to be the Sun. He asked his Highly Esteemed Colleague what he thought about the possibility of looking for starlight in daytime. Clearly this would be necessary since starlight emanated from a far-away point in space, and if it could be seen in daylight at an angle that makes it pass close to the Sun, bending might be detected if one knew the expected position of the star. One could then compare the expected position of the rays with their observed position, affected by bending because of their close passage to the Sun, and thus establish the existence of the phenomenon. Einstein wanted to know whether astronomers had a way of seeing stars in daylight and finding a star that appears near the Sun's position in the sky.

In early 1913 Einstein wrote Freundlich, again thanking him for his very interesting letter and for his great devotion to the search for proofs of the theory. He included titillating details of his continuing work on the extended concept of relativity and sprinkled the letter with questions clearly aimed at arousing his younger colleague's excitement for the project. Einstein's letter gives the flavor of his frantic quest for a finished theory. He writes in strong terms about his feelings about competing theories—those of Abraham, Mie, and Nordström. Gunnar Nordström (1881–1923), a Finnish physicist, did some ingenious work on Einstein's field equations. The development of these equations by Einstein and Grossmann ran into trouble because of dependencies of parameters. Nordström's idea was to try to develop an alternative theory of general relativity where the speed of light, c, does not depend on a field introduced in Einstein's equations. In his letter to Freundlich, and in following correspondence, Einstein brings out the frantic nature of his quest. This is a fantastic theory, Einstein writes about Nordström's work, but it has a small probability of being right. If Nordström's

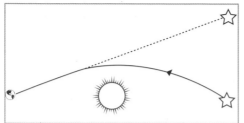

Above: The bending of light due to the mass of the Sun (highly exaggerated). We project the path of light back to the apparent, shifted star position since our instinct tells us that light travels in a straight line. Einstein predicted a shift about double that of Newton and measurements of the shift indicate that Einstein's theory of gravity is correct.

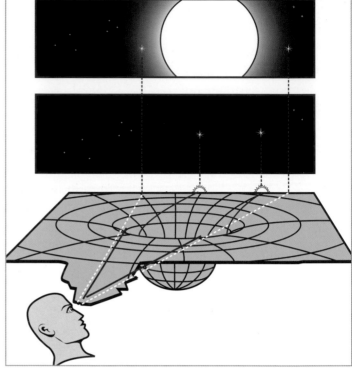

Right: Bending of light by the Sun is depicted (again exaggerated) as a consequence of the curvature of space near the Sun. Upper rectangle: stars when the Sun is eclipsed by moon appear as indicated. Lower rectangle: where those same stars really are in the sky when the Sun is elsewhere in sky.

theory is correct, Einstein writes, then there will be a redshift due to gravitation, but no bending of light. Einstein is desperate, therefore, to find a way of detecting whether light rays get bent in the gravitational field of massive objects: such a test would show whether Einstein—or Nordström—is right. His language leaves no doubt about Einstein's competitiveness. He is convinced that he alone is correct with his (not yet complete) theory.

It is in this letter, written at an uncertain date in early 1913, that Einstein first mentions an eclipse of the Sun. Earlier, in 1912, Einstein seemed to think that starlight could be observed in daylight as it passed by the Sun. Sometime between late 1912 and early 1913, he and Freundlich apparently reached the conclusion that there was no such possibility. At some point it must have occurred to one of them that the phenomenon of a total eclipse of the Sun would offer a perfect venue for the experiment. During a total eclipse it is indeed daytime and the Sun is up, but stars should be observable as well because of the darkness afforded by the Moon's shadow. So while nature did not give us a big enough Jupiter, it did give us this wonderful phenomenon, which occurs roughly every couple of years somewhere on Earth and allows us to view stars as well as the exact position of the Sun in the middle of the day.

As soon as he realized this point, Einstein felt great excitement. In a letter he tells Freundlich that he had read in an American journal that several optical systems should be used together in an eclipse in order to see stars around the sun. Here he says that this seems reasonable to his "layman's brain." But in subsequent correspondence Einstein is anything but a layman when it comes to astronomy. Evidently the great theoretician had come to the conclusion that his theory alone would not be worth much without physical verifications. He seems to have taught himself a great deal about astronomy in a relatively short period of time. In many letters he occupies himself with very technical questions about the actual process of designing a system to view an eclipse and carefully preparing the photographic plates necessary for making the photographs of stars in the vicinity of the Sun.

On August 2, 1913, Einstein repeats his conviction: "Theoretically we came to a result—I am quite convinced that the rays of light get bent. I am exceptionally interested in your plan to observe stars in daytime." He then continues with a long dissertation on small particles suspended in the atmosphere and their possible effects on visibility, the quality of the pictures which might be taken, and other technical astronomical points. He then explains to Freundlich "When operating with an optical system, you must get the whole Sun in the picture, together with the part of the sky that interests us—all on one plate. But it has been recommended that two optical systems be used together. What is not clear is how the two pictures will be used together. I would like very much to hear from you what you think about this and any other methods." It seems that Einstein was so determined that everything work right that he was not going to leave to astronomers the details of their everyday work.

And the theoretician was not working on the light bending problem alone. In the same letter he continues saying he is very curious about Freundlich's research on double stars. Freundlich's idea was to look at systems of double stars orbiting each other. Then, if one could somehow estimate the masses of the two stars in a pair as well as their radial velocities around each other, possibly one could detect the gravitational redshift predicted by Einstein's general relativity as the light from one star passes by the other star. This, unfortunately, proved an experimental dead-end. Neither Freundlich's work nor that of others would lead to any result. The phenomenon would be finally detected in an experiment done at Harvard University in the 1960s. But pursuing this particular goal, Freundlich would make serious calculation errors that would eventually irritate Einstein. The letter concludes with the statement that if such an experiment should somehow lead to a detection of a difference in the speed of light (rather than its frequency—as manifested by a redshift), then "the entire relativity theory inclusive of gravitation theory would be false." In closing, he says how happy he would be to finally see Freundlich when he comes to Zürich with his bride on their honeymoon.

By the next letter, sent by Einstein from Zürich on October 22, 1913, they had already met in Switzerland and talked extensively about the problem of detecting the putative bending of light rays from distant stars as they pass by the Sun. After the perfunctory "Herr Kollege," Einstein writes: "I thank you very much from the heart for your extensive news and the deep interest you have shown in our problem." Apparently Freundlich had tried to obtain pictures taken by other astronomers of eclipses past and attempted to discern any images of stars near the shadow of the Sun. In all of these attempts he had failed. It is not hard to understand why this happened. While the Sun itself is completely hidden during a total solar eclipse, the same is not true for the Sun's corona. Bright tongues of fire extend from the hidden Sun to large distances around the circular dark shadow of the moon hiding the body of the Sun. Stars in the area of the corona are very difficult to discern, and detecting their shifts would require an experiment designed specifically for this purpose. Unfortunately no one had performed this experiment before, since no one had perceived a need to detect shifts in stars' positions near the Sun.

From the rest of the letter it becomes clear that the idea of the eclipse was Einstein's and not Freundlich's. In fact, Freundlich's careless nature—as would be evidenced later when he would make elementary errors in computing mass estimates of doublestars—comes through in this early context. Einstein spends considerable time in his letter countering Freundlich's apparent contention that detection of starlight shifts near the Sun could be done in daytime without a total solar eclipse. Einstein explains patiently that he had asked local astronomers whether such an endeavor would be possible and was answered with a resounding "no."[3]

By December 7, 1913, Einstein and Freundlich had agreed on the venue for the experiment to verify the bending of light rays around the Sun: an expedition should he mounted to go to the Crimea to observe the total solar eclipse predicted for August 1914. They were now deep into the details of the expedition, and the remaining question was how to finance it. Having heard from Freundlich that all was arranged, a complete plan of how to travel to Russia and from there to the Crimea, how to use the telescopic equipment he had devised, how to take pictures of the Sun and the surrounding sky during the eclipse, and how to compare the pictures with those of the same area of the sky taken at night with the stars near the Sun in their usual places, Einstein immediately contacted Planck. He asked him for help in obtaining financial support to try to prove the part of general relativity he felt he had already developed.

It seems that the Prussian academy, however, was not excited enough about the project to fund it. In his letter of December 7 to Freundlich, Einstein says that Planck was interested in the problem, but if the academy would not allocate funds, he, Einstein, would spend his own meager savings on the venture. Einstein, apparently in frustration and anger at not being able to get funding, underlined in his letter: "I will not write to Struve." Hermann Struve was the director of the Royal Observatory at Potsdam. Einstein had hoped to secure funding for the project from the Observatory, but apparently he had been rebuffed. Then he added: "If nothing works then I shall pay from my own small savings at least the first 2,000 marks. So please order the necessary plates and let's not lose time because of the question of money."

Einstein's 1913 letter to George Ellery Hale, director of the Mount Willson Observatory, inquiring if the gravitational bending of light by the Sun could be observed in daylight.

And then suddenly things began to happen, and nothing could stop the course of events—both of science and of history. On April 6, 1914, Einstein and his family moved from Zürich to Berlin. With Haber's help he found a flat, but within a short time Mileva and Albert separated and she took the children with her and returned to Switzerland. Einstein then moved to a bachelor's apartment and seems to have adjusted to the change, even though it was quite painful as he was very attached to his two sons. He reacquainted himself with relatives in Berlin and found one of them, Elsa Einstein, a cousin, especially pleasant and began to develop a close friendship with her. Within five years, after a divorce from Mileva, the two were married.

On July 2, 1914, Einstein was made a member of the Prussian academy. At 34, he was by far the youngest. All the others were scientists of long standing and more advanced ages. From reports of his conversations with colleagues while still in Zurich before being notified of the honor soon to be bestowed upon him, we

The Eddington Expedition

Eddington's 1919 announcement that the deflection of starlight confirmed the theory of general relativity received a mixed reception. The scientific community, aware of the difficulties involved, was cautious. But the British press, in need of new heroes, immediately catapulted Einstein to worldwide fame.

In the middle of World War I, Eddington had received a copy of Einstein's 1916 paper on general relativity from Wilhelm de Sitter at Leiden University. It was possibly Eddington's enthusiasm that persuaded the Astronomer Royal, Sir Frank Dyson, to begin preparations to send two expeditions to test Einstein's theory during the solar eclipse of May 29, 1919. At the time no one knew when the war would end and whether the expedition would in fact be possible.

During the war, Eddington, a Quaker, declared himself a pacifist and escaped internment only by agreeing to lead the eclipse expedition. He was as opposed to the hatred of all things German that pervaded Britain during the war, as Einstein had been to the nationalist fervor that swept Germany. His sup-

port of Einstein's theory helped prevent a post-war breakdown of scientific communication. In 1919, Eddington wrote to Einstein, regarding the eclipse expedition results, "It is the best possible thing that could have happened for scientific relations between England and Germany."

The armistice of November 11, 1918, left just enough time for Eddington's team to reach Principe, an island off the west coast of Africa for the anticipated eclipse. Another team set out for Sobral in Brazil.

May 29 dawned, accompanied by torrential rain in Principe. The rain stopped at noon and about 1:30 P.M. the Sun emerged briefly. During the eclipse, Eddington took 16 photographs but complained that clouds interfered with the star images.

In fact, only two photographs were barely usable. From them Eddington calculated a star-beam displacement between 1.31 and 1.91 arcseconds, compared to Einstein's predicted 1.75 arcseconds and Newton's 0.87 arcseconds. The Sobral team's clearest 8 plates gave an average deflection of 1.98 arcseconds but 18 other lower quality views averaged to 0.86 arcseconds.

The results were not overwhelmingly in Einstein's favor, but Eddington's conviction that relativity was right led him to give more weight to images with the larger star beam deflection.

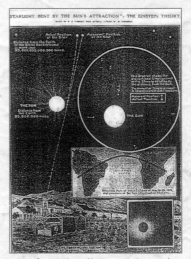

Page from the *Illustrated London News*, November 22, 1919.

At a special meeting on November 6, 1919, Dyson officially declared to the Royal Society that Einstein's general theory of relativity was confirmed. The November 7 London *Times* reported: "REVOLUTION IN SCIENCE: New Theory of the Universe—Newtonian Ideas Overthrown." — PH

know that Einstein did not care much for the distinction. Still, he gave a good speech to the membership, thanking them for the honor and for the freedom membership in the academy would now give him to pursue his research full time. As a member of the academy he would not have to worry about teaching duties or other obligations and would he able to devote his entire time to his research.

From correspondence with colleagues we know that Einstein liked living in Berlin and the new status his position afforded him. He was now also in a position to pursue funding for the experimental project with renewed vigor.

Where Einstein had failed while still in Zürich, Freundlich had in the meantime met with some partial success: Director Struve had (grudgingly) agreed to allow Freundlich to take on the eclipse project, but without allocating any observatory funding. Now in Berlin and a member of the academy, Einstein went to work hard on the problem of money. Finally, the academy granted the project 2,000 marks—the money Einstein stood ready to commit of his own savings—and earmarked the funds for the conversion of scientific instruments for the purpose of observing the eclipse, and also for the purchase of the necessary photographic plates. There was still the need for 3,000 marks for travel and freight costs to the Crimea. In one of the many twists of fate in the story of the eclipse, the money came from what in hindsight looks like an unlikely source.

Gustav Krupp (1870-1950) was a German tycoon whose arms manufacturing firm could by that time have been held indirectly responsible for many massacres, including that of the Armenians by the Turks, who used Krupp arms. In 1918, Krupp designed special long-range guns for the purpose of shelling the civilian population of Paris from a distance of seventy-four miles, which would kill 256 Parisians.[4] It was Krupp money that allowed Hitler to campaign against the Versailles Treaty, and in 1933 gave Hitler the necessary votes to gain a majority in the Reichstag and gain absolute control in Germany. The firm supplied the Nazis in World War II, and was an indispensable tool for the Nazi horror. In 1914, Gustav Krupp contributed 3,000 marks to support the expedition to look for proof for Einstein's theory of general relativity.

As the time of the expedition drew near, Einstein became more and more excited, agitated and withdrawn. His biographer Ronald Clark tells about Einstein's frequent visits to the Freundlich family during the tense period before the planned eclipse expedition. Einstein apparently did not leave anything to chance as he kept his expedition leader, Freundlich, constantly in sight. He often brought his work with him when visiting at the Freundlich house, where, before dinner was over, he would push back his plate and start writing equations on top of his hosts' expensive tablecloth. Freundlich's widow would tell Clark many years later that she regretted not keeping the tablecloth, as her husband had suggested, since it would probably be worth a lot.[5]

It seems that with time Einstein harbored a combination of two feelings. First, he was very anxious about the results of the coming eclipse expedition. He was now a scientist of some renown: his special theory of relativity had been reasonably well-received by the scientific community—although it still had opponents. The general theory, still in its infancy, was getting attention from other scientists and receiving both fierce competition and a fair amount of skepticism. His colleagues at the academy were all older and mainstream in their views and career paths, and among them Einstein was and would continue to be an outsider. He even had to force himself during this period to abandon his natural inclination for careless dress and to don respectable attire as appropriate to his new status. Einstein was desperate for positive proof that his outlandish theory about space and time and gravitation was correct. At the same time, Einstein was becoming more and more confident about the validity of his theory. In a letter to his good friend Michele Angelo Besso, Einstein wrote: "I no longer doubt the correctness of the whole system, whether the observation of the eclipse succeeds or not. The sense of the thing is too evident." Ironically, Einstein was wrong. As fate would have it, Freundlich would go off to the Crimea looking for light deflection of an amount only half that actually existing in nature. But as Freundlich was about to leave, Einstein had already made up his mind that if the results of the crucial test were to be negative, the experiment would be faulty—not his theory. This would be one of the strongest examples of how in certain cases, to a theoretician, the equation designed to describe nature takes on a life of its own and is viewed as so elegant and so divine that reality takes second place to the formula.

On July 19, 1914, Erwin Freundlich left Berlin with two colleagues, one of them a technician with the famous German lens manufacturer Carl Zeiss. Within a week's travel they reached the town of Feodosiya in the Crimea and prepared their equipment for the eclipse. Freundlich brought four different cameras as well as telescopic equipment to maximize the chances of taking at least one good picture showing clearly stars in the vicinity of the Sun during the eclipse. The German team met another from Argentina, also there to photograph the eclipse for other purposes. Interestingly, the Argentine team came to try to capture on film Vulcan—a hypothesized small planet near the Sun believed to exist because of small systematic aberrations that had been detected for decades in the orbit of the planet Mercury. Vulcan and its putative orbit near the Sun were believed to be responsible for the perihelion problem of Mercury. In another bizarre twist of fate, it would be Einstein's general theory of relativity, which the German team came to test, that would eventually solve the perihelion problem. The shifting of the orbit of Mercury is not due to another planet. None exists. It is due to the effects of the Sun's gravitational field on the planet which lies so close to it. And within a few short years, it would he Freundlich himself who would compile a long list of historical astronomical observations of the orbit of Mercury that, together with general relativity, would solve the problem. Now the two teams, the one from Argentina and the German one, shared information and techniques, as well as equipment in tense anticipation of the Sun's disappearance for a precious two minutes on August 21. [Editor's note: See sidebar on Mercury on page 125]

But all this time, history was taking another course—one far away from the direction of science and knowledge. Three weeks before Freundlich's departure from Berlin to the Crimea, Archduke Franz Ferdinand, the heir to the throne of the Austro-Hungarian empire, was visiting the city of Sarajevo, the capital of the province of Bosnia, which had been annexed by the Empire. The Serbian foreign minister took the unusual step of warning the archduke about the visit, advising him that there had been Serbian agitation in the capital and that perhaps now was not a good time for a visit. Franz Ferdinand was undeterred. On June 28, as the archduke's motorcade was carrying him to a ceremony in town hall, a bomb was hurtled at the archduke's car. It exploded, but the archduke and his wife, the Duchess of Hohenberg, were unhurt. But the plot against the Austro-Hungarian prince was thicker—other conspirators waited for him farther down his road. A second bomb was thrown at the archduke, but did not explode. Continuing on its way, the motorcade came to a point where the archduke's car had to pull back and change direction. At that moment, the third conspirator, a nineteen-year-old student named Gavrilo Princip, pulled out a pistol and fired at Franz Ferdinand and his wife, killing them both. The conspirators belonged to a terrorist group called Black Hand, which was actually opposed to the Serbian government. The group's aim was to achieve independence of the south Slavic peoples from the Hapsburg empire. The murder sent shock waves around the world. The storm clouds that would bring the First World War began to thicken over Europe. It seems, however, that despite the anger about the attack voiced both by the Hapsburgs and by their ally, the German kaiser in Berlin, the organizers of the Freundlich expedition to the Crimea were completely oblivious to the political implications of the situation, and never considered that war with Russia might be an imminent possibility. While these mighty forces of war were about to be unleashed, Freundlich and his crew were calmly preparing to watch an eclipse, set to happen on August 21, in Russian-controlled territory.

The kaiser was participating in a regatta in the harbor of Kiel when a note about the assassination of Archduke Franz Ferdinand was thrown onto his yacht folded into a gold cigarette case. The kaiser was enraged and decided to return summarily to Berlin. His ambassador in Vienna suggested that a mild punishment be meted out to the Serbs, but Wilhelm II was implacable. He was determined that the Serbs must be "disposed of—and soon." Public opinion in Germany backed him, and on July 4, the German ambassador to Britain advised Lord Haldane that he was very worried about the developing situation and the distinct possibility of war. Britain advised moderation and hoped for peace. It had nothing to gain from war, and everything to lose. But this was not the case for Germany and Austria-Hungary. The kaiser was especially belligerent because of his feelings about Russia. He was convinced that Russia must be stopped or it would dominate Europe and threaten German hegemony on the continent. The kaiser offered strong support for Emperor Franz Josef in trying to avenge the death of his son on the Serbians—this despite a report of the results of the investigation of the murder of the archduke, which found no involvement by the Serbian government.

On July 23, 1914, the Austro-Hungarian empire issued an official ultimatum to Serbia. The document was unique in the history of nations, for by it Austro-Hungary attempted to dictate to Serbia what to do on an internal as well as external level. There were fifteen demands in the ultimatum ranging from requiring the Serbian government to prohibit all anti-Austrian propaganda within its borders to insisting on the inclusion of Austrian officials in the committee set to investigate the murder. The alternative to full compliance by Serbia was war. The Serbs agreed to all but one of the conditions, but instead of agreeing to negotiate, Franz Josef mobilized his forces and prepared to attack. The Serbs counted on strong Russian support for its Serbian allies, and Germany stood ready to aid the Austro-Hungarians. The Russian czar, hoping to avert a war, made a proposal on July 27 that would have the Austro-Hungarians and the Serbs talking to each other to negotiate an agreement, but the Austrian government rejected the proposal out of hand. With the nations of Europe taking sides with the Russians and Serbs or with the Austro-Hungarians and Germans, if an armed conflict were to erupt, it was clear that it would not be contained and would become a world war.

Early on August 1, Czar Nicholas appealed for the second time to the German kaiser that their long friendship should prevail and prevent bloodshed between their two nations, but Wilhelm wouldn't budge. However, he hoped to confine the conflict to a war in the east only, without an attack on France and the Low

Countries. But his generals already had plans in place for a western front as well. Later that day, German troops crossed the border into Luxembourg and took the village of Trois Vierges. Trying to limit the war, the kaiser ordered his troops to cross back into Germany, but within hours changed his mind and sent the German army back into Luxembourg and on their way to Belgium. In the evening of the same day, August 1, King George V sent frenzied telegrams from London to Berlin and St. Petersburg in a last-ditch effort to stop the impending First World War. But these attempts were all in vain. Late that evening, the German ambassador to Russia went to see the Russian foreign minister at his palace in St. Petersburg and presented him with the German declaration of war on Russia.

As war broke out, the German team led by Freundlich found itself deep inside enemy territory. Since the Germans carried sensitive optical equipment, the Russians could well suspect them of being spies. In the first days of August, 1914, Freundlich's team was arrested. The team members were kept as prisoners of war. On August 4, Einstein, sick with anxiety, wrote to his friend Paul Ehrenfest: "My dear astronomer Freundlich will become a prisoner of war in Russia instead of being able to observe the eclipse. I am worried about him." The prisoners of war were taken from the Crimea to the city of Odessa, where they were kept for almost a month. But fortuitously the Germans had just captured a group of high-ranking Russian officers. The Russians were eager to exchange prisoners of war and the Prussian academy intervened with the German government and secured the release of Freundlich and his colleagues for the Russian officers. By September 2, Freundlich was back in Berlin. But Einstein's hopes for verification of his theory through observation of the eclipse were dashed.

While Freundlich spent the rest of the war years in Berlin, even working part of the time for Einstein, the relationship between the two soured. Einstein's point of view on it is expressed in a letter he wrote Freundlich on September 10, 1921, saying: "I don't think it would be helpful for us to see each other. I am glad that we are understanding each other more (compare it with 1914). For the whole change in our relationship we can be thankful to the English." Einstein apparently blamed Freundlich, who risked his life and freedom for Einstein's theory, for the misfortunes of war, and it seems that his rancor would last for five years, until an Englishman would succeed where Freundlich was not allowed to tread.[6]

But as fate would have it, by that time, 1919, Einstein's theory of general relativity would be complete and he would have corrected his error about the true magnitude of the deflection of light rays by the Sun's gravity. This correction happened on

November 18, 1915, when Einstein announced the expected bending of a light ray just grating the edge of the sun as 1.75 arcseconds, twice the amount he had predicted in 1914. A philosophical question comes up: What would have happened had history allowed Freundlich to carry our his experiment—detecting a shift of 1.75 arcseconds (plus or minus an experimental error) instead of 0.87 arcsecond as Einstein had predicted (actually, Einstein had made an additional, arithmetic error, obtaining the value 0.83 arcsecond). Would general relativity then be judged correct or false?

It should be pointed out that the smaller quantity, 0.87 arcsecond (when computed without the math error) corresponds to the shifting of light when a light ray is considered a particle and when one therefore uses the old Newtonian theory. It was the true incorporation of relativity that would lead to a doubling of the value. So quite possibly, had Freundlich's experiment worked, general relativity may not have been considered proven by the scientific community. Perhaps Einstein should have just waited patiently to complete his theory before seeking confirmation, and not blamed his loyal astronomer.

Einstein's ungrateful attitude toward Freundlich manifested itself in many ways over the following years, and one can only feel compassion for the astronomer who had risked so much for Einstein and believed in a theory that had aroused so much skepticism in the scientific community of the time.

Over time, Einstein's dismissive, and at times insensitive, attitude toward the astronomer becomes apparent in his letters. Gone are the "Highly Esteemed Mr. Colleague" addresses in letters Einstein sent him until the failed attempt of 1914. They are replaced by a simple "Dear Freundlich." And in an undated letter of 1917, Einstein writes: "Yesterday Planck spoke with Struve about you. Struve cursed you. You don't do what he expects you to do. Planck thinks that the best solution for you is to get a job teaching theoretical astronomy, and he thinks that you'd have a very good chance of getting one. I think he is right as one shouldn't put all hopes in getting an observatory job. Best regards, A. Einstein."

Einstein continued his correspondence with Freundlich over the years. It seems that he was often asked to help Freundlich find a job or publish a paper. From the tone of letters it is clear that Einstein now felt an important member of the German academic elite, often dropping the name of his renowned friend Planck. Freundlich seems to have been unable to hold positions, and in one letter of 1919 Einstein wrote: "I think a lecturer at the university would he a good position, but it is not easy to get. Don't let this give you gray hair, but enjoy your vacation. All will come to an end. Your nerves are frayed and without a layer of bacon to protect your head. Best regards to you and your wife from our mutual friends, A. Einstein." In another letter, Einstein says he will recommend that the academy accept Freundlich's work if Freundlich would answer six technical questions Einstein put to him. On March 1, 1919, by a remarkable coincidence, Einstein wrote Freundlich that he had just read a clear and enjoyable exposition of the work of the English astronomer Arthur Eddington. Unbeknownst to Einstein, Eddington was at that moment about to embark on a trip to an island off the coast of equatorial Africa to watch an eclipse of the Sun and try to detect the bending of star light to prove Einstein's theory of general relativity.

References

1. Actually, the exact place and time of Einstein's first meeting with Freundlich remain a mystery. Ronald Clark, who bases some of his claims about the Freundlich-Einstein relationship in his biography on a conversation with Mrs. Kathe Freundlich, claims that the two men met for the first time in Zurich in 1913 (*Einstein: The Life and Times*, New York:Avon, 1984, p. 207). However, in a letter to Michele Besso, dated Prague, March 26, 1912 (Document 377, *The Collected Papers of Albert Einstein*, M.J. Klein, et al., editors, Princeton University Press, 1993, Vol. V), Einstein tells his friend that he is soon going to Berlin to meet Planck, Nernst, Haber, and "an astronomer." This astronomer is very likely Freundlich, since he is the only astronomer with whom Einstein had any dealings during this time. Einstein's notebook containing appointments during his stay in Berlin that year has been recovered and studied by Jurgen Reno and his colleagues at the Max Planck Institute for the history of Science in Berlin. While the notebook contains important astronomical ideas as well as the names and times of meetings with various people, no mention of Freundlich is made there. However, the most convincing piece of evidence is a 1935 letter from Leo W. Pollak to Einstein (Document 11–180 of the Einstein Archives in Jerusalem) in which he says that he introduced the two men in 1911.
2. The collection is kept at the Pierpont Morgan Library in New York.
3. This feat is still not possible today. Even during an eclipse, detecting the bending of light requires a complicated procedure.
4. Martin Gilbert, *A History of the Twentieth Century*, Vol. 1, New York: Morrow, 1997, p. 490.
5. Ronald W. Clark, *Einstein: The Life and Times*, New York: Avon, 1984, p. 222.
6. This letter clearly proves that Einstein's relationship with Freundlich started to deteriorate when Freundlich's eclipse expedition failed. Other researchers of the Einstein-Freundlich relationship apparently missed this point. In a recent book, *The Einstein Tower*, Stanford University Press,1997, pp. 137–8, Klaus Hentschel claims that the relationship began to sour in 1921 when Freundlich attempted to obtain money for one of Einstein's manuscripts, which enraged the latter.

Above: Book burning in Germany on May 10, 1933. Singled out for destruction by the Nazis were the works of Kafka, Freud and Einstein.

Above: Cartoon of Einstein burning a book labeled "Hitler Terror." The caption reads, "A writer's fame, it seems, is also relative."

Right: This is thought to be the last photograph of Einstein in Germany, taken on December 1, 1932, by a passer-by who recognized him. Einstein was to leave Germany forever a few days later.

Einstein's Germany

———o———

by Fritz Stern

There was nothing simple about Einstein, ever. His simplicity concealed an impenetrable complexity. Even the links to his native Germany were prematurely ambiguous. At a time when most Germans thought their country a hospitable home, a perfect training ground for their talents, Einstein was repelled; in 1894, as a fifteen-year-old, he left Germany and became a Swiss citizen. Twenty years later, a few weeks before the outbreak of the Great War, he returned to Germany and remained for eighteen years of troubled renown, years in which he appreciated what was congenial and opposed what was antipathetic in Germany. Long before Hitler, he felt unease. He could joke about his multiple, if uncertain, loyalties—the better perhaps to hide his feelings. In 1919, at the moment when fame first engulfed him, he explained in a letter to the London *Times*: "Here is yet another application of the principle of relativity for the delectation of the reader: today I am described in Germany as a 'German servant,' and in England as a 'Swiss Jew.' Should it ever be my fate to be represented as a *bête noire*, I should, on the contrary, become a 'Swiss Jew' for the Germans and a 'German savant' for the English."

His fame, his capacity for homelessness, and the degradation of his country made Einstein a citizen of the world, seemingly detached from Germany. But I believe that his early encounters with Germany, his hostility to its official culture, shaped his public stance. My deliberately ambiguous title is meant to suggest that Einstein's Germany was both real and imaginary—that he had his own perception of reality. The German experience haunted Einstein to the very end, as it haunted so many of his generation later. It was the text of his political-moral education, the background against which he came to mold his unorthodox views and play his controversial public role.

In Einstein's time, Germany was the promise and later the nemesis of the world, the country that had a decisive bearing on world politics and where, for a moment that seemed a lifetime, the moral drama of our era was enacted. At certain critical moments, Einstein and even his closest colleagues described radically different responses. I believe this diversity will help to complicate our understanding of Germany, and this will be desirable, because Germany's past has often been treated with didactic simplicity. Einstein and Germany: they illuminate each other....

It is generally said of Einstein that he revolutionized modern physics and natural philosophy and that his genius had no equal save that of Newton's. But I shall not—and I could not—deal with what was central to him. I shall deal with the public figure, with the first scientist-hero to appear in the Western world. I shall concentrate on the thoughts that were important to this public figure who placed his scientific fame at the service of his moral indignation. The genius hovers in the background, and the occasional partisan in the foreground. In doing this, I was mindful of what Lionel Trilling has said:

> Physical science in our day lies beyond the intellectual grasp of most men... This exclusion of most of us from the mode of thought which is habitually said to be the characteristic achievement of the modem age is bound to be experienced as a wound given to our intellectual self-esteem. About this humiliation we all agree to be silent; but can we doubt that it has its consequences, that it introduces into the life of mind a significant element of dubiety and alienation which must be taken into account in any estimate that is made of the present fortunes of mind?

I felt this exclusion the more as I came to realize the intensity of the aesthetic joy that Einstein and his colleagues found in their discoveries, as their correspondence exemplifies. We are shut out from that knowledge and from that particular beauty. Lionel Trilling was abundantly right in calling this exclusion an unacknowledged wound.

Exclusion from substance was compounded by my more or less accidental familiarity with some of the men around Einstein and, as a child, with the fringes of that milieu. From time to time I shall allude to some of these personal ties, which added puzzlement and poignancy to my efforts to understand even a part of Einstein's world. I read in Einstein's unpublished correspondence with the historian's habitual hope that the archives would yield some nuggets to shock or prod the mind; the letters were marvelously human, but Einstein remained elusive and enigmatic. The search has been fascinating and disheartening—and has fully borne out what a friend said at the very beginning: Einstein is the hardest person to say anything about. His own friends found him inscrutable, and not even their love of him offered a firm bridge of understanding.

In analyzing the scientific ideal that some historians cherished, Richard Hofstadter once said: "The historian is quickly driven to a kind of agnostic modesty about his own achievement. He may not disparage science, but he despairs of it." Einstein would have agreed with this judgment, though drawn different conclusions from it. The one time I met him—in 1944, while an undergraduate at Columbia—he inquired after my plans, and I told him I was in a quandary, not knowing whether to continue my original purpose, which was to study medicine and thus follow in the footsteps of a father, two grandfathers, and four great-grandfathers, or to switch to history, an old interest turned into a new passion by the power of my teachers. To Einstein this was no quandary. Medicine, he said, was a science (which I doubt) and history was not—though it is significant that I cannot remember his rather harsh words about history. I chose not to follow his advice, but I will confirm his view that history is not a science, that it is an approximation of a time and space that we knew not.

At the risk of risible compression, let me recall some of the characteristics of modern Germany, particularly those that would have impinged on Einstein's life and thought. He was born in the decade of Germany's unification, and he died a decade after its dissolution. The 1870s were a heady and extravagant time for a country whose historic experience had been defeat and division. For centuries, Germany had been a geographic expression, everybody's battlefield, Europe's anvil on which other nations forged their destinies. In the unbroken annals of defeat, Prussia had been the sole exception; and Prussia had evolved its own ethos of frugal duty, rectitude, and obedience. It was also, as Mirabeau had pointed out at the end of the eighteenth century, not as other countries, a state with an army, but an army with a state. In the Napoleonic era, even Prussia collapsed, but political impotence had its compensations: in the shadow of defeat, the Germans created a great literary and philosophical culture and a national identity based initially on intellectual–aesthetic, not political, achievement. That culture enshrined as a moral imperative the cultivation of the self and education—at least for the elite—as the prescribed path to self-formation. There were always two strains to this conception: the ideal of the harmonious being, the rational, aesthetically literate humanist on the one hand and the demonic, inexplicable, mysteriously creative genius on the other. Einstein fitted both categories. By piously and pedantically trying to inculcate the rational, German schools often encouraged the yearning for the irrational. By the mid-nineteenth century, it was widely believed that Germany had a special vocation for learning.

Unification under Prussian aegis, achieved in battle, directed by Bismarck, codified in a constitution that preserved the privileges of a governing elite—that kind of unification was a celebration of force and a denial of earlier hopes of freedom. As Nietzsche warned, this triumph could destroy the German spirit, drown it by the worship of practicality and power. The new Reich, rapidly industrializing, exuded power. But the country became still harder to govern; new social cleavages appeared next to old regional and religious divisions. Nationalism and militarism were a means of providing cohesion, of overcoming a sense of *unfulfilled* unity. Bismarck's Germany was an authoritarian state of uncertain viability, but it was also a government of laws, a haven of constitutionality as compared with primitive, autocratic Russia, a country without torture, callous and sentimental, rigid, efficient, hardworking, bent on achievement.

Whatever the shortcomings of the new Reich, Bismarck's generation had seen the fulfillment of the great national ideal. The next generation—epitomized by the young emperor, with his dreams and delusions that could never banish his anxieties—thirsted for its own glory, for its own imprint on history. Germany had become a giant in the center of Europe: it had the best army, the strongest economy, the most efficient industry. But what was its vocation, its purpose? In a celebrated phrase, Max Weber warned that unification would be "little else than a piece of folly which was committed by the nation in her old days, and which, in view of its costly nature," should not have been embarked upon at all unless Germany would now take the next step and become a world power.

What was it that Germany sought after 1890? It sought what every aspiring nation in Europe had sought before: recognized greatness, a measure of hegemony. Perhaps the Germans were more frantic in their search for greatness, but then their day had come late. They wanted grandeur, as others had wanted it before; they needed to exorcise centuries of dependency. Europe's competition for greatness, which involved more than political dominion, its ethos of heroic striving, was the very hallmark of its exacting and triumphant civilization. Europe had always been a crucible of genius. Should the Germans—the originators of the Faustian myth—restrain their will and not try for collective preeminence?

Let me cite one more example of this exhortation to greatness. Few scholars in Germany were as critical of the nation's development and of Bismarck's character as was the great historian Theodor Mommsen; few railed as much against the servility and political nonage of their fellow citizens, few combatted anti-Semitism as vigorously as he did. But he too felt the grip of greatness, and he too preached sermons of duty, as did the professoriate throughout Europe. I believe the call to greatness gnawed at Germans more deeply than at others. Witness Mommsen's rectorial speech:

> Of course we are proud of being Germans and we do not disguise it. Of all the boasts, none is more empty and less true than the boast about German modesty. We are not at all modest and we do not want to be modest or appear to be. On the contrary, we want to continue to reach for the highest in art and science, in state and church, in all aspects of life and striving, and we want to reach for the highest in everything and all at once. There is no laurel wreath which would be too magnificent or too ordinary for us…we think it normal that our diplomats as our soldiers, our physiologists as our sailors stand everywhere in the front rank… But even if we content ourselves in no way to be content, we are not therefore blind… in research and instruction… there is no standing still… if you don't go forward, you stay back and fall behind.

Even in Mommsen, then, we find this call for greatness as the only alternative to decline.

The contradictions of imperial Germany have often been noted. Economic giant, master of the disciplined society, model of technical proficiency, nurturer of talent—and yet a nation that recklessly defied prudence in dealing with foreign nations, a governing class that suffered from the paranoid fear that Germany was threatened by subversion at home and encirclement abroad. This alternation between presumption and anxiety grated heavily on some Germans and on many foreigners.

In Germany as elsewhere, the generation before 1914 was prodigious in talent and achievement; it was then that Germany attained a preeminent place in the natural sciences. German universities, which had long thought scientific studies a secondary concern, suddenly discovered that their scientists had won world renown. Universities had not been inhospitable to talent—provided it came in politically respectable, male, preferably Protestant guise. Any deviation from the norm had to be paid for by a superabundance of talent and, on the whole, was admissible only in the newer and politically neutral fields, such as medicine and the natural sciences. In those fields, achievement was more easily measurable and more immediately useful. In medicine and physics, in particular, the barriers against Jews began to be breached early.

It is notoriously hard to account for creativity. Was the German flowering a result of Germany's having the highest rate of literacy and the highest per capita expenditure for public education? Was it that German industry decided early on to support scientific research with particular largesse? Was the system of the master and the apprentices, the professor and his school, productive not only of dependency and exploitation but also of a special bond? Did success have anything to do with the fact that German scientists had a particularly austere view of their profession, so much so that one of the early Nobel laureates, Wilhelm Ostwald, spoke of all the grief and loneliness that a true scientist must endure," because every important discovery must be paid for by a human life. No doubt there was much suffering, acknowledged and unacknowledged, but there was also an extraordinary measure of camaraderie and high spirits.

In our century, scientific achievement, whatever its causes, can be measured by the incidence of Nobel Prizes won. From the inception of that prize to the rise of Hitler, Germans garnered a larger share of prizes than any other nationality, about 30 percent. In some fields the share was higher still; of these German Nobel Prizes, German Jews won nearly 30 percent; in medicine, 50 percent. Germans and Jews collected a disproportionate share; and although Harriet Zuckerman has recently demonstrated how complicated the notion of disproportionateness is, it does seem clear that Germans and Jews shared a certain immodesty in talent.

The prominence of German Jews also says something about their place in culture, about the milieu in which they worked, which mixed, perhaps uniquely, hospitality and hostility—and perhaps both were needed for this extraordinary achievement. I have tried to suggest that Germans had a veneration for learning, a yearning for greatness, a lingering insecurity. German Jews shared these traits and found further sustenance for them in their own distinct past. Jews did not foster talent; they hovered over it, they hoarded it, they nearly smothered it. Elsewhere I have pointed out that the rise of German Jewry is one of the most spectacular leaps of a minority in the social history of Europe, but their new prominence was painfully precarious and recalled Disraeli's desperate boast to young Montefiore: "You and I belong to a race that can do everything but fail." It is impossible to talk about Einstein's Germany without talking of German-Jewish relations, and to this theme I shall return later.

I have tried to suggest some of the contradictory aspects of German culture. At this point we may be more familiar with the darker sides, with possible portents of later disaster. Einstein seemed peculiarly attuned to these portents; his friends, as we shall see, relished the virtues of German life. Perhaps our own I. I. Rabi said it all when he remarked to me the other day that he had found German culture "brutal and brilliant," that he had come to post-1918 Hamburg "knowing the libretto but learning the tune." In the first third of this century German institutes of learning and research orchestrated many voices into one tune of discovery. The German contribution to our civilization was immense, and the greatness that eluded Germans in politics they realized in the realms of science and of art.

Let me now turn to Einstein and his experiences with Germany. We know little of his early life. He was no child prodigy; rather, his reticence in speaking for the first three years, his difficulty with learning foreign languages, and his mistakes in computation have been a source of endless comfort to the similarly afflicted or to their parents, though affinity in failure may not suffice for later success. He went through a brief but intense religious phase, the end of which, he said, left him suspicious of all authorities. His parents, secularized Jews, had little to do with his intellectual development; an uncle fed his mathematical curiosity. His father was an amiable failure, mildly inept at all the businesses he started. In 1894, his parents went to Italy to start yet another business, leaving the fifteen-year-old Albert behind in a well-known Munich gymnasium. The authoritarian atmosphere and the mindless teaching appalled him. There is more than a hint of arrogance about the young Einstein, and hence it does not strain credulity to believe that his teacher exclaimed: "Your mere presence spoils the respect of the class for me." He was a rebel from the start.

Encouraged by his teachers' hostility, he decided to quit school and leave Germany. His unsuccessful career facilitated his later fame in Germany: Erik Erikson has rightly referred to "the German habit of gilding school failure with the suspicion of hidden genius." It is often said that Einstein left school because he objected to its militarism. I find this unpersuasive: Bavarian militarism? I would suppose that there might have been stifling Catholicism, insolent, thoughtless authoritarianism, a repulsive tone—all of which would have sufficed to discourage a youth like Einstein. I suspect Einstein left so precipitously in order to escape serving in the German army; by obtaining Swiss citizenship in time, he could do so without incurring the charge of desertion. His first adult decision, then, was to escape the clutches of compulsion—and the image of Einstein as a recruit in a field-grey uniform does boggle the mind. He left Germany without regrets. His first encounters with that country had not been happy.

There followed the obscure and difficult years in Switzerland, the failures, the marginal existence, the Zurich Polytechnic, and, finally, the security of the patent office in Bern. From there in 1905 emerged the four papers destined to revolutionize modern physics and cosmology. They were published in the *Annalen der Physik*, and Max Planck was the first man to recognize the genius of the unknown author. The international scientific community took note as well, and Einstein finally received his first academic appointments. In 1914, while he was a professor at the Zurich Polytechnic, two German scientists appeared, Walter Nemst and Fritz Haber, in order to offer him an unprecedented position: salaried membership in the Prussian Academy of Sciences so that he would not have to teach, though he would have a chair at the university as well. When Nernst and Haber left, Einstein turned to his assistant, Otto Stern, and said: "The two of them were like men looking for a rare postage stamp." The remark was perhaps an early instance of that self-depreciatory humor, that modesty of genius.

As a native Swabian, Einstein found Prussian stiffness uncongenial; the gentler, less strident rhythm of southern Germany or Switzerland was more to his taste. He began his new German life in April 1914 with some trepidation at his "Berlinization," as he called it. Berlin was the world's preeminent center of the natural sciences, and Planck, Haber, and a dazzling array of talent rejoiced at having this young genius at the head of their circle. Three months later the war shattered the idyllic community. Einstein had returned to Germany in time to see the country seized by the exaltation of August 1914, when almost all Germans were gripped by an orgy of nationalism, by a joyful feeling that a common danger had at last united and ennobled the people.

The intoxication passed; the business of killing was too grim to sustain the unbridled enthusiasm of August 1914. The elite rallied to the nation—as it did elsewhere too. In the fall of 1914, ninety-three of Germany's best-known scientists and artists, including Planck, Haber, and Max Liebermann, signed a manifesto that was meant to repudiate Allied charges of German atrocities, but by tone and perhaps unconscious intent argued Germany's complete innocence and blamed all misfortunes and wrong-doing on Germany's enemies. The manifesto of the ninety-three has often been seen as a warrant for aggression, as a declaration of unrestrained chauvinism. I suspect it was as well the outcry of people to whom the outside world mattered and who intuitively sensed that the Allies would come to cast Germans as pariahs again. Some of the ninety-three probably hoped for continued respect across the trenches—and signed a document that had the opposite effect. It was not the last time that Germans confirmed the sentiments they set out to deny. With but few exceptions, intellectuals everywhere joined in this chorus of hatred and in the cry for blood. So did the guardians of morality and the servants of God, the priests who sanctified the killing as an act of mythical purification. In time, some of the ninety-three turned moderate—or perhaps they remained the patriots they had been—but others passed them on the right, in the nation's wild leap to pan-German madness. Einstein was alone and disbelieving. The war that was to politicize everyone as the cause of universal grief politicized him as well. Before 1914 he had never concerned himself with politics; his very departure from Germany had been a youthful withdrawal from the claims of the state. Now, for the first time, he ventured forth from his study, convinced of the insanity of the war, shocked by the ease with which people had broken ties of international friendship and mutual respect. A pacifist asked him to sign a counter-manifesto addressed to Europeans, demanding an immediate, just peace, a peace without annexations. It was the very first appeal he ever signed. It was never published—for want of requisite signatures. Somewhat later he joined a tiny group of like-minded democrats and pacifists. In November 1915 the Berlin Goethebund asked for his opinion about the war, and he sent a message with this rather special ending: "But why many words when I can say everything in one sentence and moreover in a sentence which is particularly fitting for me as a Jew: Honor Your Master Jesus Christ not in words and hymns, but above all through your deeds."

His work remained his central passion. But intermittently he forsook his work in order to bear witness in an unpopular cause for what he took to be right. He had been a pacifist and a European of the first hour, never touched by the frenzy that ravaged nearly all. Convinced of Germany's special responsibility for the outbreak and the continuation of the war, he hoped for its defeat.

To understand Einstein's isolation, one must look at the responses of his friends and colleagues. Fritz Haber, for example, became the very antithesis of Einstein. Haber, Einstein's senior by nineteen years, was a chemist of genius, a born organizer, and in wartime an ardent patriot. Without Haber's process for fixing nitrogen from the air, discovered just before the war, Germany would have run out of explosives and fertilizers in the first six months of the war. During the war, he came to direct Germany's scientific effort; in 1915 he experimented with poison gas and supervised the introduction of the new weapon at the Western Front. In order to operate within a military machine that had no understanding of the need for a scientist, he received the assimilated rank of colonel. He relished his new role; the marshaling of all one's talents and energies in a cause one believes in and in the shadow of danger—that is a heady experience. Einstein, the lonely pacifist who had come to feel his solidarity with Jews, and Haber, the restless organizer of wartime science and a converted Jew—the contrast is obvious. For all their antithetical responses, Haber and Einstein remained exceptionally close and, on Haber's side, loving friends. Haber's life was a kind of foil to Einstein's, and it encompassed the triumphs and the tragedy of German Jewry. I shall return to him because his relations with Einstein were so important—and because he happened to have been my godfather and paternal friend of my parents.

Einstein had been horrified at the beginning of the war, but I doubt that even he could have imagined the full measure of disaster: the senseless killing and maiming of millions, the starving of children, the mortgaging of Europe's future, the tearing of a civilization that appeared ever more fragile. For what? Why? Einstein blamed it on an epidemic of madness and of greed that had suddenly overwhelmed Europe—and Germany most especially. The old German dream of greatness had turned into a nightmare of blind and brutal greed. During the later phases of the war, Einstein was again totally absorbed in his work, but whiffs of hysteria would reach him—and always from the German side, I doubt that he knew of the excesses on the other side.

Einstein had been right about the war. At its end, many felt as he had at the beginning. The war was a great radicalizing experience, pushing most people to the left and some to a new, frantic right. If there had been no war, bolshevism and fascism would not have afflicted Europe. The war discredited the old order and the old rulers; antagonism to capitalism, imperialism, and militarism appeared everywhere. Lenin's Bolsheviks offered themselves as the receivers of a bankrupt system; bolshevism was a speculation in Europe's downfall. Liberal Europeans pinned their hopes on Woodrow Wilson, but that hope faded in the vengeful spirit of Versailles. The logic of events had brought many Europeans to share Einstein's radical-liberal, faintly socialist, thoroughly internationalist views.

For a short time Einstein had hopes for Germany. Defeat had brought the collapse of the old and the rise of a new, democratic regime, as he had expected. He supported the new republic, and in November 1918, at the height of the German Revolution, cautioned radical students who had just deposed the university rector: "All true democrats must stand guard lest the old class tyranny of the right be replaced by a new class tyranny of the left." He warned against force, which "breeds only bitterness, hatred and reaction," and he condemned the dictatorship of the proletariat in what was the first of his occasional bitter denunciations of the Soviet Union as the enemy of freedom. At other times and in different contexts, he would sign appeals of what we have come to call "front organizations."

We now come to a fateful coincidence in the rise of the public Einstein. In March 1919 a British expedition headed by Arthur Stanley Eddington had observed the solar eclipse. In November it was announced that the results confirmed the predictions of the general theory of relativity. It was in London that the President of the Royal Society and Nobel Laureate, J. J. Thomson, hailed Einstein's work, now confirmed, as "one of the greatest—perhaps the greatest of achievements in the history of human thought." Somehow it seemed as if Einstein's achievement would revive the old international community of science. The world listened. Almost overnight Einstein became a celebrated hero—the scientific genius, untainted by war, of dubious nationality, who had revolutionized man's conception of the universe, newly defined the fundamentals of time and space, and had done so in a fashion so recondite that only a handful of scientists could grasp the new mysterious truth.

The new hero appeared, as if by divine design, at the very moment when the old heroes had been buried in the rubble of the war. Soldiers, monarchs, statesmen, priests, captains of industry—all had failed. The old superior class had been found inferior; *Disenchantment* was the proper title for one of the finest books written about the war. "Before 1914," Noel Annan has asserted, "intellectuals counted for little"; after the war, and in a sense in the wake of Einstein, they counted for more. Einstein now became a force, or at least a celebrity, in the world.

After 1919, he appeared more and more often as a public figure. His views were continually solicited, and he obliged with his ideas about life, education, politics, and culture. He had a special kinship with other dissenters from the Great War; like Bertrand Russell, Romain Rolland, and John Dewey, he became what the French call *un homme de bonne volonté*. His views—rational, progressive, liberal, in favor of international cooperation, condemnatory of the evils of militarism, nationalism, tyranny, and exploitation—these views described as well a cast of mind characteristic of Weimar intelligentsia.

The intellectuals of Weimar—and this needs to be said at a time when Weimar is often portrayed as some sort of Paradise Lost—were a shallow lot in their moralizing politics. Their views often seemed utopian and simplistic, pious and fiercely polemical by turns. They were cynical, as Herbert Marcuse once put it to me about himself, because they knew how beautiful the world could be. They lived in a world peopled by George Grosz caricatures and three-penny indictments of bourgeois falsehood. It is perhaps too simple to say that they lived off the bankruptcy of the old order, but they did rather revel in the crudity of their opponents. It is not good for the mind to have dumb, discredited enemies. The real strength of Weimar were clusters of talent: Heidelberg around Max and later Alfred Weber; Göttingen in mathematics; the Bauhaus and the Berlin circles.

Einstein stood above these progressive intellectuals, in consonance with them, but usually more complicated, less predictable, and always more independent than they. But he too was a theorist without a touch of practical experience. Einstein offered his prescriptions the more readily because he had been so overwhelmingly right when the multitudes had been wrong. By 1919 he had not only overthrown the scientific canons of centuries; he had also defied conventional wisdom and mass hysteria in wartime. His views were often deceptively simple; they were not as naive as has often been alleged nor quite so profound as admirers thought. There was no reason to think that a scientific genius would have special insights into other realms. He had reflected on some issues and felt strongly on others; for the rest, it became clear that genius is divisible and can be compartmentalized.

Einstein's views and prescriptions were unassailably, conventionally well intended, but they often lacked a certain *gravitas*, a certain reality—in part, I think, because he approached the problems of the world distantly, unhistorically, not overly impressed by the nature or intractability of the obstacles to ideal solutions. He was not a political thinker; he was a philosopher, moralist, prophet, and the travails of the world would prompt him to propose or support social remedies. Sometimes these remedies would be blueprints of utopia addressed to people who had lost their footing in a swamp and were sinking fast.

At a much later time, in fact at a moment when Einstein had attacked the Nazi government, Max von Laue questioned whether the scientist should deal with political issues. Einstein rejected such considerations: "You see especially in the circumstances of Germany where such self-restraint leads. It means leaving leadership to the blind and the irresponsible, without resistance. Where would we be if Giordano Bruno, Spinoza, Voltaire, and Humboldt had thought and acted this way?" Einstein's models were instructive, and Laue pointed out that they were not exact natural scientists and that physics was so remote as not to prepare its practitioners for politics in the same way that law or history did. On that letter, Einstein simply scribbled—don't answer.

Like so many thinkers of the 1920s, Einstein underestimated the force of the irrational, of what the Germans call the demonic, in public affairs. That is what so ill-prepared them for an understanding of fascism. In their innocence they thought that men were bribed to be fascists, that fascism was but frightened capitalism; in its essence, it was something much more sinister and elemental. In his social commentary, Einstein left out the very thing he once called "the most beautiful experience we can have: the mysterious."

What gave his views exceptional resonance was the magic of his person and his incomparable achievement. He was taken by many as a sage and a saint. In fact, as I have said before, he was an unfathomably complex person. In the complexity of nature he found simplicity; in the complexity of his own nature, the principle of simplicity ranked high. Indeed, it was his simplicity, his otherworldliness, that impressed people. His clothes were simple, his tastes were simple, his appearance was meticulously simple. His modesty was celebrated—and genuine, as was his unselfishness. How many scientists or academics ask for a lower salary than offered, in the absence (let it be quickly added) of private means? He was a lonely man, indifferent to honors, homeless by his own admission, solicitous of humanity, and diffident about his relations with those closest to him. At times he appeared like a latter-day St. Francis of Assisi, a solitary saint, innocently sailing, those melancholy eyes gazing distractedly into the distance. At other times he was playing with the press, finding himself in the company of the famous and the powerful despite himself.

In some ways, I believe, he came to invest in his own fame, perhaps unconsciously to groom himself for his new role. He lectured in distant lands, "a traveller in relativity." In 1921, after his first visit to the United States he said: "The cult of individuals is always, in my view, unjustified.... It strikes me as unfair, and even in bad taste, to select a few [individuals] for boundless admiration, attributing superhuman powers of mind and character to them. This has been my fate, and the contrast between the popular estimate of my powers and achievements and the reality is simply grotesque." This admiration would be unbearable except that "it is a welcome symptom in an age which is commonly denounced as materialistic, that it makes heroes of men whose goals lie wholly in the intellectual and moral sphere.... My experience teaches me that this idealistic outlook is particularly prevalent in America." He knew that he had become a hero—and was endlessly surprised by it. In 1929 he described himself as a saint of the Jews. He played many roles by turns, each, I think, completely genuinely; he was a simple man of complex roles.

In the simplicity and goodness that were his, I detect, perhaps wrongly, a distant echo of his encounters with German life. Could one imagine a greater contrast between his German surroundings and himself, between people so formal in their bearing, so attentive to appearance, so solicitous of titles, honors, externals, and himself? Did the insolence of office, the arrogance of the uniform push him into ever greater idiosyncratic informality? Was not his appearance a democratic rebuke to authority?

In the immediate postwar era, Einstein was friendly to the governments of Weimar and appalled by the vindictiveness of the Allies, who seemed to have caught what he had thought was a German disease. In all his public stands he had what Gerald Holton has called a "vulnerability to pity," and in the early 1920s he had a fleeting moment of pity for Germany. He refused to leave it at the time of trial. For years he was an uncertain member of the League of Nations' International Commission on Intellectual Cooperation, intermittently resigning when he thought the Commission too pro-French, too *Allied*. He hoped to restore an international community, Germans included. In the end he asked Fritz Haber to take his place. Successive German governments regarded him as a national asset, perhaps the sole asset in a morally and materially empty treasury. They saw in his travels and in his fame the promise of some reflected glory. But his own hopes gradually faded. He had warned Walter Rathenau against assuming the foreign ministry; Jews should not play so prominent a role. When rightwing assassins—widely hailed in Germany as true patriots—killed Rathenau, Einstein had reason to fear for his own life. The inborn servility of the Germans, he thought, had survived the successive shocks of 1918.

Immediately after the war and at the beginning of his popular fame, Einstein embraced several causes. Having embraced them, he would often embarrass and repudiate them as well. He was the antithesis of an organization man. Unstintingly he would help individuals and chosen causes, but I doubt that he would listen to them. He remained a detached theorist who saw the nature of the world wantonly violated. But at times his commandments contained visionary practicality. A pacifist during the war, he now became Germany's most prominent champion of organized paci-fism. He hated militarism—blindly, as its defenders loved it—blindly. He condemned "the worst outgrowth of herd life, the military system … I feel only contempt for those who take pleasure marching in rank and file to the strains of a band. Heroism on command, senseless violence and all the loathsome nonsense that goes by the name of patriotism—how passionately I despise them!" This, surely, is exemplary of the spirit of the1920s, formed by the experience of the first war and soaked in the we-they antithesis that precluded understanding. It precluded the understanding that had led William James to plead for a moral equivalent of war, for something prac-tical that would make peaceful use of the old martial virtues. Einstein insisted that "the advance of modem science has made the delivery of mankind from the menace of war… a matter of life and death for civilization as we know it." But Einstein did not grapple with the psychological issues, with people's desire for danger and com-radeship. In his exchange with Freud about the nature of war he acknowledged that "the normal objective of my thought affords no insight into the dark places of human feeling and will." For Einstein, war was a disease, a disorder planted by men of greed, to be abolished by men of good will through the creation of international sovereignty or through a revolutionary pacifism, that is, through the refusal of men to bear arms in peace or war. He called for resistance to war; but in 1933, almost immediately after Hitler's assumption of power, he renounced pacifism altogether—to the fury of his doctrinaire followers. In fact, he urged the Western powers to prepare themselves against another German onslaught.

His second great cause was Zionism, which he seems to have embraced during the war. By November 1919 the *Times* of London referred to him as an "ardent Zionist... keenly interested in the projected Hebrew University at Jerusalem and [he] has offered to collaborate." By the early 1920s he became a public advocate of Zionism—to the surprise and likely dismay of many of his colleagues. Assimilated Jews must have found this reminder of Jewish apartness painful; internationalists would have boggled at the implied argument for a new national community. But Einstein had come to feel a sense of solidarity with Jews, especially with Jewish victims of discrimination, and he seemed to believe in the existence of an ineradicable antagonism between gentiles and Jews, especially between Germans and Jews—with the fault by no means all on one side. Hence his view that Jews needed a spiritual home and a possible haven. He specifically cited the discrimination that talented Jews from Eastern Europe and from Germany suffered at German universities.

In 1921 Chaim Weizmann persuaded Einstein to join him on a trip to the United States to raise money for the projected Hebrew University in Jerusalem. For Weizmann, Einstein's support was critical; for Einstein, his visit to Jerusalem in 1923 was a deeply moving experience. Still, there were conflicts. Einstein railed against the mediocrity of the American head of the university; he saw him as a creature of the crass American-Jewish plutocrats for whom Einstein had contempt even as he helped to lighten their financial burden. He quarreled publicly with Weizmann over the policies of the Hebrew University and repeatedly threatened to withdraw his sponsorship. He urged a Palestinian presence that would promote, not injure, Arab interests. In 1929, at the time of major attacks on Jewish settlements, he again pleaded with Weizmann for Jewish-Arab cooperation and warned against a "nationalism à la prussienne," by which he meant a policy of toughness and a reliance on force:

> If we do not find the path to honest cooperation and honest negotiations with the Arabs, then we have learned nothing from our 2000 years of suffering and we deserve the fate that will befall us. Above all, we should be careful not to rely too heavily on the English. For if we don't get to a real cooperation with the leading Arabs, then the English will drop us, if not officially, then de facto. And they will lament our debacle with traditional, pious glances toward heaven, with assurances of their innocence, and without lifting a finger for us.

Weizmann replied instantly, at the height of the Palestinian violence, with a four-page handwritten letter. He expounded his views, which were somewhere between

Against Einstein

Einstein's theories, and his support for internationalism, pacifism and the Weimar Republic made him a target for the radical right in post World War I Germany. A lecture series opposed to relativity and sponsored by the "Working Group of German Scientists for the Preservation of Pure Science" became popular in 1920 and sparked anti-Semitism. Philipp Lenard—who ironically conducted the photoelectric experiments for which Einstein's explanation won him the 1921 Nobel Prize—attacked "Jewish physics."

In 1931, a book appeared in Germany called *100 Authors against Einstein*. The book's thesis was that Einstein must be wrong because so many people thought so! The book was reviewed—and dismissed—in *Die Naturwissenschaften* with the reviewer concluding, "One can only hope that German science is not shown up by such depressing rubbish again."

By 1933, officially sanctioned anti-Semitism was on the rise as boycotts of Jewish businesses and professionals became frequent. Einstein heard of these while in the United States and decided not to return to Germany. He also resigned from the Prussian Academy of Sciences. His supporter Max Planck was unhappy with Einstein's decision, saying that it increased the difficulties for the Jews who remained. At the time, Planck thought that it was possible to prevent the National Socialists from taking over the Academy and using it for propaganda Ultimately, and tragically, he was to fail.

During World War I Planck had been swept up in the German nationalism that Einstein abhorred. After the war, however, he realized his mistake and managed to strike a balance between German nationalism and scientific internationalism. He worked for the Weimar Republic for the benefit of science, but eventually recognized the pernicious nature of the National Socialists towards science and society. Sadly, his son was executed for his part in a failed attempt to assassinate Hitler.

After Einstein's resignation, the Academy secretary publicly charged Einstein with slandering Germany, adding that the Academy had no cause to regret Einstein's resignation. At a subsequent meeting, this decision was approved, although Max von Laue insisted that the record show that no scientific members had been consulted. Planck regretted the decision, but did not oppose it. The Academy that rallied behind the new German government in 1933 was a far cry from the Academy that had wooed Einstein to Berlin in 1914.
— PH

Above and Below: Anti-Semitic cartoons celebrating Einstein's departure from Germany. **Left:** The Einstein Tower, a solar observatory designed by Erich Mendelsohn and built in 1921 near Potsdam, Germany.

Der Hausknecht der Deutschen Gesandtschaft in Brüssel wurde beauftragt, einen dort herumlungernden Asiaten von der Wahnvorstellung, er sei ein Preuße, zu heilen.

Zionist extremists and the irenic Einstein—who, in the meantime, had criticized the Jewish stance publicly. Weizmann pointed to the recalcitrance of the Arab leaders, their fanaticism, their inability to understand anything but firmness. He pleaded with Einstein to cease his injurious attacks on the Zionists. Of course they would negotiate in time, Weizmann insisted, but "we do not want to negotiate with the murderers at the open grave of the Hebron and Safed victims." Einstein remained skeptical. Weizmann, desperate to retain his support, had written to Felix Warburg a year earlier: "There is really no length to which I would not go to bring back to our work the wonderful and lovable personality—perhaps the greatest genius the Jews have produced in recent centuries and withal so fine and noble a character."

At the time of the greatest need for a Jewish home in Palestine, immediately after Hitler's seizure of power, Einstein formally broke with the Hebrew University and with Weizmann. The correspondence between the two men suggests all the intractable issues about Jewish-Arab relations, all the differences between the safe outsider and the practical statesman. In April 1938, Einstein resigned his position and again warned against a "narrow nationalism." Once again Weizmann explained that at the moment when five million Jews faced, as he put it, "a war of extermination," they needed the support of the intellectual elite of Jewry, and not, by implication, public criticism. Einstein was not an easy ally. To some he must have appeared as a man of conscience and of unshakable principle; to others, as an uncompromising fanatic in purity, impervious to practical exigencies. As Robert Oppenheimer put it in his memorial lecture: "He was almost wholly without sophistication and wholly without worldliness...There was always with him a wonderful purity at once childlike and profoundly stubborn."

It would be hard to imagine three causes less pleasing to the bulk of the German professoriat than liberal internationalism, pacifism, and Zionism. Unlike many academics, Einstein took education with the utmost seriousness—and academics with magnificent irreverence. He had great faith in the possibilities of primary and secondary education; at one point he said that if the League of Nations could improve primary education, it would have fulfilled its mission. His ironic contemplation of universities found expression in private letters. He once complimented his close friend Max Wertheimer, the Gestalt psychologist: "I really believe there are very few who have been so little harmed by learning as yourself." In 1924 he wrote: "In truth, the University is generally a machine of poor efficacy and still irreplaceable and not in any essential way improvable. Here the community must take the point of view that the biblical God took towards Sodom and Gomorrah.

For the sake of very few, the great effort must be made—and it is worth it!"

Einstein's success—the enormous acclaim, especially abroad at a time when most German scientists were still banished from international meetings—caused much ill will at home. His opinions enraged the superpatriots. Some physicists condemned the fanfare surrounding the dubious theory of relativity; one fellow laureate attacked it as "a Jewish fraud." To anti-Semites, Einstein became a favorite and obvious target. The waves of hatred spilled from the streets into the lecture halls, and Einstein's occasional and sometimes ill-considered deprecations only made things worse.

Germany frightened him again. His hopes for the Weimar Republic had dimmed. As early as 1922, his life was threatened. He traveled even more than before, but still he refused handsome offers from Leiden and Zurich, universities with which he had the closest ties. An American university also dispatched a letter of invitation, hitherto buried in the archives, dated 26 February 1923; the timing was delicately perfect because Germany found itself in the throes of foreign occupation and rampant inflation.

> Dear Professor Einstein:
>
> The duties and occupations of the professorship will be precisely what you wish to have them. Our aim will be to make it easy and convenient for you to pursue your personal studies and course of reflection without onerous academic or public burdens of any kind. The annual salary of the Professorship is $10,000 or about 40,000 gold marks. I wish to put this invitation before you on behalf of [Columbia] University with all possible urgency, and to beg you to accept it in the interest of science, of international comity, and, I trust, of the most successful prosecution of your own personal work.
> (Signed) Nicholas Murray Butler

Alas for Columbia, Einstein declined, though pleased to have been asked "by your magnificent university." He stayed despite his misgivings about Germany; he stayed because Berlin in the 1920s was the golden center of physics; he stayed because proximity to Planck, Laue, Haber, and others was a unique professional gift, because, as he wrote Laue in 1928: "I see at every occasion how fortunate I can call myself for having you and Planck as my colleagues." In 1934 he wrote Laue that "the small circle of men who earlier was bound together harmoniously was really unique and in its human decency something I scarcely encountered again." In 1947 he wrote Planck's widow that his time with Planck "will remain among the happiest memories for the rest of my life."

The unpublished correspondence among these men suggests even more than a professional tie. The letters bespeak a degree of humane collegiality, a shared pleasure in work, as well as a delicacy of sentiment, a candid avowal of affection, which in turn would allow for confessions of anguish and self-doubt, of melancholy as well as high spirits. They spoke of joys and torment, in close or distant friendship, in an enviable style. The letters also breathe a kind of innocence, as if science was their insulated realm, nature the great, enticing mystery, and one's labors of understanding exclusively an intellectual pursuit, remote from social consequences. Such clusters of collaboration and of friendship have always existed, I suppose, and they have made life better and infinitely richer. Germany may have had a special knack for breeding them.

Einstein's Germany included gentiles and Jews, working together in extraordinary harmony. And still it can be stated categorically that none of the Jewish scientists escaped the ambiguity, the intermittent hostility, that Jewishness produced in Imperial and Weimar Germany. Neither fame nor achievement, neither the Nobel Prize nor baptism offered immunity. Passions were fiercer in Weimar, in that cauldron of resentments. Most official barriers against Jews had been lowered, but new fears and hatreds came to supplement old prejudices. Three incidents may illustrate the uncertain temper of the time. In 1921, Haber begged Einstein not to go to America with Weizmann, on the ground that Germans would take amiss his travels in Allied countries with Allied nationals at the very time when the Allies were once again tightening the screws against Germany. To persuade Einstein, Haber warned that German anti-Semites would capitalize on his seeming desertion and that innocent Jewish students would be made to suffer. Anti-Semitism, rampant as it was, need not be goaded; Einstein's warning to Rathenau originated in a similar apprehension. Or take another incident. In 1920, a well-known physicist opposed the appointment of the later laureate Otto Stern: "I have high regard for Stern, but he has such a corrosive Jewish intellect."

Or consider this last example. In 1915 the king of Bavaria, confirming the Nobel laureate Richard Willstaetter's appointment to a professorship, admonished his minister: "This is the last time I will let you have a Jew." Ten years later, discussing with his colleagues a new academic appointment, Willstaetter proposed a candidate. A murmur arose: "another Jew." Willstaetter walked out, resigned his post, and never entered the University again, the unanimous pleas of his students notwithstanding. For the next fourteen years he had daily, hour-long telephone calls with his assistant so that she could conduct the experiments in a laboratory that he would no longer enter. A man of conscience and of courage, someone who did not blink at the reality of anti-Semitism. But his stand in 1924 was his undoing a decade later. A devoted German, but now no longer a civil servant, he assumed that the Nazis would leave untouched a private scholar. He believed that some Jews had contributed to this new storm. He could not comprehend the radical newness of the phenomenon. In February 1938 he wrote my mother urging her that one ought not to leave Germany without the most careful reflection. He himself refused exile until the aftermath of Crystal Night forced him into it.

I cite Willstaetter's example among many precisely because of its contradictory nature: awareness of anti-Semitism could cloud one's perception of Nazism. If anti-Semitism had always existed, then perhaps Nazism was but an intensification of it. It is not uncommon these days to hear summary judgments about German Jewry, about their putative self-surrender, their cravenness, or their opportunism. These judgments often have a polemical edge; they are likely to do violence to the past and to the future: the myth of yesterday's self-surrender could feed the delusion of tomorrow's intransigence. If our aim is to understand a past culture, we must note that German-Jewish scientists thought Germany their only and their best home, despite the anti-Semitism that crawled all around them. They may have loved not wisely but too well, and yet their sentiments are perhaps not so much an indictment of themselves as a tribute to the appeals of Germany. We owe that past no less than what we owe any past: a sense of its integrity.

Let me hasten to the denouement. In 1932 Einstein left Germany provisionally, with the intention of returning to Berlin for one semester each year. Hitler's accession to power the next year changed all that. Einstein immediately denounced the new regime, and in response he was extruded from the Prussian Academy, his books were burned, his property seized. The first Nazi decrees for the purification of the universities would have allowed some Jews to maintain their positions. Einstein's non-Aryan friends spurned such sufferance and resigned. German physics was decimated, and a few remaining masters battled to defend some shreds of decency, some measure of autonomy. Laue once wrote Einstein that in teaching the theory of relativity he had sarcastically added that it had of course been translated from the Hebrew. Even such jokes—to say nothing of Laue's eulogies of Jewish colleagues—aroused Nazi wrath. The Nazis proscribed the very mention of Einstein, even in scientific discussions. They would have wished him to be an unperson.

For most, exile was hard; the habits of a lifetime are not easily shaken. For others, as the physicist Max Born put it, "a disaster turned out to be a blessing. For there is nothing more wholesome and refreshing for a man than to be uprooted and replanted in completely different surroundings." Resiliency was a function of age and temperament. For Haber, exile was a crushing blow and led to a final irony in his relations with Einstein. By mid-1933 he wrote to Einstein that as soon as his health would allow it, he would go to Palestine, but in the meantime he begged Einstein to patch up his public quarrel with Weizmann. Einstein replied at length: "pleased… that your former love for the blond beast has cooled off a bit. Who would have thought that my dear Haber would appear before me as defender of the Jewish, yes even the Palestinian cause. The old fox [Weizmann] did not pick a bad defender." He then lashed out against Weizmann and concluded:

> I hope you won't return to Germany. It's no bargain to work for an intellectual group that consists of men who lie on their bellies in front of common criminals and even sympathize to a degree with these criminals. They could not disappoint me, for I never had any respect or sympathy for them aside from a few fine personalities (Planck 60% noble, and Laue 100%). I want nothing so much for you as a truly humane atmosphere in which you could regain your happy spirits (France or England). For me the most beautiful thing is to be in contact with a few fine Jews—a few millennia of civilized past do mean something after all."

The German patriot Haber died a few months later in Basel, enroute to Palestine. And Einstein found a refuge at the Princeton Institute under conditions not dissimilar from what the Prussian Academy had offered him twenty years earlier. For as Erwin Panofsky has said of the Institute for Advanced Study, it "owes its reputation to the fact that its members do their research work openly and their teaching surreptitiously, whereas the opposite is true of so many other institutions of learning."

Einstein's public life continued to be dominated by his fear of Germany. He warned the West against a new German onslaught. He abandoned the pacifism he had so fervently espoused and in 1939 signed the famous letter to President Franklin D. Roosevelt urging the administration to prepare the United States because Germany might develop nuclear fission for military purposes. In the winter of 1945, when Germany was desolate in defeat and when the Morgenthau spirit, if not the plan, had a considerable grip on American thinking, a fellow laureate and old friend, James Franck asked Einstein to sign a manifesto of exiles that would appeal to the United States not to starve the German people. Einstein vowed that he would publicly attack such a plea. The German police—and he said this eleven months after the war—were still killing Jews in the streets of Germany; Germans had no remorse, they would start another war. Franck pleaded with him that to give up all hope for a moral position in politics would be tantamount to a Nazi victory after all. But Einstein, who had signed so many appeals that he himself once said he was not a hero in no-saying, scathingly rejected Franck's plea. For him, genocide was Germany at its most demonic; after Auschwitz he could muster no magnanimity. Even the righteous could not redeem the "country of mass murderers," as he called Germany. He rebuffed Laue's plea to help a young German physicist. He knew that Planck, who lost one son in the first war, had lost another whom the Nazis murdered because of his participation in the plot against Hitler. The serene Einstein, always the champion of the rights of the individual against the collectivity, now proclaimed the principle of collective guilt. At that moment, of course, the world shared Einstein's horror at German inhumanity. But in him the violence of sentiment, the total absence of that vulnerability to pity, puzzles, for it shows how desperately deep and all-consuming had been his antipathy to Germany.

Even his postwar laments about America, his horror at McCarthyism, were shaped by his image of Germany. America, he believed, was somehow following the path of Germany. The world of politics he saw through German eyes—always

But let me end on a different note. Greatness in any guise is not in vogue today, not in my discipline and not in our culture. Historians feel that it is now the turn of the forgotten, both for reasons of retroactive justice and for heuristic purposes. The argument is compelling, the feeling comprehensible, though there is an implicit deprivation involved, for as Einstein in a rather German formulation put it: "the example of great and pure individuals is the only thing that can lead us to noble thoughts and deeds." We are uncomfortable even with the rhetoric of greatness, devalued as it so often has been. I would simply say that I find it inspiriting to look upon great peaks, as from an alpine village, and contemplate the distant mountain —old, awesome, unattained and unattainable, mysterious.

It is often asserted that a culture must be judged by its treatment of minorities and deviants; a student of the German past would find this a cogent and, indeed, irrefutable argument. It is a necessary, but not a sufficient criterion. A culture must also recognize, recruit, and, in a sense, form talent; it must know how to coax talent into achievement. This too is a test of its virtue and of its instinct for survival. These are responsibilities that speak most directly to our universities, to every university.

PART TWO

EINSTEIN'S EARLY SCIENCE

ANNALEN
DER
PHYSIK.

BEGRÜNDET UND FORTGEFÜHRT DURCH

F. A. C. GREN, L. W. GILBERT, J. C. POGGENDORFF, G. UND E. WIEDEMANN.

VIERTE FOLGE.

BAND 17.

DER GANZEN REIHE 322. BAND.

KURATORIUM:

F. KOHLRAUSCH, M. PLANCK, G. QUINCKE,
W. C. RÖNTGEN, E. WARBURG.

UNTER MITWIRKUNG
DER DEUTSCHEN PHYSIKALISCHEN GESELLSCHAFT
UND INSBESONDERE VON
M. PLANCK

HERAUSGEGEBEN VON

PAUL DRUDE.

MIT FÜNF FIGURENTAFELN.

LEIPZIG, 1905.
VERLAG VON JOHANN AMBROSIUS BARTH.

The title page of *Annalen der Physik,* Volume 17, 1905, in which Einstein's first relativity paper appeared. Having already published in this journal, Einstein had acquired the privilege to publish as many papers as he wished in its pages. He could therefore publish five papers in quick succession during that year. (Preceding page: Einstein as a bureaucrat in the Swiss Patent Office.)

How Can We Be Sure That Albert Einstein Was Not a Crank?

by Jeremy Bernstein

From time to time I entertain myself with the following fantasy: The year is 1905. I am a professor of physics at the University of Bern. For many years, I have been teaching, probably from the same set of notes, respectable courses based on what is for me the familiar and comfortable physics of the nineteenth century. I teach the mechanics of Newton, the relatively modern theories of electricity and magnetism of James Clerk Maxwell, along with good solid nineteenth-century thermodynamics. I believe that atoms exist although I am troubled occasionally by the question that, around the turn of the century, Ernst Mach asked Ludwig Boltzmann: "Have you seen one?" All in all, it is a good, comfortable life. Then, with no warning at all, a series of physics papers begins arriving in the mail. They carry the return address of the Swiss National Patent Office in Bern. The covering letter identifies their author as a patent examiner—a technical expert "third class"—of whom I have never heard. He does not even have a doctoral title. Upon browsing through the papers, I discover that this doctorless unknown is claiming—using totally unfamiliar kinds of reasoning—that essentially all of the physics I have been teaching is wrong. Not just wrong in a few minor details, but fundamentally wrong. What would my reaction be? What should it have been? In short, how could I then have known that the author of these papers—the twenty-six-year-old Albert Einstein—was not a crank?

Before responding to this question, let me note that, as a matter of historical fact, at least one contemporary individual did receive these papers in more or less the way I just described. He was a young man named Conrad Habicht. Einstein had met him in 1901 in Schaffhausen, in a private school where Einstein had gone to teach since no proper university would hire him. Habicht was studying mathematics and later taught it in high school. But in 1903, he, Einstein, and a third young man named Maurice Solovine founded an entity in Bern they called the "Olympia Academy." Solovine had answered an ad in a Bern newspaper—placed there by Einstein—offering physics lessons for three Swiss francs an hour, and in this way became Einstein's lifelong friend. The so-called academy met every week to discuss intellectual matters—principally philosophy. In later life, Einstein recalled that, when he founded the theory of relativity, their readings of Hume had as much influence on him as anything else. On one of Einstein's birthdays, Habicht made an attempt to introduce him to caviar. The experiment came to naught since on that particular evening Einstein was lecturing to the "academy" on Galileo's principle of inertia and was so absorbed that he paid no attention to what he was eating. In 1905, the "academy" broke up. Its three members went their separate ways, but continued to communicate by letter. That year Einstein sent Habicht a letter that began, "I promised you four papers…" It was these four papers that laid the foundations of twentieth-century physics. The only one of the four that Einstein considered, in his words, "very revolutionary" led eventually to the discovery of the quantum theory. The last of the papers—"employing," as he said, "a modification of space and time"—created the theory of relativity. One can only wonder about Habicht's reaction to this letter.

Now back to my fantasy. It is the year 1905, and I, Jeremy Bernstein, have just received a copy of this letter and the four papers contained therein. Not having any hint of who this Einstein might be—after all Habicht *knew* who he was—what will I do with them? Will I even bother to look at them at all? Most physicists, and especially those of us who write about science for the general public, receive crank papers all the time, which we routinely throw away without reading or, at least, without reading carefully. How can we be sure that we have not thrown away a paper written by another Einstein? How can we know? What criteria can we use? As I describe my answer to these questions, I think that you will see that they apply to science and not to art. They will not, I think, help you to decide what to make of Marcel Duchamp's *Nude Descending a Staircase*, James Joyce's *Finnegans Wake*, or John Cage's *Four Minutes and Thirty Three Seconds*. If you have been exposed to the last, you will know that during the specified time interval, a pianist sits at a piano in front of an audience and does nothing. I think, however, my criteria will help you to distinguish between a theory of relativity and an idea for making a perpetual-motion machine. I also think that exploring this matter will take us deep into the nature of science itself.

I would propose two criteria to help us distinguish between crank science and the real thing: "correspondence" and "predictiveness." The term *correspondence* I have stolen from Niels Bohr, who used what he came to call the "correspondence principle" to help him construct the first quantum theory of the atom. He began working on it in 1911 when he was a visitor at the University of Manchester. Bohr pictured the atom as consisting of a tiny, massive, positively charged nucleus (the atomic nucleus had been discovered by Bohr's mentor at Manchester, Ernest Rutherford, a year or two earlier) around which circulate the negatively charged electrons of much smaller mass. Bohr's novel suggestion was that these electrons are allowed to circulate only in selective orbits—"Bohr orbits," we call them. The picture of these circulating electrons has become one of the defining symbols—logos—of the atomic age. If one studies the Bohr orbits for, say, hydrogen, one discovers that the various ones closest to the nucleus have distinctive energies. The energy difference between the electrons in one orbit and those in another is substantial. According to Bohr, when this energy difference is given up as an electron transits from orbit to orbit, it reappears as the energy of the light quanta given off by the atom. These orbital transitions account for the beautiful spectra that are characteristic of these elements. For the orbits in which the electron is far removed from the nucleus, however, the energies merge into each other, creating a kind of continuum of energies. The quantum nature of the electron's motion becomes unimportant. We say that the motion has become "classical," in that, for all practical purposes, it obeys the laws of classical physics. That there is a well-defined classical limit to Bohr's theory is crucial. Otherwise, we could not understand why quantum effects are imperceptible to us. Nothing in our daily experience suggests that the orbits we move in are restricted in any way by the quantum theory. If we whirl a stone around our heads, the size of the circle is up to us. It is true that there are Bohr orbits for this situation as well, but they are so close together that we are never aware of them. The melding of quantum and classical concepts in a well-defined limit is Bohr's correspondence principle. I would like to use it in a more general sense. I would insist that any proposal for a radically new theory in physics, or in any other science, contain a clear explanation of why the precedent science worked. What new domain of experience is being explored by the new science, and how does it meld with the old?

To me, this is such an important benchmark for distinguishing real science from its imitations that I would like to illustrate it with another example. I would like to compare some of the proposals in Einstein's 1905 relativity paper with a hypothetical paper I might well also have received in that year in which the author claims to have discovered a perpetual-motion machine. This marvelous device, the author assures us, will, once started, continue to operate indefinitely without any additional infusion of energy. (A cynic might say that this is a fairly decent description of academic life.) Let me contrast this with one of the outré proposals in Einstein's paper: namely, that the mass of an object *increases* with its velocity. In fact, Einstein's claim goes even further: when an object approaches the speed of light—186,320 miles a second—its mass becomes *infinite*! When I first learned of this as a freshman at Harvard in 1948, I thought it the most astounding thing I had ever heard of—although, if pressed, I would have had a hard time explaining what was meant by the term *mass*.

At first sight, this proposal of Einstein's seems to be entirely mad. We know that mass is a measure of the resistance of an object to an attempt to change its momentum. Nothing in our common experience gives us any hint that this resistance has anything to do with the speed at which the object is traveling. Why, then, am I not allowed to dismiss Einstein's paper out of hand as a violation of the generalized correspondence principle—as crank science? If we study it, we notice that all the novel relativistic effects—the ones that defy common sense—vanish in a world in which the velocity of light is infinite as opposed to simply being very large. The order of magnitude of these effects depends on the ratio of the velocity in question to that of light—squared in the case of the increasing inertial mass. The familiar pre-Einsteinian world is one in which light is transmitted instantaneously. The correspondence principle, applied here, means that we can make a smooth transition from the relativistic to the classical worlds by considering velocities that are very small compared with that of light.

One may well wonder how large these relativistic effects are in practice. It is amusing to contemplate what sort of speeds were available in 1905 for terrestrial vehicles. I believe that the fastest speed anyone had traveled around the time of the publication of Einstein's paper was about a hundred miles an hour. This is the speed an expert racer can get using one of the special sleds—luges—that run the Cresta ice chute in Saint-Moritz. On my first and only try, I, as an absolute novice, got going at fifty miles an hour. Experts were doing a hundred easily. The ratio of a hundred miles an hour to the speed of light is about 1 in 10,000,000. But the relativistic correction to the mass goes as the square of this: that is, in this case the correction is an effect of one part in ten raised to fourteen powers. It is little wonder that these effects play no role in our daily experience. It takes all the refinements of a physics laboratory working with high-speed particles to show them up.

Newton, Einstein and Mercury

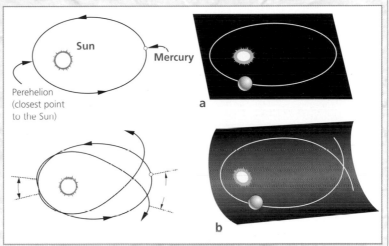

Left: Mercury's revolution around the Sun is not a perfect ellipse, as predicted by Kepler (top), but (bottom) one that slowly rotates as its perihelion advances. **Right:** Mercury's whirling orbit becomes clear when one views it not in flat, Newtonian space (a), but in curved spacetime (b).

Although Newton's and Einstein's ideas were revolutionary, they were conceived within the background of scientific theories of their day. Newton built on Kepler's third law, published in 1619, which related the time it took planets to orbit the sun with their mean distance from the sun. The result was Newton's universal law of gravitation, which related the force between objects to their masses and the distance between them: the larger the mass the greater the force; and the farther the distance between them, the less the force. The law states that gravitational attraction is a property of all matter—all particles of matter in the universe attract all other particles.

Among the successes of Newton's universal law of gravitation were the predictions of the existence of the planets Neptune and Pluto. In the mid nineteenth century astronomers had observed that the orbit of Uranus did not precisely follow the path predicted by Newton's law. This gave rise to speculation that another planet must be perturbing Uranus' orbit. A French astronomer, Urbain Jean Joseph LeVerrier, (1811-1877) made the calculations that led to the discovery of Neptune in 1846.

In 1845 LeVerrier had also predicted that according to Newton's gravitational law, the anomalous orbit of Mercury—the procession (spinning) of the ellipse of its orbit—must be due to another, as yet, unknown planet. But the missing planet was never found and the discrepancy in the orbit of Mercury was still a mystery when Einstein was developing his ideas about gravity.

Einstein saw gravity as the bending of four dimensional spacetime by large masses. Out of all the planets in our solar system, he suggested, this effect would be felt most strongly by Mercury, the planet closest to the large mass of the Sun. In 1917, Einstein calculated how the warp of spacetime near the sun would affect Mercury's orbit and his predicted results exactly matched those measured by astronomers.   —PH

Einstein himself was fully aware of the importance of correspondence. Indeed, of the four papers he sent to Habicht, he begins the one that he described as "very revolutionary" with a discussion of how correspondence applies to his new and radical views on light. Traditional optical phenomena, Einstein notes, were wonderfully well described by a theory in which light is thought of as a smooth wave.

But, he argues, these phenomena involve the average behavior of light over time. However, when light is emitted and absorbed, as in the Bohr atom, this is an instantaneous process, and the old ideas need not apply to it. This was the first hint of the duality between particles and waves that is the heart and soul of the modern quantum theory. Einstein's paper, as radical as it was, clearly spelled out why it was not in conflict with what one already knew.

Let me contrast this with the way a typical crank paper—say, on perpetual motion—presents itself. Cranks tend to use a lot of exclamation points and capital letters, but that is the least of it. To a professional physicist, it is immediately clear that any perpetual-motion machine violates what is known as the second law of thermodynamics. This law states that in any realistic process entropy—the measure of disorder—increases. Machines stop because the production of entropy by friction slows them down. Of course, like any law of physics, the second law of thermodynamics must, in the last analysis, be based on experience. But all our experience, academic life aside, is consistent with its validity. At the very least, the crank must explain why his device (all the crank papers I have so far seen have been written by men) alone of all the machines ever constructed violates this law. What has everyone overlooked? Where is the correspondence? Whoever actually takes the trouble to read these papers will find that this question never occurs to the crank. The crank is a scientific solipsist who lives in his own little world. He has no understanding nor appreciation of the scientific matrix in which his work is embedded. I would gladly read a paper on perpetual motion which began by explaining why we had all been overlooking something about entropy, something that makes a correspondence with the science we know. In my dealings with cranks, I have discovered that this kind of discussion is of no interest to them. If you find a specific flaw in their machine, they will come back the next day with a new design. The process never converges.

The second criterion that genuine science should satisfy is predictiveness. Real scientific ideas cry out to be tested in laboratories. Einstein's 1905 papers are, as I will discuss, full of predictions. This, too, is in complete contrast to the typical crank paper. Papers on perpetual-motion machines only predict that a given machine will run perpetually. One wonders how even *that* is supposed to be tested. (There are, of course, crank papers that predict the end of the world, which—considering the way things are going—may give one pause.) I would like to discuss two of the predictions in Einstein's 1905 relativity paper. They are fascinating to me because one of them was actually tested and found to be in disagree-

Five Nobel Prize winners at a dinner party for Millikan, hosted by von Laue in Berlin, 1931. From left to right: Walter Nernst, Albert Einstein, Max Planck, Robert Millikan and Max von Laue.

ment with the theory; and the other, if it had been tested, would also have been in disagreement with the theory. The reasons are quite different and tell us a good deal about the scientific method.

I have already noted that one of Einstein's predictions was that mass will increase with velocity—an absurdly small effect, as I have said for terrestrial vehicles. But even in Einstein's day, there were objects that moved with speeds comparable to that of light. In 1896, the French physicist A. H. Becquerel discovered that uranium is radioactive. By the turn of the century, he had identified one component of this radioactivity as consisting of electrons. These had been identified a few years earlier by the British physicist J.J. Thomson. Becquerel's radioactively emitted electrons are rather energetic and move with speeds comparable to that of light. If Einstein was right, these electrons should show, by modifications in their trajectories, the dependence of their masses on their velocities.

Around the turn of the century, the German physicist Walter Kaufmann began a series of experiments designed to measure the mass of the electrons emitted—

The Annus Mirabilis—Einstein's Good Year

The year 1905 stands out as one of the most remarkable in the history of science, comparable only to Newton's "annus mirabilis"—miracle year—of 1666, when Newton devised the calculus and laid the foundation for all of Newtonian physics. While working as a clerk in the Swiss Patent Office, the 26-year-old Einstein published five papers in **Annalen der Physik** (Annals of Physics) that established him as an important voice in modern physics. Contrary to popular belief, however, these were not Einstein's first papers published in serious scientific journal; nor was he pursuing only physics as a pastime. He was determined one day to secure a university position and saw these publications as furthering that goal.

The first paper was published in March 1905. It dealt with the nature of light and explained a phenomenon known as the photoelectric effect. In spite of the many experiments conducted in the nineteenth century that indicated that light was a wave phenomenon, it had been observed that light had the ability to dislodge electrons from matter as if the light was composed of particles. Einstein revived the notion promoted by Max Planck that light consisted of discreet "quanta" of energy and this explained the observed bevavior of electrons. It was for this paper (and not for relativity) that Einstein was awarded the Nobel Prize for 1921.

The second paper was published a month later and it consisted of a determination of the dimensions of atoms based on the phenomenon known as Brownian motion. Named after botanist Robert Brown, who noticed that particles moved seemingly randomly when suspended in a liquid, Brownian motion gave an indication of the exact number and size of atoms (or molecules) in a solution. It was the first proof of the reality of atoms, and is considered a brilliant display of mathematical physics.

Just eleven days later, Einstein published a second paper on Brownian motion, this time relating the behavior of suspended particles to the energy of atoms, based on the kinetic theory of heat. These two papers opened up new avenues of mathematical research in kinetic theory that would be pursued for decades.

The fourth Einstein paper published in 1905 was entitled, innocuously, "On the Electrodynamics of Moving Bodies," but it contained the revolutionary introduction of the basic principles of special relativity. From the premise that light has the same speed for all observers moving uniformly relative to one another, Einstein derived the way space and time are measured differently by these observers. Later, Einstein would claim that the paper

Above: Einstein receiving the Max Planck Award from Planck in 1929
Left: Stamp issued to commemorate Einstein's annus mirabilis.

was the culmination of ten years of wondering about a question that had puzzled him as a child: What would a person see if he could ride atop a beam of light?

The fifth paper was more a postscript to the fourth. In it, Einstein derived his famous equation $E=mc^2$ from the fact that all physics must be essentially the same for all observers moving uniformly relative to one another. If that's so, then the energy of an object in motion must still be present in the reference frame in which the object is at rest. The original paper did not present the equation in just this form, but it was massaged into that familiar form by physicists like Max Planck, whose inclusion of relativity in his lectures did much to ensure Einstein's acceptance. — HR in this case, by the electron decay of radium. His idea was to force the electrons to follow curved paths in a magnetic field. By measuring the curvature, he could indirectly measure the mass. Measuring the mass of an electron by putting it on a scale is hopeless: it takes 10—raised-to-30 powers' worth of them to weigh 1 pound.

The idea that the mass of the electron might depend on its speed was actually in the air even prior to Einstein. A contemporary German theorist named Max Abraham had proposed a theory—now long forgotten—that predicted such a dependency but one different from Einstein's. As far as I know, the first reference to Einstein's 1905 relativity paper in a physics journal is to be found the following year in the same journal, *Annalen der Physik*, where Einstein had published. It is in Kaufmann's paper which states categorically that his experimental results disagree with Einstein's prediction but do agree with those of Abraham. Einstein's reaction to this paper was extremely interesting. He dismissed the experimental results. He was certain that they *had* to be wrong since he had more confidence in the overall coherence of his theory than he did in the experiments. A decade later, it turned out that he was right when the experiments were shown to be wrong. Einstein clung to this Platonic view of the relation between theory and experiment for the rest of his life. A remarkable instance occurred in 1919, a few years after he had produced his scientific masterpiece—the general theory of relativity and gravitation. One of the predictions of that theory is that light can be bent by gravity. An extreme example of this is where, in the interior of a black hole, the light is bent back on itself and never gets out. The idea of curved light rays was first carefully tested in 1919 during an eclipse of the sun. Astronomers were able to observe light from stars that passed close to the rim of the occulted sun. The small shift found agreed almost exactly with Einstein's prediction. When the news—which made Einstein an international celebrity—came to Berlin, he happened to be with a young student named Ilse Rosenthal-Schneider. She later reported what happened: "When I was giving expression to my joy that the results coincided with his calculations, he said quite unmoved, 'But I knew that the theory is correct'; and when I asked what if there had been no confirmation of his prediction he countered, '*Da könnt mir halt der liebe Gott leid tun—die Theorie stimmt doch*' 'Then I would have been sorry for the dear Lord—the theory is correct.'"

The second example of a prediction in Einstein's relativity paper is also remarkable because it is wrong. He states it in so many words: "Thence we conclude that a balance clock at the equator must go more slowly, by a very small amount, than a precisely similar clock situated at one of the poles under otherwise identical conditions." This is Einstein's colorful way of describing one of the most disturbing features of the theory of relativity: what is known as "time dilatation." The theory predicts that a clock in motion keeps time at a slower rate than an identical clock at rest. My teacher Philipp Frank used to say apropos of this, "Travel and live longer." Time dilatation effects depend on the ratio of the square of the velocity in question to that of light. We now have atomic clocks so accurate that this prediction can be tested directly. I suspect that when Einstein put the bit about clocks on the equator and clocks on the pole into his paper, he was making a bit of a joke. A clock on the pole does not rotate with the earth, while a clock on the equator is moving at about a thousand miles an hour. Using the 1905 version of relativity—we call it the "special" theory of relativity as opposed to the "general" theory—predicts that the equatorial clock lags its polar twin by about a hundred and two billionths of a second per day! Einstein was then living in the nation of watchmakers—but billionths of a second? It is probably just as well that this experiment was not carried out in 1905 because it would have given a null result in total disagreement with Einstein's prediction. What Einstein did not know in 1905 was that two effects act on the clock: the velocity effect which he discussed, and a gravitational effect which he did not fully understand until a decade later. The two effects conspire in the following way. The experiment takes place at sea level. Sea level is determined by balancing the force of gravitation which holds the sea down, with the centrifugal force that is trying to eject it from the spinning earth. The net force at sea level is zero. If one puts this into Einstein's formula for the time dilatation, one finds that there is a time dilatation but that it is independent of latitude. The clock at the pole runs at the same rate as the clock at the equator. This was actually tested in 1977 when the physicist Carroll Alley of the University of Maryland led a team that flew three cesium clocks from Washington, D.C., to the Thule Air Force Base in Greenland. The clocks are also affected by the motion of the flight itself. But when that is subtracted out, the resulting null result is in perfect harmony with Einstein's combined special—and general—relativity predictions.

What would Einstein have made of this disagreement with his 1905 paper had he been confronted with it then? Assuming that he did not simply dismiss it, as he had the Kaufmann mass experiments, he would have come to the conclusion that something was missing from his original theory. Perhaps he would have been driven by experiment instead of by purely aesthetic considerations to the discovery of the general theory of relativity. Einstein never believed that constructing fundamental theories in physics is a matter of simple extrapolation from experiment. It involves, he felt, a creative leap. He once wrote that "the creative principle resides in mathematics. In a certain sense, therefore, I hold it true that pure thought can grasp reality, as the ancients dreamed." He felt that if the equations are beautiful enough they have to be right. This is one of the reasons many of his contemporaries had such a hard time accepting his work. Not only were the results counter intuitive, but they had been arrived at by completely unfamiliar thought processes.

A case in point is the matter of the "luminiferous aether." This ghostly medium made its way back into the physics of the nineteenth century because experiments performed at the beginning of the century had persuaded physicists that light consisted of waves. The ether, or "aether," had first found its way into physics at the hands of Newton's contemporaries who could not accept the idea that gravitation propagated across empty space. Newton thought that light consisted of particles—an idea that went out of fashion after experiments performed in the nineteenth century seemed to show that it was a wave phenomenon. Beams of light could be made to interfere with each other, even pass through each other—something that, it was thought, only waves could do. But the only sorts of wave motion people were familiar with, were waves that oscillate in a medium such as water or air. The electromagnetic theory of light propagation invented by Maxwell in the middle of the century allowed, however, for the possibility of light waves propagating in a vacuum. This notion was too much for Maxwell and his contemporaries, so they invented a medium—the aforementioned aether—which, as someone later observed, became the subject of the verb "to oscillate." As Maxwell himself put it in 1865, "We have therefore some reason to believe, from the phenomena of light and heat, that there is an aethereal medium filling space and permeating bodies, capable of being set in motion, and of transmitting that motion from one part to another, and of communicating that motion to gross matter so as to heat it and affect it in various ways." Indeed, a cottage industry developed among theoretical physicists and applied mathematicians in the nineteenth century which produced more and more complex models of the aether. The models became so complicated and even self-contradictory that it is a wonder the enterprise lasted as long as it did. In its defense, I should point out that while it was going on, a number of the mathematical tools we still use in theoretical physics were created. No doubt, as a knowledgeable professor of theoretical physics in Bern in 1905, I would have been familiar with the latest aether literature; probably some of my work would have been devoted to aether-related problems. Imagine, then, how I would have reacted to the one and only reference to the aether found in Einstein's 1905 relativity paper. It is in a sentence that occurs in the third paragraph and reads in part, "The introduction of a 'luminiferous aether' will prove to be superfluous because the view here to be developed will [not] require an 'absolutely stationary space'." There it is! My life's work dismissed as superfluous in a single sentence written by a twenty-six-year-old who did not even have an academic job. No wonder I would have been inclined to dismiss Einstein as a crank.

How, then, was Einstein's early work actually received—before he was famous—by his contemporaries? This is a fascinating subject in its own right, and much has been written about it—country by country. All things considered, I think the report card is surprisingly good. After being rejected in 1907 by the University of Bern where he had applied to be a *Privatdozent*—a kind of privately paid instructor—he was accepted a year later when he produced the thesis report he had failed to produce the previous year. This was fairly typical of the young Einstein. When it came to such formalities, he did pretty much what he felt like doing. A year later—1909—he was made an associate professor at the University of Zurich. What is more remarkable, that same year, he was given the first of his many honorary degrees—this one by the University of Geneva. I was so struck by this that I decided to investigate it further. I was fortunate to have the aid of Professor Pierre Speziali, a historian of science now retired from the University of Geneva. It was Professor Speziali who uncovered and put together the magnificent collection of letters exchanged for over fifty years between Einstein and his best friend Michele Besso. Professor Speziali informed me that the proposal to grant Einstein this honorary degree came from a professor of physics at the University of Geneva named Charles-Eugene Guye. In Zurich, he had taught a course in electricity and magnetism that Einstein had taken while an undergraduate prior to the turn of the century. In 1900, Guye received an appointment in Geneva. He read Einstein's relativity paper in 1905 and was immensely impressed by it. Indeed, in 1907 he began a series of experiments to confirm Einstein's relation between mass and velocity. It was these experiments—completed in 1915—that gave the most precise verification of this aspect of the theory. Professor Speziali informed me that the original documents concerning Einstein's honorary degree—including the citation—appear to have been destroyed.

As one might imagine, the comprehension and acceptance of Einstein's early work broke down along generational lines. Physicists—at least the good ones—of about Einstein's age understood almost immediately the importance of at least some of what he had done. Leopold Infeld, who came from Poland to work in the

1930s with Einstein at Princeton, wrote:

> My friend Professor Loria told me how his teacher Professor Witkowski in Cracow ... read Einstein's paper [on relativity] and exclaimed to Loria, "A new Copernicus has been born! Read Einstein's paper." Later when Professor Loria met Professor Max Born at a physics meeting, he told him about Einstein and asked Born if he had read the paper. It turned out that neither Born nor anyone else there had heard about Einstein. They went to the library, took from the bookshelves the seventeenth volume of *Annalen der Physik* [now a collector's item worth many thousands of dollars!] and started to read Einstein's article. Immediately Max Born [who was a few years younger than Einstein] recognized its greatness.

It is ironic that twenty years later, when Born introduced the now generally accepted probability interpretation of the quantum theory, Einstein became its most powerful opponent. Indeed, many people have argued that Einstein's refusal to accept the quantum theory itself bordered on crankiness. Another young physicist who immediately understood the importance of Einstein's paper was Max von Laue. (He, like Born, was later, in 1914, to receive the Nobel Prize.) Von Laue was then in Berlin, but Einstein's paper made such an impression on him that he made a special trip to Bern just to meet him. He may well have been the first professional physicist Einstein actually met after his student days. In the 1930s, von Laue, who remained in Germany, showed extraordinary courage when he publicly opposed the Nazis' attempt to disassociate Einstein's name from his theory—to turn relativity in some sort of "Aryan physics." My teacher Philipp Frank began corresponding with Einstein in 1907 and, soon after, was publishing papers on relativity.

With the older generation the situation was more complex. At one end of the spectrum were physicists who understood absolutely nothing and were proud of it. A representative of this genre was one W. F. Magie who appears to have been a professor of physics at Princeton. In 1911, when delivering the presidential address to the American Association for the Advancement of Science, he chose to express his views on relativity. He remarked, "I do not believe that there is any man now living who can assert with truth that he can conceive of time which is a function of velocity [the time dilatation discussed earlier] or is willing to go to the stake for the conviction that his 'now' is another man's past." This, despite the fact that in 1911 the special theory of relativity had almost become a part of classical physics. Indeed, that very year, von Laue had published his great monograph, the first comprehensive text on the theory—*Das Relativitatsprinzip*, a book that can still be read with pleasure.

A much more complex case is that of Max Planck. Planck, who was born in 1858 and died in 1947, would certainly be on the list of the greatest physicists of the twentieth century. He was awarded the Nobel Prize in physics in 1918 for his work on what is known as "black body radiation." This is the radiation produced in the interior of a heated metal cavity. It is also the kind of radiation left over from the photons produced in the Big Bang. His work paved the way for Einstein's radical 1905 paper on the nature of light. In this paper, as I have mentioned, Einstein postulated that light has a particulate nature. He noted that this could be tested by shining light on metal surfaces and studying how the electrons were ejected—what is known as the photoelectric effect. It was for this work that Einstein received the Nobel Prize in 1922 and not for the relativity theory, which the Swedish Academy found too speculative.

Planck was an immediate convert to the theory of relativity. In fact, von Laue, who was his assistant, first learned the theory in a series of seminars Planck gave about it soon after Einstein published his first paper. It was Planck, it appears, who published the first theoretical paper on the new theory. But he simply could not deal with Einstein's paper on the quanta. He was sure it was wrong in the same instinctive way that Einstein later decided that the probabilistic quantum theory was wrong. Indeed, Planck spent a decade trying futilely to derive his radiation formula from classical physics. Einstein spent the last fifteen years of his life trying to derive the results of quantum mechanics from classical physics. Planck's feelings about the quantum were so strong that when in 1913 he proposed Einstein for the Royal Prussian Academy of Science, he felt obliged to write that Einstein "sometimes may have missed the target in his speculations, as for example, in his theory of light quanta, [but that] cannot really be held against him." It is unlikely that Einstein knew of this when, in 1928, he nominated for the Nobel Prize Erwin Schrodinger and Werner Heisenberg, the two original creators of the modern quantum theory. To qualify his recommendation, however, Einstein felt obliged to add that "it still seems problematic to me how much will ultimately survive of [their] grandiosely conceived theories."

Wolfgang Pauli occasionally used to say that there were physics papers so bad they were not even wrong. That is the trouble with most crank papers. Of course, there are some that are *simply* wrong. But to me these are more like bad science than vintage crank. The vintage crank paper has no correspondence and no predictiveness. That is why it, and its author, are, as a rule, so hard to deal with. It is as if the crank is speaking in tongues. Furthermore, the authors of these papers do not want to be instructed: they want to be *affirmed*. All of us who have tried to work in a deep science know just how hard it is to get to the frontier—just how much devoted training is involved. Even Einstein went through this apprenticeship. The notes he took in H. F. Weber's 1887-88 lectures at the Swiss Federal Polytechnic School in Zurich still exist. They are the notes of a conscientious student with a clear understanding of the physics that preceded his own. The typical crank appears to regard all this apprenticeship as beneath his intellectual dignity. He wants to go right to the head of the class. No apprenticeship for *him*. By now, having been at it for many years, I feel that I can spot a crank paper in physics after reading a few lines. Nonetheless, I am made cautious by my fantasy. Will I be able to spot the next Einstein if he or she sends me, out of the blue, the equivalent of Einstein's four papers in the mail—art unframed. I hope so. But sometimes I wonder.

Cartoon of Einstein by Low from the supplement to *The New Statesman and Nation*, October 21,1933.

Professor Einstein

Low

Einstein's Bovine Dreams

by João Magueijo

W HEN I WAS ELEVEN, my dad gave me a fascinating book by Albert Einstein and Leopold Infeld called *The Evolution of Physics*. In its opening lines, it compares science to a detective story. Except that the challenge is not *who dunit*; it is why the world works the way it does.

As in any good mystery, the investigators are often led astray. Time and again they must backtrack, separating false clues from real ones. But finally the day comes when a picture emerges, when enough facts have been collected so that they can start applying that uniquely powerful human tool, the power of deduction, to make sense of it all. With a theory of how the mystery arose, and a little luck, they conjecture that certain facts about the case *must* be true. They then test these facts, and, they hope, solve the mystery.

A few paragraphs into that book, however, the mystery analogy is abruptly abandoned. Scientists, we learn, face a dilemma that crime fighters do not. In the mystery of the universe game, scientists never get to say "Case closed." Whether they like it or not, they are never really dealing with one mystery, but rather with one small piece of a huge, interlocked series of mysteries. More often than not, once they solve one piece of the puzzle, that solution suggests that old solutions to other parts of the puzzle are wrong, or at least require reexamination. The game of science can accurately be described as a never-ending insult to human intelligence.

Despite the "indignity" to which it subjects us, I immediately found physics fascinating. I liked in particular the manner in which the mysteries of the universe are posed: The questions asked are often superficially very simple but in reality extremely deep in meaning; they are also beautifully coated in the abstractions of thought experiments and pure logic.

But it was not until I was well into my own career as a physicist that I realized that most problems in physics are not approached in a coolly rational manner; at least not initially. Before we are scientists we are Homo sapiens, a species that, despite its pompous name, is more often driven by emotion than by reason. We don't always carefully sort out false clues and bad assumptions, nor do we limit ourselves to the most rational techniques of problem solving.

During the early development stage of a new idea we behave rather more like artists, driven by temperament and matters of taste. In other words, we start off with a hunch, a feeling, even a desire that the world be one way, and then proceed from that presentiment, often sticking with it long after data suggests we may be leading ourselves and those who trust us down a blind alley. What ultimately saves us is that at the end of the day, experiment acts as the ultimate referee, settling all disputes. No matter how strong our hunch is, and how well it is articulated, at some point we will have to prove it with hard, cold facts. Or our hunches, no matter how strongly held, will remain just that.

This is particularly true of that branch of physics known as cosmology—the study of the universe as a whole. Cosmology is not about this star or that galaxy; that study is usually called astronomy. Rather, for cosmologists, galaxies are mere molecules of a rather unusual substance that we call the cosmological fluid. It is the global behavior of this all-encompassing fluid that cosmologists try to understand. Astronomy is about trees; cosmology about the forest.

Needless to say the field is prime ground for speculation. Its puzzles have led us into an elaborate detective story, full of clues, wrong turns, deductions, and empirical facts. Unavoidably, part of the story also shows scientists relying upon hunches and speculations for far longer than most would care to admit.

Cosmology was for a long time the subject of religion. That it has become a branch of physics is to some extent a surprising achievement. Why should a system as apparently complex as the universe be amenable to scientific scrutiny? The answer may surprise you: The universe is, at least in regard to the forces at work, not that complex. It is far simpler, for instance, than an ecosystem or an animal. It is harder to describe the dynamics of a suspension bridge than those of the universe. This realization opened the doors for cosmology as a scientific discipline.

The big leap was the discovery of the theory of relativity, in conjunction with improvements in astronomical observations. The heroes of that story are Albert Einstein, the American astronomer and lawyer Edwin Hubble, and the Russian physicist and meteorologist Alexander Friedmann. Together they blended the constancy of the speed of light and its amazing implications with a larger mystery—the origins of our universe. And it all started with a dream.

WHEN ALBERT EINSTEIN was a teenager he had a very peculiar dream. For many years after he felt deeply marked by this dream, an obsession that would eventually be transfigured into deep reflections. These reflections were to change dramatically the way in which we understand space and time, and ultimately our perception of the whole physical reality surrounding us; indeed, they were to trigger the most radical revolution in science since Isaac Newton, and to bring into question the very rigidity of the space and time into which our Western culture has been embedded.

This was Einstein's dream:

On a misty spring morning, high up in the mountains, Einstein was walking on a path winding its way along a stream dropping from the snowy summits. It was no longer unpleasantly cold, but it was still crisp as the Sun slowly started to break through the mist. The birds sang noisily, their songs emerging above the gushing sounds of the tumultuous waters. Dense forests covered the slopes, broken only here and there by gigantic cliffs.

As the path descended further, the landscape opened slightly, and the dense forest started to give way to ever larger clearings and patches of grassy floors. Soon the hanging valleys came into view, and in the distance Einstein could see a multitude of fields, all bearing the unmistakable marks of civilization. Some of these fields were cultivated and divided by fences configured in more or less regular shapes. In others, Einstein could see cows grazing lazily, scattered throughout the meadows.

The Sun was now more confidently penetrating the mist, and as it did so it diluted the atmosphere into a tenuous soft focus through which Einstein started to make out details in the fields below. In these parts, it was common to divide the properties with electrified wire fences. They were quite ugly indeed. Also, most of them did not seem to be working at all. Look at all those cows happily munching away the hitherto inaccessible grass on the other side of the fence, their heads pushed through the wires in shocking disregard for private property.

As Einstein reached the nearest meadow, he

Figure 1

went to examine the electrified fence. He touched it and, as he expected, felt no shock—no wonder the cows along the fence didn't care about it. As he was playing with the fence, Einstein saw a large figure walking along the opposite side of the field. It was a farmer carrying a new battery and moving towards a shed across the field. Eventually, the farmer reached the shed, and Einstein saw him go inside to replace the expired battery. It was then that through the open door Einstein saw the man connecting the new battery, and *precisely at the same time* he did it, Einstein saw the cows jumping up away from the wire. *All at once. Exactly.* A fair amount of displeased mooing ensued (Figure 1).

Einstein kept on walking, and by the time he reached the farther edge of the field, the farmer was returning home. They greeted each other politely, following which a very strange dialogue took place, the sort of talk one only ever finds in the demented haze of dreams.

"Your cows have extraordinary reflexes," said Einstein. "Just now, I saw you switching on your new battery and, wasting no time, they all jumped up at once."

To this, the farmer looked utterly confused, and stared at Einstein in disbelief. "They all jumped up at once? Thank you for your compliment, but my cows are not in heat. I also looked at them when I switched on the new battery as I was hoping to scare the living hell out of them: I like to play pranks with my cows. For a short while nothing happened. Then I saw the cow nearest to me jump up, then the next one along, and so on, orderly, until the last one jumped up." (Figure 2)

Figure 2

It was Einstein's turn to feel confused. Was the farmer lying? Why should he lie? And yet he was sure of what he had seen: Farmer switches on new battery, first cow jumps up in the air, last cow jumps up in the air, all *exactly at the same time.* Still, no point starting an argument. And for some reason, he was starting to feel like strangling the farmer.

But then he woke up. What a moronic dream—and with cows, of all animals... And why had he felt pathetically homicidal for nothing? Better forget all this nonsense.

As with many strange dreams, however, some deeper meaning eventually clicks in the dreamer's mind, and indeed, before forgetting the dream altogether, Einstein suddenly had a flash. This was only a dream, and yet, in some sense, it did nothing but exaggerate a real feature of our world. Light travels very fast, but *not* at infinite speed, and what this seemingly innocuous dream was hinting at is that from such a simple physical property of light follows a totally crazy consequence: Time must be relative! What happens "at the same time" for one person may well happen as a sequence of events for someone else.

In fact light travels so fast that it looks infinitely fast, but that is merely a limitation of our senses. Careful experimentation promptly reveals the truth: Light travels at about 300,000 kilometers per second (or km/s). The finite speed of sound is more obvious to us because the sound's speed is far lower than light's: Sound travels at about 300 meters per second. So, shout at a cliff 300 meters away from you and two seconds later you will hear your echo: Your shout reaches the cliff in one second, is reflected by the cliff, and is returned to you as an echo in another second.

Flash light into a mirror 300,000 km away and two seconds later your "light-echo" will return, a phenomenon well known in radio communications in space, for example in lunar missions. The echo effect on a mission to Mars would be about thirty minutes: You send a radio message from Earth, it gets there travelling at the speed of light in about fifteen minutes, and the astronauts' reaction gets back to you in another fifteen minutes. Having an argument over the phone while on vacation in Mars is bound to be exasperating.

The cow dream depicts nothing but what actually happens in reality, albeit heavily exaggerated—something that we would indeed perceive with our senses if the speed of light were more like the speed of sound. In Einstein's dream, electricity propagates down the wires at the speed of light.* Hence the image of the farmer switching on the battery travels towards Einstein side by side with the electric pulse going down the wire. They reach the first cow simultaneously, and the pulse gives her a shock. Here it is understood that the cow's reaction time is nil*, so that the image of the farmer switching on the battery, the image of the first cow jumping, and the electric signal going down the wire, all now move towards Einstein side by side.

When they reach the next cow, she jumps up, and the image of her jumping up joins the parade. Now the image of the farmer switching on the battery; the image

---

* Artistic license taken here.

of the first two cows jumping up, and the electric signal going down the wire all move towards Einstein side by side. And so on, until the last cow. Hence Einstein sees the farmer switching on the battery, and all the cows jumping up, exactly at the same time. Had he put his hand on the wire, he would have got an electric shock and said "Scheisse!" precisely at the same time he saw it all happening. Einstein was not hallucinating; it *did* all happen at the same time. That is, at his "own" same time.

The farmer's point of view, however, is rather different. The farmer is subjected to what is really like a series of light-echoes reflected from cliffs/mirrors successively farther and farther away. He switches on the new battery, and that's like a man shouting into an abyss. The electric pulse travels towards the first cow, who jumps up when the pulse reaches her: That's like the shout moving towards a cliff in the abyss and reflecting off it. The image of the cow jumping up, travelling back towards the farmer, is now like an echo returning from the abyss. Thus there is a time delay between his switching on the battery and his seeing the first cow jump; that is, between his shout and its echo. The image of the next cows jumping up in the air is like a series of echoes generated from cliffs farther and farther away, and therefore they have larger and larger time delays—that is, they arrive successively in time.

And so the farmer is not hallucinating either: For him there is indeed a time delay between his switching on the battery and his seeing the first cow jump up. He then sees all the other cows jump up in succession, rather than at the same time. If Einstein had placed his hand on the wire, the farmer would have seen Einstein jumping up and releasing an expletive after all the cows had jumped.

There is no contradiction between the farmer and Einstein, nothing to argue about. The two observers are both saying what they saw; merely reflecting two distinct points of view. If light traveled at infinite speed, Einstein's dream would never have been possible. As things are, it is merely an exaggeration.

And yet, YES: There is a contradiction! Einstein's dream is telling us that there is no absolute concept of "it happened at the same time," absolute in the sense that it must be true for all observers without any ambiguity. Instead, Einstein's dream shows that time must be relative and vary from observer to observer: A set of events that all happen at the same time for one observer may happen as a sequence for another.

But is this an illusion? Or is the concept of time really something more complicated than what we are used to? In our everyday experience, if two events happen at the same time, they happen at the same time for everyone. Could this fact be only a rough approximation? Is this what Einstein's dream was trying to tell him? *Could time be relative?*

THE WORLD EINSTEIN was born into was one in which scientists believed in a "clockwork universe." Clocks would tick everywhere at the same rate: Time was believed to be the great constant of the universe. Likewise space was conceived as a rigid and absolute structure. These two entities, absolute space and absolute time, combined to provide the unchangeable framework for the Newtonian perception of the world: the "clockwork universe."

It is a world view that resonates throughout our culture. The truth of the matter is that we hate being qualitative, particularly when it concerns financial matters. We prefer to define a unit of money and then refer to the value of anything as a precise number of times that unit.

More generally the definition of units permits marrying the quantitative rigor of mathematics (that is, of numbers) to physical reality. The unit supplies a standard amount of a given type of thing; the number converts it into the exact amount we are trying to describe.

Thus the kilogram allows us to be precise about what we mean by seven kilograms of pineapples, and how much that should cost. The framework of our civilization would not exist without the concept of unit allied to the concept of number. No matter how poetic we claim to be, we love and cannot live without quantitative rigor. I have met very, very few genuine anarchists during my lifetime—and I have met some truly weird people.

This philosophy of life permeates our conception of space and time. Space is defined by means of a unit of length—the meter, say. I can then say that an elephant stands along a given road 315 meters onwards, and that means the number 315 times a rigid unit, the meter. We can thus be absolutely rigorous about the elephant's location.

If I want to map a given region on the surface of the earth, I introduce a two-fold spatial structure. I define orthogonal directions, say North-South and East-West. I can then specify exactly where something is relative to me with two numbers: the distance along the East-West dimension, and the distance along the North-South direction. Such a framework defines location precisely, and our obsession with knowing exactly where everything is has found perfect expression in the GPS (global positioning system): Any location on earth can now be labeled to absurd precision by a pair of coordinates.

Naturally, all this is a matter of convention. Australian aborigines map their land by songlines. Australia, for them, is not a one-to-one correspondence between points in the land and pairs of numbers, the coordinates of those points. Rather, their land is a set of highly twisted, multiply intersecting lines, along each of which runs a specific song. Each song relates a story that happened along that path, usually a myth involving humanized animals, contorted fables full of emotional meaning.

At once, the songlines create a complex tangle, so that a point cannot be just a unique pair of numbers; rather, it matters not only where you are (according to our conception) but also where you came from, and ultimately the whole of your previous and future path. What for us is a single point may for aborigines spawn an infinite variety of identities, because that point may be part of many different intersecting songlines. Unavoidably, this creates a sense of property and ownership that does not fit into our culture. Individuals inherit songlines, not areas of land. One cannot build a GPS that operates in songline space.

Yet Australia exists regardless. Songlines stress that to a large extent any description of space is a matter of choice and convention. We choose to live in a rigid and exact space made up of a set of locations, the Newtonian (some would say Euclidean) space.

All these considerations apply similarly to time. A clock is just something that changes at a regular rate: something that "ticks." A tick defines a unit of time. And a unit of time allows us to specify, by means of a number, the exact duration of a given event. What we declare to be a "regular" rate of changing is a matter of convention or definition. And yet, as with many conventions, this is not purely gratuitous; it allows us a simple and precise description of the physical reality around us.

So great is our confidence in our ability to time things that, since Newtonian days, the flow of time has been envisaged as uniform and absolute. Uniform by definition, absolute because why should different observers disagree over the timing of a given event?

Yes, why should they? And yet at the time Einstein had his dream, a crisis was in the making. His dream was a premonitory dream: This rigid conception of absolute space and time was about to be shattered.

ONE TEMPESTUOUS EVENING, the same cows featured earlier in Einstein's dream start displaying unambiguous symptoms of madness: for no reason whatsoever they start moving across the meadow close to the speed of light. Perhaps they became afflicted by an unusual strain of mad cow disease triggered by their earlier electrocution.

The farmer, on hearing the ensuing stampede, turns up in the field with a torch, but when the cows hear him approach, they quieten down and assemble near one edge of the field. But as soon as the farmer flashes his torch at the cows, they start moving away from him at very high speed, closer and closer to the speed of light. The farmer wonders whether his cows might be in heat after all.

But the farmer also wonders about something else. He has just flashed light at cows moving away from him at a speed fast approaching the speed of light. Does this mean that, as the cows nearly catch up with the light, they should see the light rays coming to a halt? This would be very odd indeed—imagine light having a rest. Is there such a thing as stationary light?

To answer this penetrating question, the farmer asks Cornelia, one of the brightest cows in the herd, to inform him of what she sees while running side by side with the light rays. She says she sees nothing unusual with the light the farmer has just flas-hed. It's light like any other. Indeed, Cornelia is very obliging, and just to make sure, she takes all the necessary steps towards measuring the speed of light. She uses standard techniques for measuring this speed, availing herself of clocks and rods that she carries with her. She returns a strange result: She finds that it is business as usual—light moves relative to her at 300,000 km/s.

It is the farmer's turn to feel like strangling Cornelia. By now thoroughly convinced that Cornelia comes from an English herd, the farmer decides to ask two other cows to measure the speed of the light coming out of his torch. But by this time disarray has set in, and the lamest cows are moving more slowly than the others. The cows selected by the farmer are moving away from him at 100,000 km/s and 200,000 km/s. To avoid a proliferation of stupid names, let us call them cow A and cow B (Figure 3).

Given that the farmer sees his light moving at 300,000 km/s he expects these more sensible cows to return the following results: The speed of light should he 200,000 km/s for cow A that's 300,000 minus 100,000), and 100,000 km/s for cow B (that's 300,000 minus 200,000). It's simple algebra after all. And we have all learnt it in school: Speeds just add or sub-tract (depending on their relative direction). So to obtain the speed

Figure 3

of the light ray with respect to each cow one should just subtract the speed of the cow from the speed of light, right? Or have all those cantankerous physics teachers in school been deceiving us all along, as we have always suspected?

Unfortunately, according to our standard perception of space and time, those physics teachers should he right. Let two cars move away from the same location, along the same straight road, at 100 and 200 km per hour. This means that while my clock ticks away one hour, one car travels 100 km, the other travels 200 km. What is the speed of the faster car with respect to the slower one?

Well, after one hour clearly the faster car is 100 km ahead of the slower one; that's 200 minus 100. So the speed of the fast car with respect to the slower one is 100 km per hour. It's logical enough: You subtract the distances, the time is the same, so you subtract the speeds. What could be controversial about this?

By the same token, if I flash light moving at 300,000 km/s at cows moving away from me at 100,000 km/s and 200,000 km/s, these cows should see my light moving at 200,000 km/s and 100,000 km/s, respectively.

But the cows yet again return a strange result. They both believe that they measure the speed of light relative to them as 300,000 km/s! Hence not only do they contradict the farmer's logic but they seem to contradict each other.

Should we believe the cows? Or should we believe that physics teacher? The good news is that experiment forces us to believe the cows! But that puts us face-to-face with a conundrum. What has gone wrong with our argument showing that the speeds should simply be subtracted? As things stand, what the cows did observe is totally nonsensical.

THIS STATE OF AFFAIRS was more or less the puzzle confronting scientists at the end of the nineteenth century. The experiments supporting the evidence supplied by the cows are now known as the Michelson-Morley experiments. They established empirically the constancy of the relative speed of light, regardless of the state of motion of the observer. If I walk on a train, my speed with respect to the platform adds to that of the train. Michelson and Morley found that light flashed from the moving earth still traveled at the same speed: In some funny sense, 1+1=1 in units of the speed of light. These experiments left physics with a deeply illogical experimental result, one that contradicted the *obvious* and logical dogma that speeds should always be added or subtracted.

This conundrum was solved by Einstein's special theory of relativity. Strangely enough, when Einstein proposed this theory, he was not aware of the Michelson-

Morley results. He probably owes more to his dream cows than to these experiments. We will therefore discuss Einstein's solution to this puzzle with reference to his cows.

Let us again employ the services of Cornelia and ask her to stand side-by-side with the farmer. As the farmer flashes light into the fields, Cornelia sets off in hot pursuit at 200,000 km/s. The farmer sees his light ray move at 300,000 km/s. Therefore, in one second he sees light travel 300,000 km away from him, and Cornelia travel 200,000 km away from him. He then *deduces* that Cornelia now

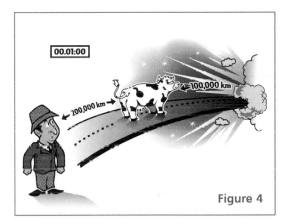

Figure 4

sees the light ray 100,000 km ahead of her, and since one second has elapsed, he thinks that Cornelia should see the light ray moving at 100,000 km/s. (Figure 4) But when Cornelia is asked to measure the speed of light, she insists that she has found it to be 300,000 km/s. What could have gone wrong?

It was here that Einstein showed his great genius and courage. He had the audacity to suggest that it could be that time is not the same for everyone. It could be that while one second went by for the farmer, only one third of a second went by for Cornelia. If that happened, Cornelia would have seen the light ray 100,000 km ahead of her, but when she divided that distance by the elapsed time she found that the result was indeed 300,000 km/s. In other words, if time goes by more slowly for observers in motion, we can explain why everyone seems to agree on the same speed of light, in blatant contradiction to what is expected from simply subtracting velocities.

But there is also another possibility. Perhaps while one second goes by for the farmer, the same happens for Cornelia, so that time is indeed absolute. Maybe it is space that is playing tricks with us. The farmer sees the light ray 100,000 km ahead of Cornelia (Figure 5) because that ray has traveled 300,000 km, whereas Cornelia has traveled only 200,000 km. But what does Cornelia see? It could be that what the farmer perceives as 100,000 km is for Cornelia 300,000 km (Figure 6). If that is so, Cornelia would also measure what she measures: One second has gone by, light is 300,000 km ahead of her according to her rods, therefore its speed with respect to Cornelia, as measured by Cornelia, is indeed 300,000 km/s.

But that would imply that moving objects should appear to be compressed along their direction of motion. Could space shrink due to motion?

These are two extreme possibilities, and of course there is a third one: a mixture of the two. It could be both that time passes more slowly for Cornelia, and that her sense of distance is distorted with respect to that of the farmer, the two effects combining to give her the same measurement of the speed of light. While for the farmer one second elapsed and the light ray is 100,000 km ahead of Cornelia, for Cornelia less time has gone by *and* the light ray is farther ahead by Cornelia's rods. In fact, when one works it all out

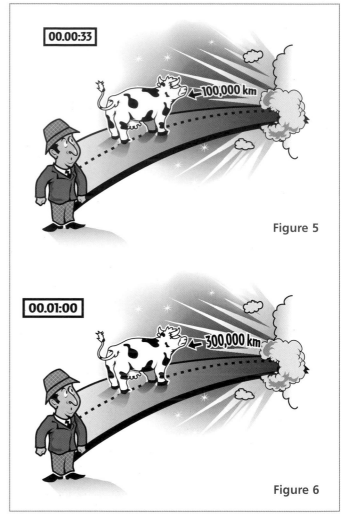

Figure 5

Figure 6

mathematically, one finds that it is indeed a mixture of these two effects that lies behind the dilemma.

This is a crazy way out. But is it true? And sure enough, the farmer soon finds that all this insanity is having an amazing effect upon his cows—they are not getting older! Because time goes by more slowly for fast-moving objects, the farmer gets older and older while his mad cows seem to grow younger by the day. A fast, mad life preserves bovine youth.

He also finds his cows distressingly compressed, nearly flattened into disks when he sees them zipping past. Movement does have a bizarre effect—time goes by more slowly, distances shrink. Of course, no one has ever tried to measure these effects with cows, but both have been observed with particles called muons, produced when cosmic rays hit Earth's atmosphere.

Clearly something had to give, in the argument leading to the subtraction of velocities. That something was the concept of absolute space and time. Einstein's cows, a.k.a. the Michelson-Morley experiments, shattered the clockwork universe, denying time and space an absolute and constant meaning. Instead, a flexible and relative concept of space and time emerged. The result is cast in what is now known as the special theory of relativity. [Editor's note: See sidebar on page 156]

WHEN ONE LOOKS at Einstein's solution to the puzzle of light, one is struck by two things: just how cranky it is—and just how beautiful. Who could have come up with such an idea? Who is this guy? One hundred years on we all know who this guy is, but if we rewind the film and look at the way the story played out in 1905, I'm afraid a rather different picture emerges.

Albert Einstein, the young man, was a daydreamer and an individualist. His school days were inconsistent. At times he would do very well, particularly in subjects he liked. At other times disaster would strike; for instance, he failed his university entry examinations the first time round. He hated the German militarism and the authoritarian nature of the education of his day. In 1896, at the age of seventeen, he renounced his German citizenship and for several years was stateless.

In a letter to a friend, the young Einstein described himself rather disparagingly as untidy, aloof, and not very popular. As is often the case with such people, he was perceived as a "lazy dog" by the sensible world (the wording comes from one of his university professors). After he graduated from university he found himself at odds with academia, and a leading professor waged a war against his ever getting a Ph.D. or a university position. Even worse, Einstein found himself at odds with the rest of the world, or, in other words, "deeply unemployed."

At the age of twenty-two, he is tragically torn. On the one hand he has the cocksure confidence of all free thinkers, and privately lets everyone know just how vacuous he feels the world of respectable attitudes to be. On the other, he has the insecurity of knowing that officially he is a no-hoper, and has to force himself to fawn upon important people in the hope of getting a job. In a letter to a famous scientist, his father paints the following portrait: "My son is profoundly unhappy about his present joblessness, and every day the idea becomes more firmly implanted in him that he is a failure in his career and will not be able to find a way back again."

Despite all efforts, Einstein never made it into academia, at least not until well after he had completed most of the work he is famous for. His early life reads like Jack London's great novel *Martin Eden*, a fact that will forever tarnish the academic world, with its endemic petty games of power and influence. Instead, after many tribulations, a friend and colleague from university days found Einstein a clerk's job at the patent office in Bern, Switzerland. The job was not well paid, but the truth is that there was hardly any work to do.

It was at his patent office desk, at the age of twenty-six, that Einstein flourished — doing little of the work he was supposed to do, while producing, among many other gems, the special theory of relativity. In a tribute to this university friend, many years later Einstein was to say: "Then at the end of the studies… I was suddenly abandoned by everyone, facing life not knowing which way to turn. But he stood by me, and through him and his father I came to Haller in the patent office a few years later. In a way, this saved my life; not that I would have died without it, but I would have been intellectually stunted."

"This guy" was therefore someone treading on the margins of society, and, in the end, happy to do so. And who else could have come up with something so apparently insane as the theory of relativity? Unfortunately, in most such cases, especially due to isolation, what comes out are actually cranky and useless ideas. On one of my shelves lie hundreds of letters providing perfect examples of that. At the end of the day we have to give the man credit—he was not just in outsider, he was Albert Einstein. Without him, the world would have been intellectually stunted.

His paper containing special relativity was promptly accepted. The editor of the journal, who made the decision, would later say that he regarded his prompt acceptance of such a loony paper as his greatest contribution to science. But did Einstein realize what he had just done?

In her old age, Einstein's sister, Maja, would recollect the months that followed in this way:

> The young scholar imagined that his publication in the renowned and much-read journal would draw immediate attention, he expected sharp opposition and the severest criticism. But he was very disappointed. His publication was followed by an icy silence. The next few issues of the journal did not mention his paper at all. The professional circles took an attitude of wait and see. Some time after the appearance of the paper, Albert Einstein received a letter from Berlin. It was sent by the well known Professor Planck, who asked for clarification of some points which were

obscure to him. After the long wait this was the first sign that his paper had been read at all. The joy of the young scientist was especially great because the recognition of his activities came from one of the greatest physicists of that time.

IN FACT, WHAT HE HAD just done was far reaching in many different ways, well beyond the introduction of a relative space and time. Relativity went from success to success, and Einstein's early misfortunes soon came to an end as the world recognized his great achievement. The implications of relativity were immense, and as I have already said, to some extent the language of physics is nowadays the language of special relativity. But this is not primarily a book on relativity, so let me just highlight what I consider to be the three most important consequences of the theory.

The first is that this constant speed, the speed of light—which is the same for all observers at all times and in all corners of the universe—is also a cosmic speed limit. This is one of the most perplexing effects predicted by the theory of special relativity, but it results logically enough from its foundation principle. The proof goes as follows: If we cannot accelerate or brake light, we also cannot accelerate anything traveling more slowly than light all the way up to the speed of light. Indeed, such a process would be exactly the reverse of decelerating light, and if it were possible, its mirror image would also have to be possible in contradiction with special relativity. Thus, the speed of light is the universal speed limit.

This fact may sound strange, but physics is often counterintuitive. Indeed, science fiction films are fond of showing spaceships breaking the speed-of-light barrier. According to relativity, it's not so much that you would get a cosmological speeding ticket; more to the point, relativity shows that you simply would not have enough power to do it, no matter what the nature of your engine.

The existence of a speed limit has a tremendous impact on the way in which we should perceive ourselves in the universe. Our nearest star, Alpha Centauri, is three light years away from us. This means that no matter what our state of technology, a round trip there will take us at the very least six years, as measured on Earth. For the astronauts, this could mean only a fraction of a second, due to the time dilation effect. So at the end of the trip there could be a six-year mismatch between the ages of the astronauts and the loved ones left behind. This could perhaps be cause for a few divorces; but nothing too serious, one would hope.

But that is our nearest star, barely around the corner in astronomical terms.

What about something more considerable on a cosmological scale? Well, let's not even be that adventurous; let's consider a trip just to the other side of our own galaxy. Even that is thousands of light years away from us. Hence, pushing technology to the very limits, a trip there and back would take several thousand years as measured on Earth. We should also make sure that the time dilation effect is such that for the astronauts this corresponds to at most a few years, if we don't want such a space mission to become an ambulatory cemetery.

But therein lies the catch! Even if I push technology to the limits and attempt such a round trip close to the speed of light, in astronaut time it is possible to cover huge distances in a few years, but they will always correspond to thousands of years on Earth. What a pointless space mission! By the time the astronauts get back to Earth, they might as well be visiting another planet. This is no longer just a matter of a few divorces: These poor astronauts would be completely severed from the civilization they had come from.

If we want to avoid such disasters we need to keep well below the speed of light, and so not stray far from home. Our range has to be much smaller than the speed of light times our lifetime—say some dozens of light years, a ridiculous figure by cosmological standards. Our galaxy is a thousand times bigger than that; our local galaxy cluster is a million times larger.

The overall image to emerge is one in which we are confined to our little corner of the universe. A bit like life on Earth if we could not move faster than a meter per century—awfully limited roaming ability. This is very depressing indeed.

A SECOND IMPORTANT IMPLICATION of the theory of relativity is the conception of the world as a four-dimensional object. Usually, we conceive space as having three dimensions: width, depth, and height. What about duration? Yes, to some extent anything also has a "time-depth," or duration, but we know that time is essentially different from space. So including time in the count or not is essentially an academic matter. Or it *was*, before the theory of relativity.

According to relativity, we find that space and time are observer dependent, that duration and length may dilate and shrink depending on the relative state of motion of the observer and the observed. But if space shrinks when time dilates, isn't it as if space were transforming into time? If that is so, then the world is indeed four-dimensional. We cannot leave time outside the count, simply because space may be transformed into time and vice versa.

Such is the perception of the world nowadays called Minkowski space-time (as in the same Prof. Minkowski who once labeled his student Albert Einstein a lazy dog). Space and time, according to relativity are no longer absolute; but a mixture of the two—"space-time"—remains absolute. It's a bit like the theorem of conservation of energy; which we all have learnt in school. There are many forms of energy; including, for instance, movement and heat. Each form is not conserved by itself, because we can transform, say, heat into motion (e.g., using a steam engine). However, the energy's grand total is conserved and always remains the same. Likewise, space and time are no longer constant, but depend on who you are talking to. Depending on the observer, duration and length may be dilated or shrunk. But the grand total, space-time, is the same for everyone.

This "space-time" picture is quite revolutionary if you think about it a bit. The basic unit of existence is no longer a point in space, but the line depicted by this point in space-time when you consider it at all times: What Minkowski called the world-line. Hence think of yourself not as a volume in three-dimensional space, but as a tube in four-dimensional space-time, a tube consisting of your volume shifting forward in time, towards eternity. In a fit of nerdy witticism the physicist George Gamow titled his autobiography "*My world-line.*"

ONE LAST IMPLICATION of special relativity I wish to highlight is the famous $E=mc^2$ equation: Energy equals mass times the speed of light squared. This is probably the best-known physics formula nowadays. How did it come about?

The argument is in fact closely related to the proof that the speed of light is a universal speed limit. A few pages ago we proved this fact *logically* (showing that if we could accelerate something up to the speed of light, then conversely we should be able to decelerate light, in contradiction with the constancy of c). This is fair enough, but *dynamically* what is the reason we cannot overtake light?

If you push something you create an acceleration, that is, a change in the speed of an object. However the larger the object's mass (colloquially, the heavier the object), the larger the force you need to produce the same acceleration. What Einstein found was that the faster an object appears to move, the "heavier" it feels (or, noncolloquially the larger its mass.) He also found that as a given object is seen to approach the speed of light, its mass is seen as becoming infinitely large. And if the mass of an object becomes infinite, no force in the universe is big enough to produce noticeable acceleration anymore. All the force in the universe is not enough to produce that little extra bit of acceleration that would push an object all the way up to, and beyond, the speed of light.

This is why the speed of light works as the cosmological speed limit. You run out of steam as you try to do something illegal: What you are pushing gets heavier and heavier so that you can't push it hard enough to break the speed of light barrier and get that cosmological speeding ticket, whether you want to or not.

And what does this have to do with $E = mc^2$? What follows is Einstein's mind at its purest, guided by disarmingly simple reasons of symmetry and aesthetics. He now notes that motion is a form of energy, sometimes called kinetic energy. If by adding motion to a body you increase its mass, it looks as if by increasing its energy (here in the form of movement) you increase its mass. But what's so special about that energy's being in the form of movement? We know that we can convert any form of energy into any other. Then why not say that by increasing the energy of a body (in whatever form), we increase its mass?

It's a brave generalization, but it has implications that should in principle be observable. Heat an object and its mass should increase. Stretch a rubber band and it accumulates elastic energy; therefore its mass should also increase. Not by much; but a little bit. And so on, for all forms of energy. Thus, in a great coup of insight, Einstein, in a three-page paper published in 1905, proposes that by increasing the energy of a body by E, its mass M should increase—by E divided by the square of the speed of light: $M = E/c^2$

The argument rests on the fact that when kinetic energy is added to a body, its mass is increased, and that for reasons of symmetry this should be valid for all forms of energy.

But then there was a real brain wave, two years later, in 1907. Einstein pushed his sense of beauty and symmetry even further, for the better—or worse—of all of us. Two years before, he had noticed that confining a relation between increases in mass and energy to energy in the form of motion spoiled unity: all energy should increase the mass of a body. But doesn't this seem to imply that energy already has a mass, or, even better, that the two are the same thing?

Identifying any form of energy with mass and vice versa seems to improve the oneness, the perfection of the theory. But then if all forms of energy carry a mass, shouldn't mass also carry energy? Shouldn't mass, in fact, he identified with a form of energy? Thus, Einstein did something horribly simple to the formula above, he rewrote it as $E = mc^2$.

It looks brutally simple and yet it is a gigantic conceptual leap. Again it is a brave generalization, but not a gratuitous one: It has observational predictions, it can be tested. When you put numbers into this formula and perform a short calculation, the implication is that inside 1 gram of matter lies dormant an energy equivalent to the explosion of about 20,000 kg of TNT.

But that's obviously wrong, isn't it? How did Einstein cope with this enormous contradiction? No worries: Einstein pointed out that we do not notice energy; but only variations in energy. I feel cold if thermal energy moves from my body into the environment. I feel my car accelerating if I push the accelerator pedal and burn fuel, thereby taking chemical energy from the fuel and converting it into motion. The tremendous amount of energy harnessed inside 1 gram of matter passes unnoticed because it is never released into the world; it's just like a huge reservoir of energy sitting inside a body, never making its presence known.

In a popular science account of this concept, written by Einstein himself, he considers by analogy a phenomenally rich man who never parts from his money. He lives modestly, and goes around spending only small sums. Thus no one knows of his large fortune because only variations in his wealth are perceptible to the world. The large energy associated with the mass of objects is very similar.

Perhaps I should remind you that while all this was happening, nuclear physics was hardly a subject. The whole concept of the energy of mass evolved from pencil and paper, and, ironically, from considerations of symmetry and beauty. Little did Einstein the pacifist know what he was about to unleash.

On August 6, 1945, Einstein's "phenomenally rich man" bestowed his grim fortune upon the world.

THE THEORY OF RELATIVITY was a massive intellectual earthquake. Today, no one disputes that relativity revolutionized physics, but it also forever changed our perception of reality, not to mention its dramatic effects on the course of twentieth-century history. So much so that nowadays everyone has heard of Einstein's theory of relativity

But Einstein was not finished yet. He soon realized that his theory was incomplete, which is why be called it "special" relativity. He therefore set out at once to find the complete, "general" theory of relativity This turned out to he even more groundbreaking and mindblowing. But the story of its discovery is not so straightforward: Teenage, dreamlike innocence was lost at this point, and Einstein's struggle for the general theory of relativity would be a very adult nightmare indeed. Looking at photographs of Einstein, taken around the time he finally finished the general theory, we see an utterly exhausted man. He has the look of someone who has just emerged from a lengthy and bloody intellectual battle.

As the nineteenth century came to a close, physicists had every reason to feel good: Newtonian physics had been applied to a wide variety of physical phenomena and electromagnetism had been seemingly completely described by Scotish physicist James Clerk Maxwell's four elegant laws. Maxwell had shown that light was an electromagnetic wave that hurtled through space. Like all waves then known, however, it was presumed that light required a medium in which to travel—a substance that would do the oscillating just as air oscilates in the case of sound waves, or water for oceanic waves. This was the ether—a substance that had no weight and offered no resistance to material objects traveling through it.

Maxwell thought it might be possible to measure the speed and direction of the "ether wind"—the flow of ether past the earth as the earth orbited the Sun, but his calculations showed that any device that could measure that speed would have to have a sensitivity of one part in a hundred million—something he regarded as technically impossible.

In 1881, Albert Michelson, a physicist who had studied in Berlin with Hermann Helmholtz (the leading physicist of his day), joined the faculty at the Case School of Applied Science (later Case Western Reserve Univer-sity) in Cleveland, Ohio, and teamed up with Edward Morley, a chemist with a facilty for designing ingenious experiments. Together, they perfected an idea Michelson had first thought of in Germany for measuring variations of the speed of light in the ether. The device was called an inteferometer and it was deceptively

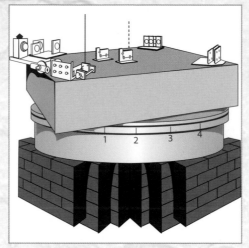

The Michelson-Morley Interferometer as it was built in a Cleveland basement.

simple in design: a beam of light was aimed at a half-silvered mirror that split it into two beams. Each half-beam would be sent in perpendicular directions to fully-silvered mirrors, and then be reunited and directed onto a screen, where interference patterns created by slight differences in the distances (or length of times) the half-beams traveled would result. Since it was impossible to measure the distances involved with such demanding accuracy, the device was floated on a pool of mercury and slowly rotated. As it spun, interference patters created by differences in the speed of the light moving with and against the ether should have been created and displayed as characteristic alternating areas of light and dark from interfering beams of light.

In 1887, the experiments were conducted—but no such variation of interference patterns could be detected. The result was a "null result"—a big zero, probably the

most surprising and perplexing null result in the history of science.

Scientists around the world suggested reasons for the null result, including suggesting that perhaps Cleveland, Ohio, was, in fact, at the center of the universe's ether. This was discounted when the experiment was peformed simultaneously in two parts of the world. In fact, variations of the experiment were to be performed many times over the next thirty years as scientists kept devising clever (and some not-so-clever) explanations for the null result. The failure of the Michelson-Morley experiment to detect the ether was so embarassing that when Michelson became the first American to be awarded a Nobel Prize in physics in 1907, no mention was made of the experiment or its result.

One explanation that was widely entertained was put forth by Irish physicist George FitzGerald and Dutch physicist Hendrik Lorentz: lengths contract in the direction of motion through space, and this compensates for the effect of the varied times of light in the Michelson-Morley apparatus. There was no reason put forth for this phenomenon, however, as an ad hoc hypothesis, it did little to alleviate the sense of crisis physicists around the world felt.

It remained for Einstein not only to explain the null result with a new theory—special relativity—but to show that the Lorentz-FitzGerald contractions are a consequence.

—HR

"What Song the Syrens Sang": How Did Einstein Discover Special Relativity?

by John Stachel

If you have read Edgar Allen Poe's "The Murders in the Rue Morgue," perhaps you remember the epigraph that Poe chose for this pioneer detective story:

> What song the Syrens sang, or what name Achilles assumed when he hid himself among women, though puzzling questions, are not beyond all conjecture.[1]

I believe that the problem of how Einstein discovered the special theory of relativity (SRT) falls into this category of "puzzling questions," that "are not beyond all conjecture."[2] Let me begin by explaining why.

When I started work on the Einstein Papers, there was already a large literature on the origins of SRT compared, say, to the rather scanty amount published on the origins of the general theory of relativity (GRT). So I assumed that the development of SRT must be fairly clear. However, I soon learned that the amount of work published on the origin of SRT and GRT are just about inversely proportional to the available primary source material. For GRT, we have a series of Einstein's papers from 1907 to 1915, capturing the successive steps of his search for the final version of the theory. In addition, there is extensive contemporary correspondence on the subject, several research notebooks, records of lectures given by Einstein during this period, not to mention a number of later reminiscences and historical remarks by Einstein.[3]

For SRT we have the paper *On the Electrodynamics of Moving Bodies*, in which the theory was first set forth in 1905 in its finished form, indeed a rather polished form (which is not to say that it bears no traces of its gestation process).

Reprinted from *Einstein ' B' to ' Z'* by John Stachel with permission of the author.

The only earlier documentary evidence consists of literally a couple of sentences to be found in the handful of preserved early Einstein letters (I will quote both sentences later). We do have a number of later historical remarks by Einstein himself, sometimes transmitted by others (Wertheimer, Reiser-Kayser, Shankland, Ishiwara, for example), which raise many problems of authenticity and accuracy; and some very late Einstein letters, answering questions such as whether he had prior knowledge of the Michelson-Morley experiment, what works by Lorentz he had read, the influence of Poincaré, Mach, Hume, etc., on his ideas; Einstein's replies are not always self-consistent, it must be noted.[4]

Yet the urge to provide an answer to the question of the discovery of SRT has proven irresistible to many scholars. It is not hard to see why: A twenty-six year old patent expert (third class), largely self-taught in physics, who had never seen a theoretical physicist (as he later put it), let alone worked with one, author of several competent but not particularly distinguished papers, Einstein produced four extraordinary works in the year 1905, only one of which (not the relativity paper) seemed obviously related to his earlier papers. These works exerted the most profound influence on the development of physics in the 20th Century. How did Einstein do it? Small wonder that Tetu Hirosige, Gerald Holton, Arthur I. Miller, Abraham Pais, John Earman, Clark Glymour, Stanley Goldberg, Robert Rynasiewicz, Roberto Torretti, *et al.,* have been moved to study this question. I shall not try to record my debts to and differences with each of these scholars, lest this survey become even longer and more tedious than it is already; but must at least acknowledge the influence of their work on my own.[5] I resisted the urge to conjecture for some years, but have finally succumbed, so I can well understand the temptation.

Contrary to my original, naive expectation, no general consensus has emerged from all this work. Given the nature of the available documentation and the difficulty of understanding any creative process—let alone that of a genius—this really is not surprising. I now believe that the most one can hope to do in discussing the discovery of SRT is to construct a plausible conjecture. Such a conjecture will be based upon a certain weighting of the scanty evidence we possess, based upon certain methodological hypotheses, as well as the imagination of the conjecturer.[6] There are bound to be differences of opinion in these matters. All one can demand is that it be made clear on what methodological hypotheses a conjecture is based, and a demonstration that the conjecture is in accord with the available evidence when the latter is weighted in accord with these hypotheses.

Let me emphasize that no such account can hope to encompass those elements of the creative process that Einstein referred as as "the irrational, the inconsistent, the droll, even the insane, which nature, inexhaustibly operative, implants into the individual, seemingly for her own amusement," for "These things are singled out only in the crucible of one's own mind." Yet one may draw courage for the type of conjecture I have in mind from another remark of Einstein: "A new idea comes suddenly and in a rather intuitive way. That means it is not reached by conscious logical conclusions. But, thinking it through afterwards, you can always discover the reasons which have led you unconsciously to your guess and you will find a logical way to justify it. Intuition is nothing but the outcome of earlier intellectual experience."

I shall discuss only this intellectual, logical side of Einstein's struggles. Before trying to reconstruct these struggles, it is well to note that his outward existence was far from tranquil during the period when he was developing SRT. While attending the Polytechnic at Zurich, thanks to the support of maternal relatives, he was plagued by the thought that he was unable to help his family, which was in dire financial straits due to constant business reverses. He was the only graduate in his section (VIA) not to get an academic post, and lived a hand-to-mouth existence for almost two years, until he got a job at the Swiss Patent Office thanks to help from a friend's father. During this period he was under severe family pressure to break with his fiancée, whom he only married in 1903 after his father's death. His first child was born in 1904, and he had to support wife and child on his modest income from the Patent Office, while his mother found work as a housekeeper. So one must not think of Einstein as a tranquil academic, brooding at leisure on weighty intellectual problems. Rather one must imagine him fitting his intellectual work into the interstices of a professional career and personal life that might have overwhelmed someone with a different nature.

The main methodological hypothesis guiding my conjecture was stated by Hans Reichenbach some time ago: ". . .the logical schema of the theory of relativity corresponds surprisingly with the program which controlled its discovery." To put it in more hifalutin' terms, also due to Reichenbach, I believe that "the context of justification" of SRT used by Einstein can shed light on "the context of its discovery."[7] This hypothesis suggests that we can learn a good deal about the development of the theory by paying close attention to the logical structure of its initial presentation in 1905, and to the many accounts of the theory that Einstein gave afterwards. Of course, I have tried not to neglect any scrap of evidence known to me, including the pitifully small amount of contemporary documentation and the later reminiscences.

But I have given special weight to Einstein's early papers, letters, and lectures, in which he sought to justify the theory to his contemporaries. Intellectually, Einstein was an exceedingly self-absorbed person, willing to go over and over the grounds for the theory again and again. These accounts, given over a number of years, are remarkably self-consistent. They provide evidence for a number of conjectures about the course of development of his own ideas, and occasionally even include explicit statements about it. I assume that by and large memory tends to deteriorate with time, and (worse) that pseudo-"memories" tend to develop and even displace correct recollections. So, a second methodological hypothesis which I shall adopt is that, in case of discrepancies between such accounts, earlier ones are to be given greater weight than later ones. Explicit remarks that Einstein makes about the discovery of SRT in the course of his later expositions must always be given great weight, but the earlier he made them the greater the weight I give to them. Of course, if some feature of Einstein's accounts remains unchanged over many years, I take this as evidence for giving such a point the most weight.[8]

It follows from these methodological assumptions that I must preface my conjectures with a brief résumé of the "logical schema of the theory of relativity" as it was first published in the 1905 paper. In this paper, as in almost all subsequent accounts, Einstein bases SRT on two fundamental principles: the principle of relativity and the principle of the constancy of the velocity of light. The principle of relativity originated in Galileian-Newtonian mechanics: Any frame of reference in which Newton's law of inertia holds (for some period of time) is now called an inertial frame of reference. From the laws of mechanics it follows that, if one such inertial frame exists, then an infinity of them must: All frames of reference (and only such frames) moving with constant velocity with respect to a given inertial frame are also inertial frames. All mechanical experiments and observations proved to be in accord with the (mechanical) principle of relativity: the laws of mechanics take the same form in any of these inertial frames. The principle of relativity, as Einstein stated it in 1905, asserts that *all* the laws of physics take the same form in any inertial frame—in particular, the laws of electricity, magnetism, and optics in addition to those of mechanics.

The second of Einstein's principles is based on an important consequence of Maxwell's laws of electricity, magnetism, and optics, as interpreted by H. A. Lorentz near the end of the nineteenth century. Maxwell had unified optics with electricity and magnetism in a single theory, in which light is just one type of electromagnetic wave. It was then believed that any wave must propagate through some mechanical medium. Since light waves easily propagate through the vacuum of interstellar space, it was assumed that any vacuum, though empty of ordinary, ponderable matter, was actually filled by such a medium, to which our senses did not respond: the ether. The question then arose, how does this medium behave when ordinary matter is present? In particular, is it dragged along by the motion of matter? Various possible answers were considered in the course of the nineteenth century, but finally only one view seemed compatible with (almost) all the known experimental results, that of H. A. Lorentz: The ether is present everywhere. Ordinary matter is made up of electrically charged particles, which can move through the ether, which is basically immobile. These charged particles, then called "electrons" or "ions," produce all electric and magnetic fields (including the electromagnetic waves we perceive as light), which are nothing but certain excited states of the immovable ether. The important experimental problem then arose of detecting the motion of ponderable matter—the earth in particular—through the ether.

No other theory came remotely close to Lorentz's in accounting for so many electromagnetic and especially optical phenomena. This is not just my view of Lorentz's theory, it was Einstein's view. In particular, he again and again cites the abberation of starlight and the results of Fizeau's experiment on the velocity of light in flowing water as *decisive* evidence in favor of Lorentz's interpretation of Maxwell's equations.

A direct consequence of Lorentz's conception of the stationary ether is that, the velocity of tight with respect to the ether is a constant, independent of the motion of the source of light (or its frequency, amplitude, or direction of propagation in the ether, etc.).

Einstein adopted a slightly—-but crucially—modified version of this conclusion as his second principle: There is an *inertial frame* in which the speed of light is a constant, independent of the velocity of its source. A Lorentzian ether theorist could agree at once with this statement, since it was always tacitly assumed that the ether rest frame is an inertial frame of reference and Einstein had "only"substituted "inertial frame" for "ether." But Einstein's omission of the ether was deliberate and crucial: by the time he formulated SRT he did not believe in its existence. For Einstein a principle was just that: a principle—a starting point for a process of deduction, not a deduction from any (ether) theory. (I am here getting ahead of my story and will return to this point later.) The Lorentzian ether theorist would add that there can only be *one* inertial frame in which the light principle holds. If the speed of light is a constant in the ether frame, it must be non-constant in every other inertial frame, as follows from the (Newtonian) law of addition of velocities.

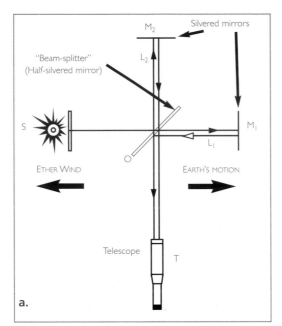 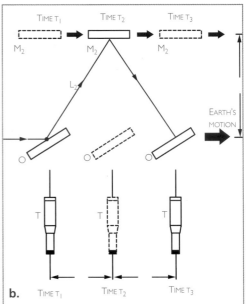

(a) A diagramatic representation of the Michelson-Morley Experiment. A beam of light from source S is split by half-silvered mirror O into two beams, L_1 and L_2, traveling along and perpendicular to the motion of the Earth through space (so the "Ether wind" is in the opposite direction). The interference pattern that shows the beams took different times is observed in telescope T. **(b)** If light really required a medium in which to propagate, L_2 would have traveled a (very) slightly longer distance than L_1 as the apparatus moved through space, giving rise to an interference pattern—but none was ever detected.

The light principle hence *seems* to be incompatible with the relativity principle. For, according to the relativity principle, *all* the laws of physics must be the same in any inertial frame. So, if the speed of light is constant in one inertial frame, and that frame is not physically singled out by being the rest frame of some medium (the ether), then the speed of light *must* be the same (universal) constant in every other inertial frame (otherwise the democracy of inertial frames is violated). As Einstein put it in 1905, his two principles are "apparently incompatible." Of course, if they really were incompatible logically or physically, that would be the end of SRT.[9]

Einstein showed that they are not only logically compatible, but compatible with the results of all optical and other experiments performed up to 1905 (and since, we may add). He was able to show their logical compatibility by an analysis of the concepts of time, simultaneity, and length, which demonstrated that the speed of light really could have the privileged status, implied by his two principles, of being a universal speed, the same in every inertial frame of reference.[10]

Now I shall begin my conjecture about Einstein's discovery of SRT. In a 1921 lecture, Einstein stated that SRT originated from his interest in the problem of the optics of moving bodies. He seems to have been fascinated from an early age by the nature of light, a fascination that persisted throughout his life. From an essay he wrote in 1895, (at age 16), we know that he then believed in the ether, and had heard of Hertz's experiments on the propagation of electromagnetic waves; but he does not show any knowledge of Maxwell's theory. In much later reminiscences, he reports that during the following year (1895–1896) he conceived of a thought-experiment: what would happen if an observer tried to chase a light wave? Could she or he catch up with it? If so, she or he ought to see a non-moving light wave form, which somehow seemed strange to him. In retrospect, he called this "the first childish thought-experiment that was related to the special theory of relativity." Reliable accounts inform us that during his second year (1897–98) at the Swiss Federal Technical Institute, or Poly as it was then called, he tried to design an experiment to measure the velocity of the earth through the ether, being then unacquainted with either the theoretical work on this problem by Lorentz or the experiment of Michelson and Morley (M-M). A precious bit of contemporary documentary evidence reinforces this later account. In a letter to his schoolmate and friend Marcel Grossmann, written in the summer of 1901 (by then both had graduated from the Poly), Einstein wrote:

> A considerably simpler method for the investigation of the relative motion of matter with respect to the light ether has again occurred to me, which is based on ordinary interference experiments. If only inexorable destiny gives me the time and peace necessary to carry it out.

At first sight, it would seem remarkable for Einstein to have written these words (which also show that he had not yet abandoned the concept of the ether), if he knew about the M-M experiment at this time.

However, while still at the Poly (i.e., before 1901) he appears to have studied Maxwell's theory (not covered in his school lectures) on his own, perhaps from the new textbook of August Föppl (which, in various reincarnations, such as Föppl–Abraham, Abraham–Becker, Becker–Sauter, has stayed in print to this day). Föppl discusses a problem which evidently made a strong and lasting impression on Einstein, since he opens the 1905 paper with a discussion of it. This is the problem of the relative motion of a magnet and a conducting wire loop. If the loop is at rest in the ether and the magnet is moved with a given velocity, a certain electric current is induced in the loop. If the magnet is at rest, and the loop moves with the same relative velocity, a current of the same magnitude and direction is induced in the loop. However, the ether theory gives a different explanation for the origin of this current in the two cases. In the first case an electric field is supposed to be created in the ether by the motion of the magnet relative to it (Faraday's law of induction). In the second case, no such electric field is supposed to be present since the magnet is at rest in the ether, but the current results from the motion of the loop through the magnetic field (Lorentz force law). This asymmetry of explanation, not reflected in any difference in the phenomena observed, must already have been troubling to Einstein. Even more troubling was the knowledge, when he acquired it, that all attempts to detect the motion of ponderable matter through the ether had failed. This was an "intolerable" (his word, about 1920) situation. Observable electromagnetic phenomena depend only on the *relative* motions ponderable matter; their explanations differ, however, depending on the presumed state of motion of that matter relative to the hypothetical ether; yet all attempts to detect this presumed motion of ordinary matter relative to the ether end in failure! He later (c. 1920) recalled that the phenomenon of electromagnetic induction compelled him to adopt the relativity principle.

In 1938 he wrote "The empirically suggested non-existence of such an [ether wind] is the main starting point [point of departure] for the special theory of relativity."[11] It is not clear when the significance of the failure of all attempts to detect the motion of ordinary matter through the ether first struck him. The letter quoted above suggests that it was after the summer of 1901. We know from a letter to another friend, Michele Besso, dating from early 1903, that he had decided to "carry out comprehensive studies in electron theory." No later than that, and quite possibly earlier, he read Lorentz's 1895 book, "Attempt at a Theory of Electrical and Optical Phenomena in Moving Bodies." Einstein surely learned about the many such failures by reading this book, since one of its main purposes was to show that such failures were compatible with Lorentz's stationary ether theory. His later comments suggest that study of this book (Einstein says this is the only work by Lorentz he read before 1905) convinced him of the essential superiority of Lorentz's approach to the optics of moving bodies; yet it also convinced him that the Lorentz theory was still not fully satisfactory. Lorentz could explain away the failure to detect motion of matter relative to the ether convincingly to Einstein in all cases but one: the M-M experiment. To explain this, Lorentz had to introduce a special hypothesis, which to Einstein seemed completely unconnected with the rest of the theory: the famous Lorentz contraction. To Einstein, such an approach was not a satisfactory way out of the "intolerable dilemma." It seemed preferable to him to accept at face value the failure of the M-M and all similar experiments to detect motion of matter relative to the ether. Taken by themselves, these negative results suggested to Einstein that the relativity principle applied to electromagnetism, while the ether should be dropped as superfluous. There has been some confusion on this important point, so I shall expand on it. Sometimes the case is presented in such a way as to suggest that it was the "philosophical concept" of the relativity of all motion, as Einstein once called it, which was the key step in his rejection of the ether. But the concept of a stationary ether, as well as of a moving ether, is quite compatible with this philosophical concept of the relativity of motion: one need only assume that motions relative to the ether in the first case, as well as relative motions of the parts of the ether in the second, have physical efficacy. The leading advocates of both the dragged-along and the immovable ether concepts, Hertz and Lorentz, respectively, both understood this and both were read by Einstein.[12]

By the time he gave up the ether concept, Einstein most likely took this philosophical conception of the relativity of all motion for granted, presumably under the influence of his early reading of Mach's *Mechanics* (around 1897). What bothered him now was that no phenomenon existed that could be interpreted as empirical evidence for the physical efficacy of the motion of ordinary matter relative to the ether, in spite of repeated efforts to find one. Yet the best available theory—Lorentz's theory—could only attempt to explain away such failures. These explanations were satisfactory, within the framework of Lorentz' theory, in almost all known cases (i.e., for all experiments sensitive only to order v/c), and Einstein even seems to have been tempted to give up what we may call his *physical* relativity principle (with no ether needed). But Lorentz's explanation of the M -M experiment seemed to Einstein so artificial that he resisted this temptation, opting for the physical relativity principle. After eliminating the ether from the story altogether, one can simply take the results of the M-M and similar experiments as empirical evidence for the equivalence of all inertial frames for the laws of electricity, magnetism and optics as well as those of mechanics. I believe Einstein gave up the ether concept and definitely opted for the physical relativity principle at least a couple of years before the final formulation of SRT, perhaps even earlier. At any rate, at some point well before the 1905 formulation at the theory, he made this choice and adhered to it thereafter.

There was a related motive for his skepticism with regard to the ether, which I shall now mention. Not only was Einstein working on problems of the optics of moving bodies, he was also working on problems related to the emission and absorption of light by matter and of the equilibrium behavior of electromagnetic radiation confined in a cavity—the so-called black body radiation problem. He was using Maxwell's and Boltzmann's statistical methods, which he had redeveloped and refined in several earlier papers, to analyze this problem. This was itself a daring step, since these methods had been developed to help understand the behavior of ordinary matter while Einstein was applying them to the apparently quite different field of electromagnetic radiation.[13] The "revolutionary" conclusion to which he came was that, in certain respects, electromagnetic radiation behaved more like a collection of particles than like a wave. He announced this result in a paper published in 1905, three months before his SRT paper. The idea that a light beam consisted of a stream of particles had been espoused by Newton and maintained its popularity into the middle of the nineteenth century. It was called the "emission theory" of light, a phrase I shall use. The need to explain the phenomena of interference, diffraction and polarization of light gradually led physicists to abandon the emission theory in favor of the competing wave theory, previously its less-favored rival. Maxwell's explanation of light as a type of electromagnetic wave seemed to end the controversy with a definitive victory of the wave theory. However, if Einstein was right (as events slowly proved he was) the story must be much more complicated. Einstein was aware of the difficulties with Maxwell's theory—and of the need for what we now call a quantum theory of electromagnetic radiation—well before publishing his SRT paper. He regarded Maxwell's equations as some sort of statistical average—of what he did not know, of course—which worked very well to explain many optical phenomena, but could not be used to explain all the interactions of light and matter. A notable feature of his first light quantum paper is that it almost completely avoids mention of the ether, even in discussing Maxwell's theory. Giving up the ether concept allowed Einstein to envisage the possibility that a beam of light was "an independent structure," as he put it a few years later. "which is radiated by the light source, just as in Newton's emission theory of light."

So abandonment of the concept of the ether was a most important act of liberation for Einstein's thought in two respects: It allowed Einstein to speculate more boldly on the nature of light and it opened the way for adoption of his relativity principle as a fundamental criterion for all physical laws. I must add a word about Einstein's use of such principles as a guide to further research. In 1919 he explicitly formulated a broad distinction between constructive theories and theories of principle. Constructive theories attempt to explain some limited group of phenomena by means of some model, some set of postulated theoretical entities. For example, many aspects of the behavior of a gas could be explained by assuming that it was composed of an immense number of constantly colliding molecules. Theories of principle formulate broad regularities, presumably obeyed by all physical phenomena, making these principles criteria ("rules of the game") that any constructive theory must satisfy. For example, the principles of thermodynamics are presumed to govern all macroscopic phenomena. They say nothing about the micro-structure or detailed behavior of any particular gas, but do constitute limitations on any acceptable constructive theory of such a gas. Any theory not conserving the energy of the gas, for example, would be immediately rejected. Since the turn of the century, Einstein had been searching for a constructive theory of light, capable of explaining all of its properties on the basis of some model, and was to continue the search to the end of his days. But, "Despair[ing] of the possibility of discovering the true answer by constructive efforts," as he later put it, he decided that the only possible way of making progress in the absence of such a constructive theory was to find some set of principles that could serve to limit and guide the search for a constructive theory.[14] There is no contemporary evidence showing when Einstein adopted this point of view (he first indicated it in print as early as 1907). I believe he had done so by 1905. The structure of the 1905 SRT paper is certainly compatible with his having done so. It is based on the statement of two such principles, deduction of various kinematic consequences from them, and their application to Maxwell's electrical and optical theory.

To return to the main thread of my conjecture, I believe that Einstein dropped the ether hypothesis and adopted his relativity principle by 1903 or 1904 at the latest. This is by no means the end of the story. It seemed that he must then drop Lorentz's version of Maxwell's theory, based as it was on the ether hypothesis. With what was he to replace it? There is good evidence suggesting he spent a great deal of effort trying to replace it with an emission theory of light—the sort of theory suggested by his concurrent researches into the quantum nature of light.[15]

An emission theory is perfectly compatible with the relativity principle. Thus, the M-M experiment presented no problem; nor is stellar abberation difficult to explain on this basis.[16]

Einstein seems to have wrestled with the problems of an emission theory of light for some time, looking for a set of differential equations describing such a theory that could replace the Maxwell-Lorentz equations; and trying to explain a number of optical experiments, notably the Fizeau experiment, based on some version of the emission theory. He could not find any such equations, and his attempt to explain the Fizeau experiment led him to more and more bizarre assumptions to avoid an outright contradiction. So he more-or-less abandoned this approach (you will soon see why I say more-or-less), after perhaps a year or more of effort, and returned to a reconsideration of the Maxwell-Lorentz equations. Perhaps there was a way of making these equations compatible with the relativity principle once one abandoned Lorentz's interpretation via the ether concept.

But here he ran into the most blatant-seeming contradiction, which I mentioned earlier when first discussing the two principles. As noted then, the Maxwell-Lorentz equations imply that there exists (at least) one inertial frame in which the speed of light is a constant regardless of the motion of the light source. Einstein's version of the relativity principle (minus the ether) requires that, if this is true for one inertial frame, it must be true for all inertial frames. But this seems to be non-sense. How can it happen that the speed of light relative to an observer cannot be increased or decreased if that observer moves towards or away from a light beam? Einstein states that he wrestled with this problem over a lengthy period of time, to the point of despair. We have no details of this struggle, unfortunately. Finally, after a day spent wrestling once more with the problem in the company of his friend and patent office colleague Michele Besso, the only person thanked in the 1905 SRT paper, there came a moment of crucial insight. In all of his struggles with the emission theory as well as with Lorentz's theory, he had been assuming that the ordinary Newtonian law of addition of velocities was unproblematic, it is this law of addition of velocities that allows one to "prove" that, if the velocity of light is constant with respect to one inertial frame, it cannot be constant with respect to any other inertial frame moving with respect to the first. It suddenly dawned on Einstein that this "obvious" law was based on certain assumptions about the nature of time always tacitly made. In particular, the concept of the velocity of an object with respect to an inertial frame depends on time readings made at two different places in that inertial frame. (He later referred to this moment of illumination as "the step.")[17] How do we know that time readings at two such distant places are properly correlated? Ultimately this boils down to the question: how do we decide when events at two different places in the same frame of reference occur at the same time, i.e., simultaneously? Isn't universal simultaneity an intuitively obvious property of time? Here, I believe, Einstein was really helped by his philosophical readings. He undoubtedly got some help from his readings of Mach and Poincaré, but we know that he was engaged in a careful reading of

Hume at about this time; and his later reminiscences attribute great significance to his reading of Hume's *Treatise on Human Nature*. What could he have gotten from Hume? I think it was a relational—as opposed to an absolute—concept of time and space. This is the view that time and space are not to be regarded as self-subsistent entities; rather one should speak of the temporal and spatial aspects of physical processes; "The doctrine," as Hume puts it,"that time is nothing but the manner, in which some real object exists." I believe the adoption of such a relational concept of time was a crucial step in freeing Einstein's outlook, enabling him to consider critically the tacit assumptions about time going into the usual arguments for the "obvious" velocity addition law. This was the second great moment of liberation of his thought. I shall not rehearse Einstein's arguments here, but it led to the radically novel idea that, once one physically defines simultaneity of two distant events relative to one inertial frame of reference, it by no means follows that these two events will be simultaneous when the *same* definition is used relative to another inertial frame moving with respect to the first. It is not logically excluded that they *are* simultaneous relative to all inertial frames. If we make that assumption, we are led back to Newtonian kinematics and the usual velocity addition law, which is logically quite consistent. However, if we adopt the two Einstein principles, then we are led to a new kinematics of time and space, in which the velocity of light is a universal while simultaneity is different with respect to different inertial frames; this is logically quite consistent. The usual velocity addition law is then replaced by a new one, in which the velocity of light "added" to any other velocity ("added" in a new sense—it would be better to say "compounded with") does not increase, but stays the same! The Maxwell-Lorentz equations, when examined with the aid of this new kinematics, prove to take the same form in every inertial frame. They are, therefore, quite compatible with the relativity principle, which demands that the laws of electricity, magnetism and optics have this property. The presence or absence of an electric or magnetic field, is then also found to be relative to an inertial frame, allowing a completely satisfactory relativistic analysis of the example of the conducting wire loop and magnet in relative motion. Within six weeks of taking "the step," Einstein later recalled, he had worked out all of these consequences and submitted the 1905 SRT paper to *Annalen der Physik*.

This does not imply that Lorentz's equations are adequate to explain all the features of light, of course. Einstein already knew they did not always correctly do so—in particular in the processes of its emission, absorption and its behavior in black body radiation. Indeed, his new velocity addition law is also compatible with an *emission* theory of light, just *because* the speed of light compounded with any lesser velocity still yields the same value. If we model a beam of light as a stream of particles, the two principles can still be obeyed. A few years later (1909), Einstein first publicly expressed the view that an adequate future theory of light would have to be some sort of fusion of the wave and emission theories. This is an example of how the special theory of relativity functioned as a theory of principle, limiting but not fixing the choice of a constructive theory of light.

Here I shall end my conjectures on how Einstein arrived at SRT. To briefly re-capitulate, I believe that the first principle, the relativity principle, recapitulates his struggles with the mechanical ether concept which led finally to the first crucial liberation of his thought—the abandonment of the ether. The second principle, the principle of the constancy of the speed of light, recapitulates his struggle, once he had definitely opted for the relativity principle, first to evade the Maxwell-Lorentz theory by an emission theory; then to isolate what was still valid in the Maxwell-Lorentz theory after giving up the ether concept and abandoning absolute faith in the wave theory of light. The struggle to reconcile the two principles could only end successfully after the second great liberation of his thought: the relativisation of the concept of time. The resulting theory did not force him to choose between wave and emission theories of light, but rather led him to look forward to a synthesis of the two. This synthesis was finally achieved, over twenty years later, in the quantum theory of fields, to the satisfaction of most physicists, but ironically, never to that of Einstein.

I cannot ask you to accept my conjectures after all of my warnings at the outset of this paper, but will be content if you say "Se non è vero, è ben trovato," "If it isn't true, it's well contrived."

Notes

1. Poe is quoting Sir Thomas Browne's *Hydrotaphia*.
2. A preliminary question is raised by my use of the word "discovery." Is it better to speak of the discovery or the "creation" of a theory like SRT? "Discovery" suggests the finding of some pre-existent, objective structure, as when we say "Columbus discovered America." "Creation" suggests an individual, subjective act, as when we say "Tolstoy created *Anna Karenina*." Neither word seems really appropriate to describe what goes on in the scientific endeavor. Einstein apparently preferred the word "Erfindung" (invention) to describe how scientific theories come into being. Speaking of Mach, Einstein says: "Er meinte gewissermessen, dass Theorien durch *Entdeckung* und nicht durch *Erfindung* entstehen."

WHAT SONG THE SYRENS SANG

Einstein-Besso Correspondance (Hermann, Paris 1972), p. 191, dated January 6, 1948.

3. In the study of the discovery of GRT, therefore, one may hope to formulate conjectures which can he either confirmed or refuted. For example: A study of Einstein's published papers and private correspondence between 1912–1915 convinced me that the standard explanation for his failure to arrive at the correct gravitational field equations until the end of this period—namely, his presumed lack of understanding of the meaning of freedom of coordinate transformations in a generally covariant theory and the ability to impose coordinate conditions that this freedom implied—could not be correct (see "Einstein's Search for General Covariance, 1912-1915," presented at the Ninth International Conference on General Relativity and Gravitation, July 17, 1980). On the basis of his study of a research notebook of Einstein from the early part of this period, John Norton was able to prove that Einstein already was aware of the possibility of imposing coordinate conditions on a set of field equations, and indeed had used the harmonic coordinate conditions (see John Norton, "How Einstein found his field equations: 1912-1915," *Historical Studies in the Physical Sciences*, 4, 253 (1984) For reasons discussed in the text, one cannot hope to confirm or disconfirm most conjectures about the origins of SRT.

4. For a survey of this material for the period up to 1923, see J. Stachel, "Einstein and Michelson: The Context of Discovery and the Context of Justification," Astron. Nachricht. 303, 47 (1982). Unless otherwise noted, quotations from Einstein are cited from this paper, which gives the full references.

5. See Arthur 1. Miller, *Albert Einstein's Special Theory of Relativity* (Addison-Wesley, Reading 1981), which contains references to his earlier papers as well as those of Holton, Hirosige arid many others: Abraham Pais, *'Subtle is the Lord...' The Science and the Life of :Albeit Einstein* (Oxford Univ. Press, New York, 1982); Stanley Goldberg, Understanding Relativity (Birkhäuser, Boston 1984): Roberto Torretti, *Relativity and Geometry* (Pergamon, Oxford 1983). Earman. Glymour and Rynasiewicz have not yet published a full account of their views; I thank them for making available copies of several preprints on this subject.

6. A popular epigram among historians runs: "God is omnipotent, but even He cannot change the past. That is why He created historians."

7. See the reference in footnote 4 for the source of the citations from Reichenbach. If my thesis here is correct, this argues against the still widely held view that these two contexts should be rigorously separated. But in this paper I shall not elaborate on the wider issue.

8. For example, Einstein's statements of the second principle of SRT, the light principle remained remarkably consistent throughout his lifetime. Indeed, an apparent exception in the printed text of his article "What is the theory of Relativity?," published originally in English translation in the *Times* of London in 1919, proved to be based upon an incorrect transcription of his manuscript.

9. Much of the anti-relativity literature, which still continues to grow in volume if not in weight, is based on attempts to show that the two principles are indeed logically incompatible.

10. Sometimes (e.g., by Pais and Goldberg), this consequence of Einstein's two principles is asserted to be his second principle. This is incorrect factually (Einstein's account of the second principle is one of the most consistent features of his discussions of SRT over the years—see footnote 8). and disturbing for several reasons: (a) it makes it impossible to explain why Einstein refers to the two principles as apparently contradictory. There is no contradiction apparent between the relativity principle and this deduction from it; (b) it is logically defective, since the two principles would no longer he logically independent, as they are in Einstein's formulation; (c) most important for present purposes, this formulation deprives us of important clues to Einstein's reasoning that led to the development of SRT.

11. Einstein to Max Talmey, June 6, 1938. The German text reads: "Die empirisch suggerierte Nichtexistenz einer solchen bevorzugten 'Wind-Richtung' ist der Haupt-Ausgangspunkt der speziellen Relativitatstheorie."

12. Hertz said: "... the absolute motion of a rigid system of bodies has no effect upon any internal electromagnetic processes whatever in it, provided that all the bodies under consideration, including the ether as well, actually share the motion" (Electromagnetic Waves, p. 246). Lorentz said: "That one cannot speak of the absolute rest of the ether, is self-evident indeed, the expression wouldn't even have any meaning. If I say for short, the ether is at rest, this only means that one part of this medium is not displaced with respect to the others and that all perceptible movements of the heavenly bodies are relative movements with respect to the ether." Versuch, p. 4 (1895).

13. He was not alone in transferring statistical methods from ordinary matter to radiation. Planck had already done so, but Einstein did not see the relation of his work to Planck's until after publishing his first paper on the subject.

14. See Albert Einstein, *Autobiographical Notes* (Open Court, LaSalle 1979), pp. 48 (German text) and 49 (English translation).

15. One such piece of evidence, not cited in my earlier paper (see footnote 4), has only recently come to light. It occurs in the most complete review of SRT that Einstein ever wrote. It was prepared in 1912 but never published, and is still in private hands. Luckily, a copy has come into the possession of the Einstein Archive. In it, Einstein explains at some length the difficulties that are encountered (and presumably these are the ones he had encountered), if one tries to explain the results of the Fizeau experiment on the basis of an emission theory of light combined with the relativity principle and Galilei-Newtonian kinematics, [See *The Collected Papers of Albert Einstein*, vol. 4. *The Swiss Years: Writings* 1912–1914 (Princeton University Press, Princeton 1995), Doc, 1, "Manuscript on the Special Theory of Relativity," pp. 32–36].

16. The earliest explanation of stellar abberation had been based on the emission theory.

17. Abraham Pais has mentioned this in describing his conversations with Einstein.

Einstein's Dilemma

by Shimon Malin

1. "God Does Not Play Dice"

Einstein's conviction that quantum mechanics is not a fundamental theory of nature was a consequence of his paradigm, his firm beliefs about the nature of reality.

"It must have been around 1950," writes Einstein's biographer Abraham Pais. "I was accompanying Einstein on a walk from The Institute for Advanced Study [in Princeton] to his home, when he suddenly stopped, turned to me and asked me if I really believed that the moon exists only if I look at it."[1]

When he realized the fallacy of Mach's philosophy, Einstein became a realist. (Mach was not a realist. For him, the elements of reality, i.e., sensations, did not point to a "real" world beyond themselves.) It was obvious to Einstein that there is an objective world, whose existence is independent of acts of consciousness. He was not, however, a naive realist. He did not believe that the world we perceive is the world as it is. He appreciated Kant's finding that the world as we perceive it, the phenomenal world, is largely a creation of our own minds. He believed, however, that characteristics of the independently existing reality can be discovered through science and that scientific theories, or conceptual models, are our only access to reality.

In addition to being a realist, Einstein adhered to the tenet of locality: Nothing can propagate faster than the speed of light. As we will see, locality is a consequence of Special Relativity. The Einsteinian world-view was thus based on "local realism," the combination of realism and locality. In addition, Einstein was a firm believer in strict causality, also called "determinism," which claims that there is no randomness in nature, that the present and future states of the universe, down to the smallest detail, are the result of the state of the universe in the distant past. Einstein expressed this belief in his well-known phrase, "God does not play dice." As we will see, strict causality is a claim that quantum mechanics denies. According to quantum mechanics the laws of nature are statistical; individual events in the present and the future are not entirely determined by the past.

Einstein's adherence to determinism made the gulf between his views and Bohr's views unbridgeable. For Bohr, quantum mechanics was the exciting, newly discovered fundamental theory of nature. He believed that one had to accept all its implications and fashion one's world-view in accordance with them. For Einstein, a non-deterministic theory, such as quantum mechanics, could not possibly be a fundamental theory of nature. Quantum mechanics had to be accepted, because it worked, but this acceptance was grudging and tentative. One had to keep looking for the really fundamental theory of nature, a theory that would show quantum mechanics to be a limiting case, a statistical approximation.

Realism and strict causality are distinct ideas. One can be a realist and still allow for a measure of randomness in the determination of events. But Einstein refused to take this route. This stubborn attachment to strict causality puzzled many of his fellow physicists. In other arenas Einstein showed an uncanny capacity to give up cherished notions. His own discoveries introduced major paradigm shifts. Furthermore, the statement "God does not play dice" is problematic: The doctrine of strict causality implies a God that does not allow for randomness, and in doing so it also implies that after the initial act of creation God does absolutely nothing in relation to the world—hardly an attractive concept of the Divine!

There are weighty arguments in favor of realism. It appeals to common sense, and, in the context of science, it works, at least up to a point. But the assumption of strict causality can be viewed as arbitrary. Why was Einstein so attached to it? Why did he steadfastly refuse to give it up in spite of the presence of randomness as an essential feature of quantum mechanics, a theory whose spectacular success he fully realized?

The present essay expresses my own thoughts about this issue. To begin, strict causality was just one feature of Einstein's world-view, a view that was largely based on the findings of Special Relativity. As we will see, these findings and strict causality are interrelated. To see the interrelationships, we proceed to discuss Einstein's Special Theory of Relativity. We will come back to the issue of strict causality in the last section of this essay.

2. What Is Real?

The answer to the question "What is real?" reflects the world-view of a civilization, a culture, or a system of thought. In the dominant philosophical systems of the ancient Greeks, as well as in most religions, the answer to the question "What is real?" is given in terms of "levels of being." Some elements are more real than others, and the relationships among the different elements form a distinct hierarchy.

For Plato "the Good" is the supremely real. The Forms, which occupy a lower level than the Good, are eminently real, and the sensible world, the objects in space and time reported to us through the senses, have only a shadowy kind of reality. This idea was refined by the neoplatonists in the second and third century AD. For Plotinus, the greatest of the neoplatonists, the sequence begins with "the One" and continues through the Nous (the Universal Mind or Intelligence), the World Soul, our individual souls, and nature and ends with the sensible world. The sensible world is the least real of all—its reality is a borrowed one.

In Western religious systems God is at the top of the ladder. Below Him are high beings, such as archangels and angels; the hierarchy continues with humanity, then animals, plants, and finally inanimate matter. The idea of hierarchy applies to individuals as well: In both Platonic and religious thinking the soul is higher, more real than the body.

But Greek philosophy contains another strand of thought, materialism. Materialism can be traced back to Leucippus and Democritus in the fifth century BC. It was passionately promoted by Epicurus in the second century BC, and eloquently expressed by the Roman poet Lucretius around 55 BC. Lucretius stated the essence of the materialistic doctrine as follows:

> All nature as it is in itself consists of two things—bodies and the vacant space in which the bodies are situated and through which they move in different directions... nothing exists that is distinct both from body and from vacuity.[2]

Throughout the Middle Ages this doctrine was considered heresy par excellence, and Epicurus was assigned a place of honor in Dante's hell. With the rise of modern science in the seventeenth century, however, materialism was ready for readoption.

3. The Mechanistic Universe

> It seems probable to me that God in the Beginning formed Matter in solid, massy, hard, impenetrable, movable Particles, of such Sizes and Figures, and with such other Properties, and in such Proportion to space, as most conduced to the end for which he formed them; and as these primitive Particles

being Solids, are incomparably harder than any porous Bodies compounded of them; even so very hard as never to wear or break in pieces; no ordinary Power being able to divide what God himself made one in the first Creation.

This statement of Newton's reveals a concept of reality that contains two levels, God and nature. Since God created nature, His reality is supreme, certainly higher than the reality of the nature He created. But, as Pierre Simon de Laplace and others noted in the following centuries, the concept of "God" was logically extraneous to the Newtonian paradigm. The paradigm did not need a God to create or sustain it. Though Newton himself was deeply religious and devoted more time and energy to the study of theology than he did to the study of physics, the system he introduced was the foundation of a world-view that was not only materialistic but mechanistic: nature as a perfect clockwork, a reality which contains just the clockwork itself, without a designer or a craftsman who built it.

Once the unnecessary hypothesis of God's existence was dispensed with, the stage was set for conceiving of the universe as a mere collection of particles moving in space. The immutability and supremacy of God were replaced by the immutability and supremacy of the laws of nature—Newton's laws. In the physics of the eighteenth and nineteenth centuries the clockwork-universe paradigm reigned unchallenged. Physicists, as individuals, may or may not have believed in God, but the working paradigm of physics was the atheistic, materialistic one.

The materialistic concept of nature was modified during the second half of the nineteenth century, when it had become clear that the austere Newtonian universe, composed only of particles and void, would not do. This modification did not change its basic mechanistic character. It merely added to nature a new type of ingredient, called a "field." The need for this addition arose as follows:

The Newtonian framework included the problematic notion of "action at a distance." For example: The earth is here, and the moon is there. How does the earth exert a gravitational pull on the moon? Well, it just does, acting at a distance. But those who found the notion of "action at a distance" disturbing proposed another explanation: Assume that the earth, merely through its massive presence, generates a "gravitational field" in the space around it. This field mediates the gravitational relationship between the earth and the moon. The moon is pulled toward the earth not because it feels the faraway presence of the earth but because it feels the immediate presence of the earth's gravitational field right where it is.

From the point of view of eighteenth-century physics the concept of the gravitational field is not essential. If one accepts the awkward notion of action at a distance, one need not invoke the concept of a gravitational field. When we consider the complex arena of electromagnetic phenomena, however, the concepts of "electric fields" and "magnetic fields" prove indispensable. In fact, the crowning achievement of nineteenth-century physics is a set of equations, formulated by James Clerk Max-well, which delineates the intricate relationships between these two types of fields, as well as their relationships with the electric charges and currents that give rise to them. Most of the technological inventions of the past century from electric generators and radio to television and computers, are based on these equations.

What is an electric field? Fundamentally, it is a potentiality to influence the motion of electrically charged particles. If an electric field is present in a certain region in space, and a charged particle happens to be there, it will feel a force. And if no particle is present at that region? Well, in that case nothing happens. That is why the field itself is essentially only a potentiality.

This brief account of the addition of "fields" to the mechanistic universe is conceptual rather than historical; it is written with the benefit of hindsight. The historical development was not as simple as the above paragraphs indicate. In the seventeenth and eighteenth centuries there were numerous attempts to avoid the notion of action at a distance by explaining the force of gravity as a transmission of forces through contacts among the particles of a hypothetical omnipresent subtle form of matter, called "ether." In the nineteenth century the ether was invoked once again in order to avoid the introduction of "fields" as a new type of entity by explaining the effects of electric and magnetic fields in terms of material particles. Numerous models were constructed to explain these effects as results of stresses and strains in the ether. These models ran into difficulties; the properties that the ether had to be endowed with to account for the observed phenomena were simply hard to believe. All of these models were discarded once Special Relativity appeared on the scene, because Special Relativity turned out to be inconsistent with the assumption of the existence of an omnipresent ether.

Toward the end of the nineteenth century the mechanistic paradigm of classical physics reached its high point: The real entities in the universe are particles and fields, existing and changing according to Newton's laws of mechanics and Maxwell's laws of electromagnetism. To explain phenomena such as free fall, the tides, or the magnetism of the earth, we have to show how it arises as a configuration of particles and fields subject to these laws. In principle all phenomena are explainable within this framework; hence chemistry is reducible to physics, biology to chemistry, and psychology to biology. And consciousness, which is hardly significant in the immense scheme of the largely inanimate universe, is but one specific item of study within psychology.

4. What Is Time?

The mechanistic paradigm of the late nineteenth century was the one Einstein came to know when he studied physics. Most physicists believed that it represented an eternal truth. But Einstein was open to fresh ideas. Inspired by Mach's critical mind, he demolished the Newtonian concepts of space and time and replaced them with new, "relativistic" concepts.

The full extent of Einstein's thought and achievements lies outside of the scope of this essay. We will explore in detail, however, his first major breakthrough, his discovery of the *relativity of simultaneity*. This discovery revolutionized the understanding of space and time, put the question "What is real?" in a new light, and had a bearing on the issue of whether or not "God plays dice."

The question "What is real?" is intimately related to the question "What is time?" This is especially true in the currently held paradigm of Western civilization: For Plato "the Real" was the timeless Forms, or Ideas, such as Beauty, Love, or the Number Three. For us, however, "the real" is transient things and events, such as tables and chairs, lightning bolts, or our bodies. Hence our concept of "the real" is intertwined with our concept of time.

Well, then, what is time? St. Augustine begins his investigation into the nature of time with a perceptive comment: "I know well enough what it is, provided that nobody asks me; but if I am asked what it is and try to explain, I am baffled." He goes on to argue and question as follows: The past does not exist, because it is already past; the future does not exist, because it hasn't happened yet. Only the present is real. The present, however, has no duration; it is merely the demarcation line between past and future. And yet we do have an awareness of periods of time: We have an awareness of something taking a long time, and something else taking only a short time. How is such awareness possible? If that which exists, namely, the present, has no duration, how can we be aware of "a long time"? How can we be aware of something that does not exist? Augustine's response to the question is an insight into the nature of time. As we experience "a long time," he writes, "It is not future time that is long, but a long future is a long expectation of the future; the past time is not long, but a long past is a long remembrance of the past." St. Augustine concludes: "It is in my own mind, then, that I measure time. I must not allow my mind to insist that time is something objective."[4]

St. Augustine's lucid analysis is a forerunner of the deep, if obscure, analysis of Kant, an analysis that led to the conclusion that time is a "category" of the human mind. The paradigm we currently live by regarding the relationship between time and "the real" is still informed by St. Augustine's insight: Neither the past nor the future is real; only the present is. But there is a crucial difference between our view and St. Augustine's. Augustine's basic premise is that the only reality is God. This basic premise is not shaken by the discovery that time is unreal. By contrast, God is not a part of the scientific world-view. And so, if the past and the future do not exist, we are left to conclude that *"the real world" is the world as it is at the present moment. It follows that the real world has no duration.*

Furthermore, if this is indeed the case, *we are never aware of the real world*. We can't possibly be aware of the real world because it always takes time, however short, for the carriers of sense impressions (light, sound) to reach our bodies. Even in the case of immediate perceptions, such as touch or taste, it takes time for the nervous system to process the sensory input and create an experience. In our awareness of the physical world we are always behind. By the time we experience it, we experience it not as it is but as it was. Nevertheless we are clearly convinced (and this conviction is part of our world-view) that in spite of this gap between existence and awareness, there is a real world out there, the world of the present moment. It consists of the collection of all the events that take place right now. Thus *the real world is a collection of simultaneous events.*

5. Special Relativity: The Basic Postulates

The conclusion of the last section will help us appreciate the significance of the paradigm shift introduced in 1905 by Einstein's Special Theory of Relativity. The theory is based on two postulates. The first, called *the principle of relativity*, dates back to Galileo and Descartes.

Imagine that you are traveling in a sealed train car; you cannot look outside. Imagine further, that the ride is so smooth that you don't feel any vibrations. When you entered this sealed car, the train was in motion, so you assume that it still is; but, after a while, you may begin to wonder: Is it still moving? This may lead you to ponder: Is there a way to determine whether the train is moving or not without leaving the car or looking out?

Well, it depends. If the car is accelerating, you will feel pushed back, and you will know that it is moving. If the train moves around a bend, you will feel pushed sideways, and you may even lose your balance—no doubt, the train is indeed moving.

Living Clocks and Relativity in Nature

The complexity of life is stunning, but the order that emerges is even more remarkable. Yucca plants have a reciprocal relationship with yucca moths. The moths lay eggs in the plant ovaries, then pack them with pollen balls to ensure seed production to feed the larvae. One species, *Yucca whipplei*, takes about seven years to flower. How, one may wonder, does a plant know when to flower?

Living or biological clocks are ancient and essential. In the last few decades, scientists have begun to understand the workings of these complex biological time keepers that initiate such diverse phenomena as the opening and closing of flowers, the annual migration of birds and the fierce seasonal mating fights of wild boars.

Internal clocks are synchronized with the daily cycle of light and darkness, phases of the moon and the seasons of the year—though not necessarily initiated by them. In 1729, French astronomer Jacques Ortous de Marian found that the opening and closing of mimosa leaves and later blue heliotrope flowers continued on cue even in a darkened room. Later, researchers found that circadian rhythms drift slightly unless reset daily by external time cues such as dawn and dusk.

Carl Linnaeus, of *genus* and *species* fame, published the names of flowers with such fixed closed and open times in *Philosophica botanica* in 1751. He suggested using these plants to make a living *Horologium florae*, or Flower Clock. Such clock gardens—

plants grouped in a circle divided into 12 sections, each containing flowers that open or close within that hour—became popular in the nineteenth century. It seems quite possible that

ed with triggering these events have been pinpointed in the plant *Arabidopsis thaliana* by inserting a luciferase firefly gene that can be made to glow, just like the firefly at

Flower clock proposed by Carl Linnaeus in 1751. Linnaeus probably never planted his flower clock, but gardeners implemented his idea a century later. Such living clocks are becoming popular again. Around the flower clock above are suggestions for a contemporary scheme.

Einstein would have seen such clock gardens in Europe.

Geneticists have now recorded patterns of gene expression in plants that follow a circadian rhythm, such as leaf and petal movements and the release of fragrances. Genes associat-

dusk if neighboring "clock" genes are turned on. Genetically modified *Aradopsis* plants could, in theory, light up at certain times of day, when different "clock" genes switch on, to make a modern glow-in-the-dark Flower Clock.

— PH

If, however, the train keeps moving at a constant speed and in a straight line, there is no way to tell. *But if there is no way to tell, even in principle, then the question "Am I moving or am I at rest?" is meaningless.* When we think about trains moving on the surface of the earth, the overwhelming presence of the earth makes it natural to assume that the earth is "really" at rest. But this is a kind of optical illusion. As far as the laws of nature go, they are the same on the platform and in the steadily moving train. And the motion is really relative. I, standing on the platform, consider myself at rest, and I see you moving at a constant velocity. But you, standing inside the train car, have every right to consider yourself at rest and think of me as moving at a constant velocity in the opposite direction. Who is "really" at rest? There is no such thing as being "really" at rest, or "really" in motion. Relative to me, you are moving. Relative to you, I am in motion. For inertial observers, i.e., observers whose velocities stay constant, that's the end of the story.

While this first postulate, the principle of relativity, makes sense, the second postulate does not. The second postulate, *the principle of the constancy of the speed of light*, states that the speed of light is the same regardless of the speed of the source which emits the light and regardless of the speed of the observer who measures it.

The value of this constant speed of light is about 186,000 miles per second. Light is really moving fast! The remarkable thing, however, is not the value, but the constancy of this value when measured under different conditions. Common sense tells us that this cannot be true: Suppose I travel on a train that moves at 100 miles per hour and throw a ball toward the front of the train at 30 miles per hour. Suppose, furthermore that the train is passing through a station without stopping or slowing down. If the stationmaster, standing on the platform of his station, measures the speed of the ball relative to him, the result of his measurement would be the sum of the speed of the train and the speed of the ball relative to the train, i.e., 130 miles per hour. The principle of the constancy of the speed of light says, however, that if I "throw" a pulse of light instead of a ball, this procedure of adding up the velocity of light and the velocity of the train yields a wrong result; the measurement of the speed of light by the stationmaster would yield the same result as the measurement of the passenger on the train—as if the train were not moving at all!

"Throwing" pulses of light could mean something as simple as holding a flashlight and flicking it on and off. The Michelson-Morley experiment of 1886, which surprised the world of physics by demonstrating this constancy of the speed of light, was much more refined, but its idea and its results are correctly illustrated by our simple example. [Editor's note: See pages 156 and 162.]

6. The Relativity of Simultaneity

Einstein came to accept this second postulate not because of the Michelson-Morley experiment but through an analysis of Maxwell's theory of electromagnetism. The speed of light appears as a constant in Maxwell's equations; and if, following the principle of relativity, one assumes that Maxwell's equations are equally valid in all frames of reference that are moving at a constant velocity, then this constant, the speed of light, must, indeed, be constant, i.e., have the same value in all such frames. Following this line of thinking, Einstein accepted the second postulate in spite of its apparent absurdity. This acceptance was an ingenious leap of faith. It allowed Einstein to take the position that both postulates, the principle of relativity and the constancy of the speed of light, are true, and then show that if we are willing to radically change the way we think about space and time, both postulates can indeed be a part of a consistent world-view. Let us assume the validity of the two postulates, he argued, and see what follows.

The first concept that came under Einstein's scrutiny was the concept of Simultaneity. How do we determine whether or not two events are simultaneous? This is easy if both events occur right here: If I saw them happening at the same time, then they did happen at the same time. But what about two events that are taking place far away from each other? What if one of them took place right here and the other on the moon? In this case, if I saw them at the same time, then, certainly, they did not occur at the same time: The event on the moon took place a second and a quarter before the event here, because it takes light that much time to travel from the moon to the earth. The definition of simultaneity must take into account the time gap between the happening of an event and the arrival of light from the event to the observer. To take this time gap into account, let us say, with Einstein, that events are simultaneous if an observer *situated halfway between their locations sees* them at the same instant.

Using such "midway observers" we can synchronize two clocks situated apart from each other. The clocks need to be set so that an observer situated midway between them will see them showing the same time. If these clocks are of identical construction, and neither breaks down, we can assume that, once set, they will continue to be synchronous as time goes on. This is all rather straightforward. Our last assumption holds true if neither clock is under the influence of gravity, or, in the language of the *General* Theory of Relativity, if both clocks are in regions of spacetime that are flat. We will limit our discussion to this case.

We are now ready to begin the process that will lead to "the relativity of simultaneity." The process involves one of Einstein's favorite intellectual journeys, a thought experiment. A thought experiment is an experiment that can be performed in principle (i.e., its performance is allowed by the laws of nature) and yet, because of the limitations of our technology, cannot be performed in practice. However, even though we cannot do the experiment, nothing prevents us from thinking about it, and such thought can lead to far-reaching conclusions, as we will see. We will describe our thought experiment, which is similar to the one proposed by Einstein, in the form of a story.

Once upon a time two astronauts, Peter and Julie, were traveling in their respective spaceships way out in space, far away from stars and planets. Both spaceships had clocks at each end; each pair of clocks was synchronized: Peter and Julie, standing in the middle of their vehicles, acted as "midway observers."

The two spaceships approached each other. They traveled in precisely opposite directions, and, because of Peter's carelessness, they almost collided head-on. Fortunately, the spaceships just missed each other as they zoomed past. The encounter, while not disastrous, was not uneventful. This is how Julie told me about it (see Figure 2.1).

"I was just standing there, in the middle of my spaceship, which was at rest, when Peter's spaceship appeared, zooming toward me at an enormous speed. I was relieved when I realized that it did not hit my spaceship, but right at the instant his spaceship was side by side with mine, something strange happened: Two pieces of space debris, two small rocks, hit our vehicles. It happened at the exact moment our ships were perfectly parallel, the back of his ship almost touching the nose of mine and the nose of my ship almost touching the back of his. It was at this precise moment that the fronts and rears of our two ships were hit

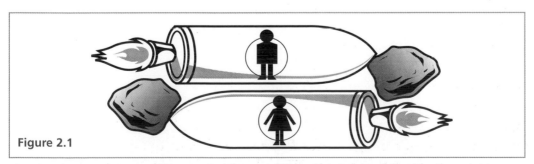

Figure 2.1

The situation at 12:00 according to Julie. The two rocks are hitting the ends of both spaceships, as Peter, standing in the middle of his spaceship, zooms past Julie, standing in the middle of hers.

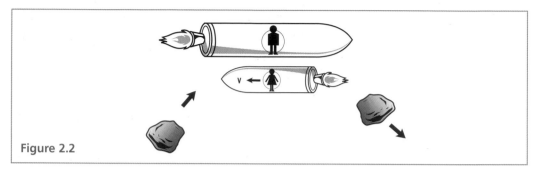

Figure 2.2

The situation at 12:00 according to Peter. One rock has already hit the front of Peter's spaceship and the rear of Julie's, while the other is about to hit the rear of Peter's spaceship and the front of Julie's, as Julie, standing in the middle of her spaceship, zooms past Peter, who is standing in the middle of his spaceship.

simultaneously by the two rocks. The rocks weren't big enough to penetrate our hulls, but they jolted our ships pretty violently—violently enough to stop both of my clocks. Later Peter told me that the rocks hit hard enough to stop his clocks too. The strangest part of it is that the jolts occurred at the exact same time!"

"And how do you know that the two jolts happened at the same time?" I asked.

"Well," Julie responded, "as I just said, when my two clocks stopped, they showed twelve o'clock. As a matter of fact, since they're still broken, they still do."

"That's strange," I responded. "I talked to Peter a little while ago, and he told me that the two jolts were not simultaneous. One of them took place a little before noon and the other a little after noon." (See Figure 2.2)

Julie pondered this piece of information for a long time. Finally she said: "What about the broken clocks at the front and rear of his vehicle? They stopped when the rocks hit. What do they read?"

When I answered, she could hardly believe her ears: *The clock in the front of his spaceship shows a fraction of a second before 12:00; the one at the rear, a fraction of a second after. According to his clocks, at 12:00 the jolt to the front of his spaceship already happened, and the other jolt hadn't happened yet.*

It was not only Julie who could not believe her ears. It is, indeed, hard to accept the relativity of simultaneity. Remember, "the world now" is what we consider real. According to Special Relativity, however, the collection of events that comprises "the world now" is different for different observers. Julie's "real world at 12:00" includes the two rocks in the acts of hitting the spaceships, while Peter's "real world at 12:00" includes neither of these events. In his "world at 12:00," the event one jolt already happened, and the other one is still in the future. What is *the really real world at 12:00* like? Einstein's message is that the concept of a really real world at 12:00 is meaningless. The relativity of simultaneity is an unavoidable conclusion of the two basic postulates of Special Relativity.

In our thought experiment the differences between Peter's and Julie's clock readings for events A and B amounted to very small fractions of a second. This is due to the fact that the two events occurred in close proximity to each other. The relativity of simultaneity can lead to very large differences in clock readings if the two events are far apart. Let us assume, for example, that the encounter between Peter and Julie took place very far from the solar system, light years away from earth. Meeting and reminiscing years after the encounter, they could have compared the time of the encounter with the times of events on earth. Peter could have said to Julie, "Remember that near collision we had? It happened on December 30, 1989, the same day my poor uncle died," "No, no," Julie would have responded, "it happened on March 4, 1990, when my nephew was born!"

Let us conclude this section with the following observation: If you compare Figures 2.1 and 2.2, you may notice something strange about the lengths of the two spaceships. In Figure 2.1 they are of equal length, while in Figure 2.2 Peter's spaceship is longer than Julie's. This is not an error. It follows from the two basic postulates of Special Relativity that length is relative! If you measure the length of spaceship (or any other object) while it is moving relative to you, the result is less than the length of the same object measured when it is at rest relative to you. This is called "length contraction."

7. An Ultimate Speed

The relativity of simultaneity is fascinating. But what does it have to do with Einstein's adherence to strict causality?

Before we respond to this question, let us introduce one more consequence of the two basic postulates of Special Relativity: The speed of light is the greatest speed that occurs in nature. Nothing can move faster than light. Light itself, and other forms of radiation, can't help but move precisely at the speed of light; this is what the second postulate tells us. By contrast, material particles and objects, such as an electron, an atom, a spaceship, or a human body, can approach the speed of light but can never reach it. And since everything in nature is either matter or radiation, nothing can move faster than light. Furthermore, since information is transmitted from one place to another by some agent, the speed of light is the ultimate speed for the propagation of signals.

The speed of light is thus the ultimate speed by which a cause can bring about an effect. If one event A is the cause of another event B, there must be enough of a time separation between them to allow a signal traveling at the speed of light from A to reach B. This amounts to a kind of separability: Events that are separated in space can be related as cause and effect only if there is enough of a time separation between them as well. This characteristic of nature is called "Einstein separability" or "locality."

8. The Eternal Present

The relativity of simultaneity is, in Alfred North Whitehead's words, "a heavy blow to the classical scientific materialism, which proposes a definite present instant at which all matter is simultaneously real."[5] What Whitehead calls "the classical scientific materialism" is the world-view we live by. The negation of our intuitive concept of the real by Special Relativity demonstrates that whatever "the real" is, it is not what we think. If "the world now" cannot be "the real world," what is? This was Einstein's dilemma.

St. Augustine was led into his investigations of the nature of time through the bothersome question "What did God do before He made heaven and earth?" "My answer," he wrote, "is not 'He was preparing Hell for people who pry into mysteries.' This frivolous retort has been made before now, so we are told, in order to evade the point of the question."[6] Contemplating the difference between time and eternity, he concludes that before Creation there was no time. Addressing God, he expresses his conclusion as follows:

> It is therefore true to say that when you had not made anything, there was no time, because time itself was of your making. And no time is co-eternal with you, because you never change; whereas, if time never changed, it would not be time.[7]

The human perspective of dividing events into past, present, and future events is irrelevant from the perspective of eternity, which is God's perspective:

> It is in eternity, which is supreme over time because it is a never-ending present, that you are at once before all past time and after all future time... Your years neither go nor come, but our years pass and others come after them, so that they all may come in their turn. Your years are completely present all at once, because they are at a permanent standstill.[8]

The seeming unreality of past and future is not really problematic for St. Augustine. The division into past, present, and future is a creation of the human mind. From God's perspective there is no such division. Hence all events, past, present, and future, have equal status as far as their reality goes. They may or may not be "really real," but the events of the present moment are not more or less real than the events of the past or the events of the future.

A similar conclusion regarding the reality of past, present, and future events follows from the relativity of simultaneity. If events that are present for one observer are past or future for another, the question of whether or not they are "real" will not be settled by checking on whether they are "now events" or not. The appeal to the perspective of God is not part of the current scientific paradigm, but Augustine's point of view regarding the reality or unreality of physical events is forced upon us by the relativity of simultaneity. Einstein resolved his dilemma by concluding that fundamentally all events, past, future, and present, are on equal footing, that they are all real. Time is an illusion.

A month before he died, Einstein received the news of the death of his life long friend Michele Besso. His letter of condolence to the Besso family contains the billowing remarkable statement:

> Now he has departed from this strange world a little ahead of me. That signifies nothing. For us believing physicists the distinction between past, present and future is only a stubbornly persistent illusion.[9]

9. The Case for Strict Causality

As you may recall from the beginning of this chapter, this long excursion into the Special Theory of Relativity was motivated by the wish to understand Einstein's commitment to the tenet of strict causality.[10] Now we are in a position to appreciate the source of this commitment.

Following the indication of Special Relativity, Einstein ascribes reality to all the events in spacetime. All events, regardless of whether they are past, present, or future events from the point of view of particular observers, are on the same footing: they are all real. The apparent unreality of the past and the future is merely the result of the way our minds process impressions.

Now, if we are sure of anything, it is that the past is fully determined. Whatever happened, happened—there is no randomness about that. But if all events are to be treated on the same footing, we must extend this ontological certainty to the present and future as well, if time is an illusion, a mere human perspective, the perspective of a limited mentality facing the vastness of the "eternal present," then, in a sense, the present and the future have already happened, just like the past.

The belief in strict determinism is, then, one element in a consistent world view that is based, on the one hand, on the belief in an independently existing reality and, on the other, on the findings of Special Relativity. According to the principles of quantum mechanics, however, the physical universe is not deterministic. This is one reason for Einstein's refusal to accept quantum mechanics as a fundamental theory of nature.

There is, however, another reason. Einstein's belief in realism included the common sense assumption that the description of a physical system requires a single model. This belief applies to all systems, large and small, even to systems that belong to the quantum domain. However, when Niels Bohr struggled to understand the quantum theory he discovered that this is not the case. Two years after Heisenberg's lecture at the University of Berlin he introduced his "framework of complementarity," according to which a full description of a quantum system involves, in general, two seemingly contradictory conceptual models. This was to be the great challenge to Einstein's paradigm: Bohr's world-view and its mainstay, the framework of complementarity.

Notes

1. A. Pais, *"Subtle is the Lord,"* p.5
2. Lucretius, *On the Nature of the Universe*, p. 39
3. I. Newton quoted in G. Holton, *Introductions to Concepts and Theories in Physical Science*, p. 298
4. St. Augustine, *Confessions* XI.14, p.264; XI.28, p. 277; XI.27, p. 276
5. A. N. Whitehead, *Science and the Modern World*, p. 118
6. St. Augustine, *Confessions* XI.12, p. 262
7. Ibid., XI.14, p. 263
8. Ibid., XI.13, p. 263
9. A. Calaprice, collector and ed., *Quotable Einstein*, p. 61
10. There seems to be some indirect evidence that toward the end of his life Einstein softened his opposition to indeterminism...[See Notes, Chapter 2, in *Nature Loves To Hide* for a full discussion]

When Time Slows Down

by Donald Goldsmith

Introduction

Einstein's theory of relativity predicts that as one travels closer and closer to the speed of light, time passes more and more slowly. That is, if we could watch a clock moving by us at nearly the speed of light, we would observe that the clock takes longer to tick off an hour than does our own stationary clock. Since we have inherited an intuitive belief that time unfolds at the same rate everywhere, such a prediction at first puzzles many people. Only by appeals to reason, experiment, repetition, and authority have astronomers and physicists been able to convince themselves, and other interested parties, that we really do observe that fast-moving clocks keep time more slowly than stationary clocks.

Amazing though it may seem, a person who makes a long journey at nearly the speed of light would return to earth with an age far less than that of a twin who remain behind on earth. The basis behind this is the experimental result elevated into a universal rule by Einstein's theory— that no matter how a source of light may be moving with respect to an observer, the observer will always find that light arrives at the same speed. This speed is denoted by "*c*" and is equal to just under 300,000 kilometers (about 186,000 miles) per second. (Astronomical distances are often measured in *light years*, the distance light travels in a year, or even in light seconds, the distance light travels in one second.)

The Speed of Light

During the 1880s, Albert Michelson and Edward Morley made observations of the light coming to us from distant stars—at times when the earth's motion in its solar orbit was directed toward the stars and at times when the motion was directed away from the stars. Despite the fact that Michelson and Morley's equipment was sensitive enough to measure the change in the speed of the star's light that would be

Reprinted with permission from *Mercury* magazine and the Astronomical Society of the Pacific (www.astrosociety.org).

expected from adding (or subtracting) the earth's orbital speed of 30 kilometers per second to (or from) the 300,000 km/sec speed of light, no such change was observed. Michelson was puzzled, Morley was puzzled, and so was everyone else in Cleveland (where the experiment was done), and throughout the world, who cared to think about what these results meant. Something was wrong with the obvious concept that light emitted from a moving source should travel at its usual speed in a vacuum, c, plus or minus the contribution from the motion of the source.

Physicists have rarely died from a lack of understanding, however, and eventually they came to believe what they could not previously understand: no matter how a source of light may be moving with respect to an observer, that observer receiving the light will always find that the light arrives at the same speed, c, not that speed plus or minus some contribution from the source's motion. Though it may seem a long jump to pass from one set of observations related to the earth's motion all the way to a general conclusion about the behavior of light, this leap appears natural once we accept the principles that the earth occupies an average position in space and that the laws of physics should be the same everywhere. We could hardly believe that some demon was rigging the show to fool two professors in Ohio (or succeeding generations of physicists who repeated their experiments with an ever-greater accuracy and still found no change in the speed of light from a moving source.)

The Theory of Relativity

In 1905, Albert Einstein pointed the way out of the disagreement between the observed facts and intuition. Einstein said that it was time for us to shed our belief in an absolute motionless space in which velocities would add and subtract from one another as we feel they should. Any belief in an absolute frame of time, where all accurate clocks would appear to run at the same rate, should also be disregarded. Instead, Einstein proposed that we start from what we observe: the measured speed of light is the same no matter what the velocity of the observer relative to the light source. Einstein drew on the mathematical formulation that preceded 1905 (but which failed to admit that the universe might truly live up to the implications of this formulation). In his epochal paper "On the Electrodynamics of Moving Bodies" he showed that if an observer watches a physical system (of particles, bicycles, or whatever) move by at a constant speed, the observer will notice that lengths in the moving system decrease, time in the moving system goes more slowly, and inertial masses (resistance to acceleration) increase, all in relation to what an observer finds for a physical system that is at rest with respect to the observer. Furthermore, the amount of the change in the lengths, the time, and the inertial masses increases as the velocity of the system relative to the observer increases. All of this works in just such a way that any observer measuring light produced by the system will always find that the light arrives at the same speed, c, fulfilling the results of the Michelson-Morley experiment. This was the point of Einstein's exposition: to fit the facts known from observation into a theoretical and mathematical framework that could be systematically generalized to apply to a wide range of possible situations.

The "special theory of relativity" that Einstein devised in 1905 predicts what an observer will see if a system moves by at a constant speed. (The general theory of relativity deals with cases of accelerating or decelerating motion.) In particular, the theory predicts that time in the system moving at a velocity, v, relative to the observer and time in the observer's own system (zero velocity relative to the observer) obey the following relation:

Time in moving system = the square root of $[1 - (v/c)^2]$ x Time in observer's system

Because no object can move faster than light—another keystone of relativity theory, and never yet contradicted by observation—the ratio v/c is always between zero and one, and so is $(v/c)^2$. Then $1 - (v/c)^2$ always has a value between 1 (for $v = 0$) and zero (for $v = c$). The square root of $1 - (v/c)^2$ likewise varies from one (if $v = 0$) down to zero (if $v = c$), and thus a given time observed in the moving system is always less than (or equal to) the time in the observer's own system.

To take a definite example, suppose that we could propel a subatomic particle at 99.5 percent the speed of light ($v/c = .995$). We would observe that time in the particle's system passes ten times more slowly than time in our own frame of reference. If we observe that such a subatomic particle has a lifetime of one one-thousandth of a second before it turns into other kinds of particles, we shall observe that its lifetime becomes one one-hundredth of a second when it moves at 99.5 percent of the speed of light relative to us. Particle accelerator laboratories like the Stanford Linear Accelerator Center routinely perform such experiments, and the results of these tests have verified the predictions of the special theory of relativity millions of times over.

Suppose that we now aim higher than the world of tiny particles and imagine that we could propel a space vehicle at 99.5 percent the speed of light towards the star Aldebaran, 32 light years away. Since the best rockets we have today can reach speeds of 1/10 of one percent of the speed of light, there clearly are energy problems to be overcome before this example becomes a reality. However, if a spaceship did travel to Aldebaran at 99.5 percent the speed of light, it would arrive there 32 years and two months later, *as measured in our time*. But for people aboard the spaceship—the moving system—time would run ten times more slowly than on earth, and we would observe that during the journey they would age not thirty-two and some years, but just about 3 1/5 years! If the space travelers returned home immediately after they arrived, at the same speed as they went to Aldebaran, they would not have aged 64 years and four months (the elapsed time on earth), but only 6 2/5 years, and would in some cases find themselves younger than their grandchildren.

The Twin Paradox

At this point, many a would-be believer in relativity theory has paused to object: Wait a minute! If the rules of the theory hold good for any observer, why don't the space travelers observe *us* on earth to be aging more slowly than *they* are? They can claim the earth is moving with respect to the spaceship and thus return to find us on earth younger than they are. This question is often restated as the "twin paradox": if one twin stays home while the other rides the rocket, which is really the younger when they meet again?

The answer is that the traveling twin really comes home younger than the twin who stayed behind. There are various ways to demonstrate the result:

(1) By assertion. This method, beloved of politicians and many teachers, reminds the reader that calculations too difficult to be reproduced here have shown conclusively that the twin who travels ages less than the twin who stays on earth.

(2) By experiment. The best test of all would be to send someone to Aldebaran and back at 99.5 percent the speed of light. Proposals for government funding offering far smaller rewards have often been submitted, but none of them would cost as much as this one. We might think that we could use particle accelerators to send particles with a known lifetime down the tubes at 99.5 percent the speed of light, turn them around, and bring them back for comparison with stationary particles. In fact, although this experiment has been done in its essential form, though not precisely as described above, there is something unsatisfying about results on the subatomic level to convince ourselves that things really work the same way on a human scale. Still, the experimental results are in, and they confirm the prediction made above.

Tachyons—Faster Than the Speed of Light?

Although Einstein's special theory of relativity implies that no object can be accelerated to the speed of light from a lower velocity, the theory does allow for particles that *always* travel faster than light. Unlike ordinary matter, these hypothetical "superluminal" (faster-than-light) particles, called *tachyons* (from the Greek for "swift"), would gain more mass as they came closer to the speed of light "from above," i.e., by *losing* velocity.

First introduced by physicist Gerald Feinberg in 1967, tachyons have strange properties because of the importance of the square root of

$(- v^2/c^2)$ in Einstein's equations. Notice that when v is greater than c, our fraction is greater than 1, which means we are dealing with the square root of a negative number—or, an imaginary number.

It turns out that if tachyons do exist, they'd have very strange properties indeed. For example, they would gain energy as they approached the speed of light and lose energy as they sped up—again, just the opposite of what happens with ordinary matter.

Tachyons have never been shown to exist in reality, and if they do exist, it is unclear whether they could inter-

act with ordinary matter. Some physicists have attempted to detect such superluminal particles by watching how cosmic rays—particles that travel through space at 99.99% the speed of light—interact with the atmosphere of Earth. Some claim to have detected such particles, but the results have been inconclusive.

Tachyons remain a fascinating possibility, often featured in science fiction, theoretically capable of transmitting information at velocities greater than the speed of light, which would (again, only theoretically) make it possible to send messages to the past.   — DG

(3) By reasoning. Although we may rightly suspect our intuition about physical reality, we should trust our powers of reason. If we believe the results of the special theory of relativity outlined above, we can proceed to draw conclusions from them: the theory all stands together or all falls together, and so far it has stood the test of experiment. So let us consider the twin aboard the spaceship and the twin at home, and see whether we can answer with certainty the question of who ends up younger when they meet again.

A Space Journey

We might first pause to agree that if the twins are not together, comparing their ages becomes very difficult because we cannot pass information instantaneously from one place to another. A twin on earth signaling to a twin on Aldebaran must wait at least 64 years for a message to go to Aldebaran and return, since nothing can move faster than light. Einstein's research helped to illuminate how our intuition glides over the problem of determining whether or not two events that occur at different places are truly "simultaneous." He showed that the key intervals (separations) to concentrate on are neither intervals of distance nor intervals of time, but rather intervals of "spacetime," defined by the following relation:

(interval of spacetime) $= \sqrt{\text{square root of } [c^2 \text{ (time interval)}^2 - \text{(space interval)}^2]}$

Although two different observers moving at different (but constant) speeds relative to some system may measure different space intervals between two events that occur in that system, the two observers will measure the same spacetime interval between the two events. Fascinating as spacetime intervals are, we can see that the twin problem is in danger of slipping out of the agile grasp of our minds unless we concentrate on the simple question of the two twins who start side by side and end up side by side once the journey is over. This means that the spacetime interval questions can be set aside in favor of a direct approach to the time interval, since the twins are not separated in space from either twin's point of view when the journey starts and when it ends. The critical point here is this: *The key distinction between the traveling twin and the twin at home is that the former must reverse direction in order to come home again.*

Suppose that each twin decides to send signals to the other that will let the other twin know how fast time is passing.[1] The twins do this by each wearing a beacon that flashes once every second—each second, that is, of time as determined in their own frames of reference. We can use the observed rates of flashing as a way of measuring how fast time seems to be passing. The same purpose would be served by measuring pulse rates or tree-ring growth rates. If we observe a beacon flashing more rapidly than once per second, this corresponds to an observation that time seems to be passing more rapidly than in our own reference frame, and if we observe a rate slower than once per second, this means we are observing time to be passing more slowly than it does for us.

Now the twin blasts off from earth and travels to Aldebaran at 99.5 percent the speed of light. How often does the twin who stays behind receive a flash from the traveling twin? The answer is not once per second, but about once every twenty seconds. There are two reasons for the increase in the time interval between flashes. First, since the spaceship moves at 99.5 percent the speed of light relative to the earth, by Einstein's theory events on the spaceship seem to occur ten times more slowly than they do on earth. Also, the twin in space is constantly receding from the twin on earth, so each flash has farther to travel than the preceding one. The spaceship recedes from earth by almost ten light seconds for every ten seconds of earth time, and this adds almost another ten seconds (of increased travel time) to the ten-second interval established by the spaceship's motion relative to the earth. Hence the interval between flashes observed on earth from the twin on the spaceship is almost twenty seconds.

What does the twin on the way to Aldebaran find for the time interval between flashes from the earth? The answer is also almost twenty seconds, and for the same reasons: the earth seems to be receding from the spaceship at 99.5 percent the speed of light, so events seem to unfold ten times more slowly on earth than aboard the spaceship. Also, each successive flash must travel ten light seconds farther, and so the once-per-second flashes emitted on earth are spaced by intervals of almost twenty seconds when they are observed aboard the spaceship.

So far the twin paradox seems to be alive and well, as each twin sees the other's flashes arriving not one second apart but twenty seconds apart. But watch closely as we compare what happens throughout the entire journey, because if you miss the details that make all the fun worthwhile, you will wind up surly as the proverbial bereaved bear at the end of the description.

First, consider the whole series of events as seen by the twin who travels to Aldebaran. On the way there, flashes arrive from earth once every twenty seconds, so the twin concludes that during this time, events on earth are unfolding twenty times more slowly than events aboard the ship. To the traveling twin, the trip to Aldebaran seems to last only about 3 1/5 years, and this twin concludes that on earth only one-twentieth of this time, or just under two months, has passed on earth during the outward journey. Now suppose that the twin reaches Aldebaran and, without pausing for so much as a melloroon, returns to earth at 99.5 percent the speed of light, always watching the flashing beacon. On the return journey, flashes arrive not once every twenty seconds, *but twenty times per second*. Why? Because now each successive flash has *less* distance to travel to reach the spaceship. For this reason, even though the flashes seem to be emitted only once very ten seconds (because the earth can be considered to be moving at 99.5 percent the speed of light relative to the spaceship), there is a reduction in the travel time from the earth to the spaceship between each flash of 10 sec x (.995) = 9.95 seconds. The net result is that the time between flashes is 10 sec – 9.95 sec = 0.05 sec, so they arrive twenty times per second. This result does not indicate conclusively that the twin on earth is now aging more rapidly than the traveling twin. It indicates rather that the spaceship's motion towards earth more than compensates for the apparent slowdown in the rate of flashing. Bear in mind that each twin knows about additional travel time, and could calculate the part that it is playing in what they see, but the end result would still be what we calculate.

So the traveling twin comes home on a journey that aboard the spaceship seems to take 3 1/5 years (just as the outward journey did), but during the return journey the traveling twin sees a flash rate from earth twenty times that on board, so the twin concludes that 20 x 3 1/5 years = 64 years 2 months[2] have elapsed on earth during the journey, and the total elapsed time on earth for the journeys out and back is 64 years and four months while the total elapsed time aboard ship is 3 1/5 + 3 1/5 years = 6 2/5 years. If you have been following this epic journey closely, you will notice that all the flashes have been observed and are accounted for aboard the spaceship.

What does the twin on earth see? While the other twin flashes onward toward Aldebaran, the twin on earth receives flashes once every twenty seconds, as we described above, because time aboard the spaceship appears to be running more slowly and the spaceship is constantly getting farther from the earth. To the twin on earth, the spaceship's outward journey appears to take 32 years and two months (32 light years at 99.5 percent the speed of light), and, furthermore, the last flash from the outward journey, emitted upon arrival at Aldebaran, will take 32 years to reach the earth. The result is that for the entire 64 years and two months needed for the traveling twin to reach Aldebaran and have the flash from Aldebaran reach the earth, the earthbound twin observes time to be running twenty times more slowly than earth time. The twin on earth concludes that the total elapsed time aboard the spaceship by the time the flash from Aldebaran arrives is 1/20 x (64 years 2 months) = 3 1/5 years.

When the spaceship has begun its return journey and the flashes from this leg arrive on earth, they come not once per second but twenty times per second, since the spaceship's motion toward earth more than compensates for the apparent slowing of time aboard ship, as we described earlier for the reciprocal situation seen aboard the spaceship. The traveling twin will return to earth just two months after the flash from Aldebaran arrives on earth, because that flash travels to earth at the speed of light (for 32 years), while the spaceship comes home at 99.5 percent the speed of light, thus taking 32 years 2 months, two months longer than the flash of light.[3] During the two months between the arrival on earth of the flash from Aldebaran and the return of the spaceship, the flashes arrive twenty times per second, indicating a passage of time twenty times faster by this standard of flash counting, so the return phase aboard the spaceship seems to take 20 times 2 months, or 3 1/5 years, as observed from earth.[4] The spaceship's entire journey as seen from earth thus appears to total 3 1/5 years + 3 1/5 years = 6 2/5 years aboard the spaceship, but 64 years 4 months have passed on earth during the round trip to Aldebaran. This agrees with what the traveling twin saw. As observed aboard ship, the entire trip took 6 2/5 years, while 64 years 4 months elapsed on earth.

Thus by counting flashes and timing their arrival, both the twin on earth and the traveling twin agree that the trip to Aldebaran caused the traveling twin to age 6 2/5 years while the twin who stayed behind aged 64 years and 4 months, for a difference of almost 60 years. The cause of this difference is the reversal of direction made by the traveling twin but not by the twin on earth. On account of this key distinction, the two twins are not equivalent: they did not have the same experience even as seen from their own small worlds. The apparent paradox of the twins has thus been resolved, thanks to the immense power of logic ranged alongside the guiding principles Einstein elucidated.

So we have the secret of longevity by earth standards: travel at velocities near the speed of light, and you'll outlive your contemporaries. It turns out (but we shall not demonstrate it here) that it's not necessary to go in a straight line; it's enough to circle round and round the earth like astronauts. The men who spent 84 days in orbit in Skylab, moving at a speed relative to ourselves of about 8 km/sec (about three one-thousandths of a percent of the speed of light), aged about one five-thousandth of a second less than they would have by staying at home. Although this gain in life probably does not outweigh the risks involved, especially since we should also dock them one five-thousandth of a second of flight pay, we can imagine that someday—should humanity survive long enough—space travelers to nearby stars could use the slowing down time to help them survive the journey and return to tell their descendants what they found.[5]

References

1. This discussion follows the example of N. David Mermin in Chapter 16 of his book *Space and Time in Special Relativity* (New York: McGraw Hill, 1986). This is probably the best book written to explain special relativity theory to the lay reader, and I could not hope to improve on Mermin's explanation of the twin paradox and its resolution.

2. I have rounded off 3.2122 years to 3 1/5 years and a factor of 19.975 to a factor of twenty; the true product is 64.1637 years.

3. When the flash from Aldebaran arrives on earth the spaceship is only two light months from earth and closing fast.

4. I have rounded off as mentioned above.

5. The slowing down of time often appears in science fiction without much care for exactitude. An exception is Poul Anderson's *Tau Zero*, where a spaceship's crew grows as old as the universe and (here license enters) survives the next "big bang."

The mathematical structure of black holes was worked out when the possibility that there actually were such things in the universe was considered remote.

Einstein's Dream*

by Stephen Hawking

In the early years of the twentieth century, two new theories completely changed the way we think about space and time, and about reality itself. More than seventy-five years later, we are still working out their implications and trying to combine them in a unified theory that will describe everything in the universe. The two theories are the general theory of relativity and quantum mechanics. The general theory of relativity deals with space and time and how they are curved or warped on a large scale by the matter and energy in the universe. Quantum mechanics, on the other hand, deals with very small scales. Included in it is what is called the uncertainty principle, which states that one can never precisely measure the position and the velocity of a particle at the same time; the more accurately you can measure one, the less accurately you can measure the other. There is always an element of uncertainty or chance, and this affects the behavior of matter on a small scale in a fundamental way. Einstein was almost singlehandedly responsible for general relativity, and he played an important part in the development of quantum mechanics. His feelings about the latter are summed up in the phrase "God does not play dice." But all the evidence indicates that God is an inveterate gambler and that He throws the dice on every possible occasion.

In this essay, I will try to convey the basic ideas behind these two theories, and why Einstein was so unhappy about quantum mechanics. I shall also describe some of the remarkable things that seem to happen when one tries to combine the two theories. These indicate that time itself had a beginning about fifteen billion years ago and that it may come to an end at some point in the future. Yet in another kind of time, the universe has no boundary. It is neither created nor destroyed. It just is.

---

* A lecture given at the Paradigm Sessions of the NIT Data Communications Systems Corporation in Tokyo in July 1991.

Relativity and Flying Clocks

Einstein's general theory of relativity predicts that gravity effectively slows time down. An earth-bound clock, for instance, should run more slowly than one ticking far from Earth's gravitational influence. The special theory of relativity predicts that clocks traveling at high speeds tick more slowly compared to clocks on Earth. By using extremely accurate atomic clocks, both these effects have now been observed and verified.

Gravity Probe A, conducted by NASA in 1976, involved launching a highly accurate maser atomic clock on a two-hour suborbital flight to explore the effect of gravity on time. The GPA rocket orbited at a height of 10,000 km (6,200 miles), where Earth's gravity is less than half its sea level value. At this height, a clock should run 4.5 parts in 10 billion faster than one on Earth. As the rocket slowed, stopped momentarily, and started falling back, scientists measured the slight differences between the onboard and Earth-based clocks. The difference confirmed Einstein's predictions to a precision of 70 parts per million.

In a famous flying-clock experiment, four cesium atomic clocks were flown around the world and back again on commercial airline flights in 1971. The experiment, designed by R.E. Keating and J.C. Hafele of the U.S.

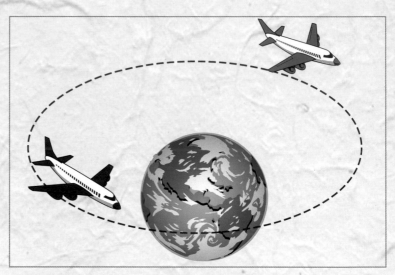

In the Hafele and Keating experiment, theory predicts that the clock flying eastward compared with a reference clock on earth should lose about 40 billionths of a second. The clock on the plane flying westward is actually moving more slowly than the clock on earth rotating 1000mph eastward, so the predicted gain for this clock is about 275 billionths of a second—exactly what was observed.

Naval Observatory, confirmed the relativistic slowing of time as predicted by Einstein's special theory of relativity.

To commemorate the 25th anniversary of this experiment, a single cesium atomic clock was flown from London to Washington and then back. Like the original flight, the 1996 flying clock experiment took two relativistic considerations into account: The speed of the aircraft and the altitude of the clock on the aircraft relative to

an observer on the ground. The overall gain for the in-flight clock was consistent with a predicted clock loss of 16.1 billionth of a second due to the aircraft's speed and a clock gain of 53 billionth of a second due to decreased gravity. Einstein would likely have appreciated the fact that the testing of relativity, in part, required the use of masers., since it was his insights that eventually led to the invention of the maser.   — PH

I shall start with the theory of relativity. National laws hold only within one country, but the laws of physics are the same in Britain, the United States, and Japan. They are also the same on Mars and in the Andromeda galaxy. Not only that, the laws are the same at no matter what speed you are moving. The laws are the same on a bullet train or on a jet airplane as they are for someone standing in one place. In fact, of course, even someone who is stationary on the earth is moving at about 18.6 miles (30 kilometers) a second around the sun. The sun is also moving at several hundred kilometers a second around the galaxy, and so on. Yet all this motion makes no difference to the laws of physics; they are the same for all observers.

This independence of the speed of the system was first discovered by Galileo, who developed the laws of motion of objects like cannonballs or planets. However, a problem arose when people tried to extend this independence of the speed of the observer to the laws that govern the motion of light. It had been discovered in the eighteenth century that light does not travel instantaneously from source to observer; rather, it goes at a certain speed, about 186,000 miles (300,000 kilometers) a second. But what was this speed relative *to*? It seemed that there had to be some medium throughout space through which the light traveled. This medium was called the ether. The idea was that light waves traveled at a speed of 186,000 miles a second through the ether, which meant that an observer who was at rest relative to the ether would measure the speed of light to be about 186,000 miles a second, but an observer who was moving through the ether would measure a higher or lower speed. In particular, it was believed that the speed of light ought to change as the earth moves through the ether on its orbit around the sun. However, in 1887 a careful experiment carried out by Michelson and Morley showed that the speed of light was always the same. No matter what speed the observer was moving at, he would always measure the speed of light at 186,000 miles a second.

How can this be true? How can observers moving at different speeds all measure light at the same speed? The answer is they can't, not if our normal ideas of space and time hold true. However, in a famous paper written in 1905, Einstein pointed out that such observers could all measure the same speed of light if they abandoned the idea of a universal time. Instead, they would each have their own individual time, as measured by a clock each carried with him. The times measured by these different clocks would agree almost exactly if they were moving slowly with respect to each other—but the times measured by different clocks would differ significantly if the clocks were moving at high speed. This effect has actually been observed by comparing a clock on the ground with one in a commercial airliner; the clock in the airliner runs slightly slow when compared to the stationary clock. However, for normal speeds of travel, the differences between the rates of clocks are very small. You would have to fly around the world four hundred million times to add one second to your life; but your life would be reduced by more than that by all those airline meals.

How does having their own individual time cause people traveling at different speeds to measure the same speed of light? The speed of a pulse of light is the distance it travels between two events, divided by the time interval between the events. (An event in this sense is something that takes place at a single point in space, at a specified point in time.) People moving at different speeds will not agree on the distance between two events. For example, if I measure a car traveling down the highway. I might think it had moved only one kilometer, but to someone on the sun, it would have moved about 1,800 kilometers, because the earth would have moved while the car was going down the road. Because people moving at different speeds measure different distances between events, they must also measure different intervals of time if they are to agree on the speed of light.

Einstein's original theory of relativity, which he proposed in the paper written in 1905, is what we now call the special theory of relativity. It describes how objects move through space and time. It shows that time is not a universal quantity which exists on its own, separate from space. Rather, future and past are just directions, like up and down, left and right, forward and back, in something called space-time. You can only go in the future direction in time, but you *can* go at a bit of an angle to it. That is why time can pass at different rates.

The special theory of relativity combined time with space, but space and time were still a fixed background in which events happened. You could choose to move on different paths through space-time, but nothing you could do would modify the background of space and time. However, all this was changed when Einstein formulated the general theory of relativity in 1915. He had the revolutionary idea that gravity was not just a force that operated in a fixed background of space-time. Instead, gravity was a *distortion* of space-time, caused by the mass and energy in it. Objects like cannonballs and planets try to move on a straight line through space-time, but because space-time is curved, warped, rather than flat, their paths appear to be bent. The earth is trying to move on a straight line through space-time, but the curvature of space-time produced by the mass of the sun causes it to go in a circle around the sun. Similarly, light tries to travel in a straight line, but the curvature of space-time near the sun causes the light from distant stars to be bent if it passes near the sun. Normally, one is not able to see stars in the sky that are in almost the same direction as the sun. During an eclipse, however, when most of the sun's light is blocked off by the moon, one can observe the light from those stars. Einstein produced his general theory of relativity during the First World War, when conditions were not suitable for scientific observations, but immediately after the war a British expedition observed the eclipse of 1919 and confirmed the predictions of general relativity: Space-time is not flat, but is curved by the matter and energy in it.

This was Einstein's greatest triumph. His discovery completely transformed the way we think about space and time. They were no longer a passive background in which events took place. No longer could we think of space and time as running on forever, unaffected by what happened in the universe. Instead, they were now dynamic quantities that influenced and were influenced by events that took place in them.

An important property of mass and energy is that they are always positive. This is why gravity always attracts bodies toward each other. For example, the gravity of the earth attracts us to it even on opposite sides of the world. That is why people in Australia don't fall off the world. Similarly, the gravity of the sun keeps the planets in orbit around it and stops the earth from shooting off into the darkness of interstellar space. According to general relativity, the fact that mass is always positive means that space-time is curved back on itself, like the surface of the earth. If mass had been negative, space-time would have been curved the other way, like the surface of a saddle. This positive curvature of space-time, which reflects the fact that gravity is attractive, was seen as a great problem by Einstein. It was then widely believed that the universe was static, yet if space, and particularly time, were curved back on themselves, how could the universe continue forever in more or less the same state as it is at the present time?

Einstein's original equations of general relativity predicted that the universe was either expanding or contracting. Einstein therefore added a further term to the equations that relate the mass and energy in the universe to the curvature of space-time. This so-called cosmological term had a repulsive gravitational effect. It was thus possible to balance the attraction of the matter with the repulsion of the cosmological term. In other words, the negative curvature of space-time produced by the cosmological term could cancel the positive curvature of space-time produced by the mass and energy in the universe. In this way, one could obtain a model of the universe that continued forever in the same state. Had Einstein stuck to his original equations, without the cosmological term, he would have predicted that the universe was either expanding or contracting. As it was, no one thought the universe was changing with time until 1929, when Edwin Hubble discovered that distant galaxies are moving away from us. The universe is expanding. Einstein later called the cosmological term "the greatest mistake of my life."

But with or without the cosmological term, the fact that matter caused space-time to curve in on itself remained a problem, though it was not generally recognized as such. What it meant was that matter could curve a region in on itself so much that it would effectively cut itself off from the rest of the universe. The region would become what is called a black hole. Objects could fall into the black hole, but nothing could escape. To get out, they would need to travel faster than the speed of light, which is not allowed by the theory of relativity. Thus the matter inside the black hole would be trapped and would collapse to some unknown state of very high density.

Einstein was deeply disturbed by the implications of this collapse, and he refused to believe that it happened. But Robert Oppenheimer showed in 1939 that an old star of more than twice the mass of the sun would inevitably collapse when it had exhausted all its nuclear fuel. Then war intervened, Oppenheimer became involved in the atom bomb project, and he lost interest in gravitational collapse. Other scientists were more concerned with physics that could be studied on earth. They distrusted predictions about the far reaches of the universe because it did not seem they could be tested by observation. In the 1960s, however, the great improvement in the range and quality of astronomical observations led to new interest in gravitational collapse and in the early universe. Exactly what Einstein's general theory of relativity predicted in these situations remained unclear until Roger Penrose and I proved a number of theorems. These showed that the fact that space-time was curved in on itself implied that there would be singularities, places where space-time had a beginning or an end. It would have had a beginning in the big bang, about fifteen billion years ago, and it would come to an end for a star that collapsed and for anything that fell into the black hole the collapsing star left behind.

The fact that Einstein's general theory of relativity turned out to predict singularities led to a crisis in physics. The equations of general relativity, which relate the curvature of space-time with the distribution of mass and energy, cannot be defined as a singularity. This means that general relativity cannot predict what comes out of a singularity. In particular, general relativity cannot predict how the universe should begin at the big bang. Thus, general relativity is not a complete theory. It needs an added ingredient in order to determine how the universe should begin and what should happen when matter collapses under its own gravity.

The necessary extra ingredient seems to be quantum mechanics. In 1905, the same year he wrote his paper on the special theory of relativity, Einstein also wrote about a phenomenon called the photoelectric effect. It had been observed that when light fell on certain metals, charged particles were given off. The puzzling thing was that if the intensity of the light was reduced, the number of particles emitted diminished, but the speed with which each particle was emitted remained the same. Einstein showed this could be explained if light came not in continuously variable amounts, as everyone had assumed, but rather in packets of a certain size. The idea of light coming only in packets, called quanta, had been introduced a few years earlier by the German physicist Max Planck. It is a bit like saying one can't buy sugar loose in a supermarket but only in kilogram bags. Planck used the idea of quanta to explain why a red-hot piece of metal doesn't give off an infinite amount of heat; but he regarded quanta simply as a theoretical trick, one that didn't correspond to anything in physical reality. Einstein's paper showed that you could directly observe individual quanta. Each particle emitted corresponded to one quantum of light hitting the metal. It was widely recognized to be a very important contribution to quantum theory, and it won him the Nobel Prize in 1922. (He should have won a Nobel Prize for general relativity, but the idea that space and time were curved was still regarded as too speculative and controversial, so they gave him a prize for the photoelectric effect instead—not that it was not worth the prize on its own account.)

The full implications of the photoelectric effect were not realized until 1925, when Werner Heisenberg pointed out that it made it impossible to measure the position of a particle exactly. To see where a particle is, you have to shine light on it. But Einstein had shown that you couldn't use a very small amount of light; you had to use at least one packet, or quantum. This packet of light would disturb the particle and cause it to move at a speed in some direction. The more accurately you wanted to measure the position of the particle, the greater the energy of the packet you would have to use and thus the more it would disturb the particle. However you tried to measure the particle, the uncertainty in its position, times the uncertainty in its speed, would always be greater than a certain minimum amount.

This uncertainty principle of Heisenberg showed that one could not measure the state of a system exactly, so one could not predict exactly what it would do in the future. All one could do is predict the probabilities of different outcomes. It was this element of chance, or randomness, that so disturbed Einstein. He refused to believe that physical laws should not make a definite, unambiguous prediction for what would happen. But however one expresses it, all the evidence is that the

Why We Believe in Black Holes

Black holes can form when stars with at least 10 to 15 times the mass of the Sun collapse in on themselves after a supernova explosion. Such a burned-out star no longer experiences outward gas pressure to counteract the force of gravity within the star, so it collapses to a singularity, or point of infinite density—an enormous mass in zero volume. In general, no light can escape from a black hole, although according to Stephen Hawking, in some circumstances this may be possible. The critical size at which stars could form black holes can be calculated from Einstein's equations. But what evidence is there that black holes really exist?

In 1964, scientist Yakov Zel'dovich, proposed a way of detecting black holes. Many stars, including our Sun, produce a wind of gas that blows from their surface. If this wind passes close to a black hole it will be pulled in at enormous speed. The increased kinetic energy increases the temperature of the gas to millions of degrees. Objects with temperatures in the range of a few hundred degrees emit infrared radiation; at millions of degrees, they emit X-rays.

To detect X-rays from possible black holes in space, NASA launched the Uhuru satellite in 1970, followed in 1978 by the Einstein X-ray telescope. Uhuru detected over three hundred X-ray stars and NASA scientists have found some promising black hole candidates, including Cygnus X-1.

In 1974, Stephen Hawking bet Kip

NASA image of the "Black Widow" pulsar. Pulsars are neutron stars — collapsed stars originally only three times the mass of the sun, now mostly neutrons. They emit radio waves that reach earth in pulses as the star rotates. Einstein's equations can be used to predict whether a supernova will result in a black hole or a neutron star.

Thorne that Cyg X-1 was not a black hole, explaining in his best-selling book, *A Brief History of Time,* that this was an insurance policy—if black holes, his life's work, did not exist, at least he would get something. The bet was for a one-year subscription to *Private Eye* magazine. In 1990, with the general consensus growing that Cyg X-1 was a black hole, Hawking conceded. Most scientists today agree that the likelihood that Cyg X-1 is a black hole is greater than 95 percent

Cyg X-1 is part of a binary system in which a black hole and companion star orbit each other. The gas cloud from the visible star is pulled sporadically into the black hole, and the emit-

ted X-rays are sometimes blocked when the companion star eclipses the black hole. Scientists now have strong evidence for about 10 binaries that they contain black holes.

In addition to black holes formed when individual stars collapse, single massive black holes (MBH) may form when stars or galaxies collide. These can be millions of times more massive than the Sun. A disc of gaseous material drawn towards and surrounding the MBH would emit huge amounts of radiation that range from infrared to gamma rays. Such radiation from the Active Galaxy NGC 4261 has been observed, making it a prime candidate for a black hole.   — PH quantum phenomenon and the uncertainty principle are unavoidable and that they occur in every branch of physics.

Einstein's general relativity is what is called a classical theory; that is, it does not incorporate the uncertainty principle. One therefore has to find a new theory that combines general relativity with the uncertainty principle. In most situations, the difference between this new theory and classical general relativity will be very small. This is because, as noted earlier, the uncertainty predicted by quantum effects is only on very small scales, while general relativity deals with the structure of space-time on very large scales. However, the singularity theorems that Roger Penrose and I proved show that space-time will become highly curved on very small scales. The effects of the uncertainty principle will then become very important and seem to point to some remarkable results.

Part of Einstein's problems with quantum mechanics and the uncertainty principle arose from the fact that he used the ordinary, commonsense notion that a system has a definite history. A particle is either in one place or in another. It can't be half in one and half in another. Similarly, an event like the landing of astronauts on the moon either has taken place or it hasn't. It cannot have half-taken place. It's like the fact that you can't be slightly dead or slightly pregnant. You either are or you aren't. But if a system has a single definite history, the uncertainty principle leads to all sorts of paradoxes, like the particles being in two places at once or astronauts being only half on the moon.

An elegant way to avoid these paradoxes that had so troubled Einstein was put forward by the American physicist Richard Feynman. Feynman became well known in 1948 for work on the quantum theory of light. He was awarded the Nobel Prize in 1965 with another American, Julian Schwinger, and the Japanese physicist Shinichiro Tomonaga. But he was a physicist's physicist, in the same tradition as Einstein. He hated pomp and humbug. and he resigned from the National Academy of Sciences because he found that they spent most of their time deciding which other scientists should be admitted to the Academy. Feynman, who died in 1988, is remembered for his many contributions to theoretical physics. One of these was the diagrams that bear his name, which are the basis of almost every calculation in particle physics. But an even more important contribution was his concept of a sum over histories. The idea was that a system didn't have just a single history in space-time, as one would normally assume it did in a classical nonquantum theory. Rather, it had every possible history. Consider, for example, a particle that is at a point A at a certain time. Normally, one would assume that the particle will move on a straight line away from A. However, according to the sum over histories, it can move on any path that starts at A. It is like what happens when you place a drop of ink on a piece of blotting paper. The particles of ink will spread through the blotting paper along every possible path. Even if you block the straight line between two points by putting a cut in the paper, the ink will get around the corner.

Associated with each path or history of the particle will be a number that depends on the shape of the path. The probability of the particle traveling from A to B is given by adding up the numbers associated with all the paths that take the particle from A to B. For most paths, the number associated with the path will nearly cancel out the numbers from paths that are close by. Thus, they will make little contribution to the probability of the particles going from A to B. But the numbers from the straight paths will add up with the numbers from paths that are almost straight. Thus the main contribution to the probability will come from paths that are straight or almost straight. That is why the track a particle makes when going through a bubble chamber looks almost straight. But if you put something like a wall with a slit in it in the way of the particle, the particle paths can spread out beyond the slit. There can be a high probability of finding the particle away from the direct line through the slit.

In 1973 I began investigating what effect the uncertainty principle would have on a particle in the curved space-time near a black hole. Remarkably enough, I found that the black hole would not be completely black. The uncertainty principle would allow particles and radiation to leak out of the black hole at a steady rate. This result came as a complete surprise to me and everyone else, and it was greeted with general disbelief. But with hindsight, it ought to have been obvious. A black hole is a region of space from which it is impossible to escape if one is traveling at less than the speed of light. But the Feynman sum over histories says that particles can take any path through space-time. Thus it is possible for a particle to travel faster than light. The probability is low for it to move a long distance at more than the speed of light, but it can go faster than light for just far enough to get out of the black hole, and then go slower than light. In this way, the uncertainty principle allows particles to escape from what was thought to be the ultimate prison, a back hole. The probability of a particle getting out of a black hole of the mass of the sun would be very low because the particle would have to travel faster than light for several kilometers. But there might he very much smaller black holes, which were formed in the early universe. These primordial black holes could be less than the size of the nucleus of an atom, yet their mass could be a billion tons, the mass of Mount Fuji. They could be emitting as much energy as a large power station. If only we could find one of these little black holes and harness its energy. Unfortunately, there don't seem to be many around in the universe.

The prediction of radiation from black holes was the first nontrivial result of combining Einstein's general relativity with the quantum principle. It showed that gravitational collapse was not as much of a dead end as it had appeared to be. The particles in a black hole need not have an end of their histories at a singularity. Instead, they could escape from the black hole and continue their histories outside. Maybe the quantum principle would mean that one could also avoid the histories having a beginning in time, a point of creation, at the big bang.

This is a much more difficult question to answer, because it involves applying the quantum principle to the structure of time and space themselves and not just to particle paths in a given space-time background. What one needs is a way of doing the sum over histories not just for particles but for the whole fabric of space and time as well. We don't know yet how to do this summation properly, but we do know certain features it should have. One of these is that it is easier to do the sum if one deals with histories in what is called imaginary time rather than in ordinary, real time. Imaginary time is a difficult concept to grasp and it is probably the one that has caused the greatest problems for readers of my book. I have also been criticized fiercely by philosophers for using imaginary time. How can imaginary time have anything to do with the real universe? I think these philosophers have not learned the lessons of history. It was once considered obvious that the earth was flat and that the sun went around the earth, yet since the time of Copernicus and Galileo, we have had to adjust to the idea that the earth is round and that it goes around the sun. Similarly, it was long obvious that time went at the same rate for every observer, but since Einstein, we have had to accept that time goes at different rates for different observers. It also seemed obvious that the universe had a unique history, yet since the discovery of quantum mechanics, we have had to consider the universe as having every possible history. I want to suggest that the idea of imaginary time is something that we will also have to come to accept. It is an intellectual leap of the same order as believing that the world is round. I think that imaginary time will come to seem as natural as a round earth does now. There are not many Flat Earthers left in the educated world.

You can think of ordinary, real time as a horizontal line, going from left to right. Early times are on the left, and late times are on the right. But you can also consider another direction of time, up and down the page. This is the so-called imaginary direction of time, at right angles to real time.

What is the point of introducing the concept of imaginary time? Why doesn't one just stick to the ordinary, real time that we understand? The reason is that, as noted earlier, matter and energy tend to make space-time curve in on itself. In the real time direction, this inevitably leads to singularities, places where space-time comes to an end. At the singularities, the equations of physics cannot be defined; thus one cannot predict what will happen. But the imaginary time direction is at right angles to real time. This means that it behaves in a similar way to the three directions that correspond to moving in space. The curvature of space-time caused by the matter in the universe can then lead to the three space directions and the imaginary time direction meeting up around the back. They would form a closed surface, like the surface of the earth. The three space directions and imaginary time would form a space-time that was closed in on itself, without boundaries or edges. It wouldn't have any point that could be called a beginning or end, anymore than the surface of the earth has a beginning or end.

In 1983, Jim Hank and I proposed that the sum over histories for the universe should not be taken over histories in real time. Rather, it should be taken over histories in imaginary time that were closed in on themselves, like the surface of the earth. Because these histories didn't have any singularities or any beginning or end, what happened in them would be determined entirely by the laws of physics. This means that what happened in imaginary time could be calculated. And if you know the history of the universe in imaginary time, you can calculate how it behaves in real time. In this way, you could hope to get a complete unified theory, one that would predict everything in the universe. Einstein spent the later years of his life looking for such a theory. He did not find one because he distrusted quantum mechanics. He was not prepared to admit that the universe could have many alternative histories, as in the sum over histories. We still do not know how to do the sum over histories properly for the universe, but we can be fairly sure that it will involve imaginary time and the idea of space-time closing up on itself. I think these concepts will come to seem as natural to the next generation as the idea that the world is round. Imaginary time is already a commonplace of science fiction. But it is more than science fiction or a mathematical trick. It is something that shapes the universe we live in.

PART THREE

EINSTEIN'S LATER SCIENCE

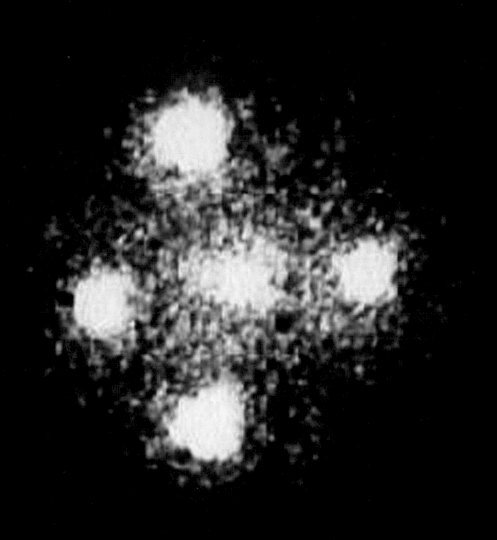

Einstein's Cross demonstrates gravitational lensing. The four outer white objects are distant galaxies that display virtually identical characteristics. Either there are four identical galaxies situated close to one another in space, or more plausibly, we are observing four images of the same galaxy that are directed towards us by the gravity of an intervening galaxy—the one in the middle.
(Preceding page: Einstein in 1933)

A Farewell Look at Gravity

by John Archibald Wheeler

Only by understanding gravity as grip of spacetime on mass, and mass on spacetime, can we comprehend even the first thing about the machinery of the world—the inertia of a particle, the motion of the planets, the constitution of a star. Without a grasp on gravity, we would not perceive the link between the fall of the nearest pebble, the orbit of Europa, the beelike buzzing of a star cluster, the two hundred million years of each tour of the Earth about the Milky Way, and the fantastic power output of the distant quasars. We would find ourselves perplexed by the doubling of the telescopic image of a distant quasar, by the expansion of the universe and the gradual slowing of this expansion, and by the primordial process that manufactured the particles and the elements. Thanks to gravity and astrophysics, we now have some understanding of every one of these processes. To gain so much from gravity, however, does not give us everything. Existence, the preposterous miracle of existence, still remains the mystery of all mysteries. Even the word *universe*, bandied about in many a book, conceals a mountain of ignorance. We don't know whether what we call "the universe" is open or closed or even whether that distinction—with growth in our understanding—may not itself fade away into thin air, into undefinability, into nonsense.

To trace out what we know today about all the applications and ramifications of gravity, the evidence for them that lies before us unexplained, the questions that dog us and the tools that we have to explore them, would take volumes. . . . [L]et us take a fleeting look at frontier topics as a happy way to say to gravity, "Farewell for today. Until tomorrow!"

The Expanding Universe

No topic will serve more simply and more quickly as a point of departure for discussing many of the processes governed by gravity than the expansion of the

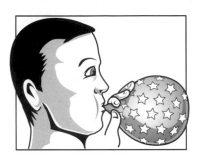

The inflation of a star-spangled balloon models the expansion of the universe. Each star finds itself at the center of an expanding pattern of stars, though there is no center for the expansion and no one star is in any position more priviledged than any other.

universe. This expansion we see in the ever widening distance between clusters of galaxies. To picture the expanding universe, we partially inflate a balloon, paste stars on its surface to represent galaxies, and blow up the balloon to a greater size. Each star sees the stars around it receding from it. Moreover, the stars twice as far away recede twice as fast.

A high-school student once asked me a question that typifies the response of many persons on first hearing about the idea of an expanding universe. "If the universe is expanding, as you say, then is everything in the universe, including ourselves, also getting bigger? If not, what takes the place of the big spaces in the universe that result from its expansion?" This student's question reflects a common misunderstanding: that is, if the distance between one cluster of galaxies and another expands, then the distance between Sun and Earth expands, the length of a meter stick expands, the diameter of every atom also expands. But if all that were true, it would make no sense to speak of any expansion at all. Expansion relative to what?

In the expanding universe, only the distance between one cluster of galaxies and another expands. Atoms don't expand. Meter sticks don't expand. We don't expand. The balloon with its firmament of stars provides a happy mental picture of the expanding universe. As the balloon expands, the distance between star and star expands, but not one star itself expands.

One deficiency of the balloon model is hard to remedy. An ant standing on its star sees itself surrounded by a mere circle of other stars, not by a whole sphere of heavenly bodies as we do. Only one parameter of direction, only one angle does the ant need to specify the direction of a particular neighbor, and one more parameter, a distance, to complete the specification of its location. The ant's world is two-dimensional. Our own world is three-dimensional. In it we require two parameters of direction, two angles, to tell our telescope which way to point to see a particular galaxy; and a third parameter, a distance, to tell how far away it is.

I have never talked with a balloon-bound ant. I do not know whether it conceives, or even can conceive, of its unbounded but closed 2-sphere rubber world to be embedded in a flat 3-space. But we, surely, need only do a little hop, skip, and jump of the imagination to think of the 3-sphere universe of galaxies to be embedded in a flat 4-space. (This flat four-dimensional *space* has nothing whatsoever directly to do with our real physical world of *spacetime*, even though that also is four-dimensional.) Almost all of this flat 4-space is totally out of our reach, quite untouchable, pure talk: the center of the embedded 3-sphere is utterly inaccessible, and so is the whole of its interior and all the endlessness outside it. Only the 3-sphere itself is real. It alone stands for the space in which we live and move, according to the Einstein-Friedmann model of a closed but unbounded universe.

I confess I have never been able, myself, to picture directly this mythical, flat, infinite 4-space and the 3-sphere universe embedded in it. In my thought I always drop a dimension. I come back to seeing myself in the world of the ant, living on the surface of a balloon, able to travel in any direction in a space that has limited extent but no limits. Einstein taught us to think of the universe in these terms, closed but unbounded. The only natural boundary condition, he argued, is *no* boundary!

If the size of the Einstein-Friedmann model universe were never to change, then in it we could proceed in any direction, go straight on with never a swerve of direction for an immense distance, and find ourselves back at our starting point. But it does change! The universe expands.

It took Einstein fifteen years to recognize and accept the expansion. His greatest hero and influence was Spinoza, who had been excommunicated from the Amsterdam synagogue in the early sixteen hundreds for denying the biblical account of a moment of creation. Where in all that nothingness before creation would the clock sit that would give the signal to start? So his reasoning ran. Based on this logic, Einstein concluded that the universe most run on from everlasting to everlasting. His great geometric theory of gravity, however, provided no natural model of the universe that would meet this requirement. Therefore, he introduced

NASA's Gravity Probe B Experiment

Einstein's general theory of relativity predicts the curvature of spacetime and an effect called "frame dragging." Gravity Probe B was launched in 2004 to try to detect these effects.

The curvature of spacetime—which is four dimensional—can be visualized in two dimensions by imagining a bowling ball in the center of a sheet of rubber, suspended above the ground. A marble placed on the rubber sheet would fall towards the bowling ball due to the dip in the rubber, rather than to an attractive force between the two spheres. Any satellite orbiting Earth should feel an effect due to Earth's curvature of spacetime. As it rotates Earth also "drags" spacetime, along with it.

Gravity Probe B was developed by NASA and Stanford University and launched on April 20, 2004. The Delta 2 rocket carried a satellite that was put into orbit 400 miles above Earth. Inside the satellite four spherical gyroscopes, each 4 centimeters in diameter, spun at 10,000 revolutions per minute. Gyroscopes rotate freely and keep their orientation so long as there are no outside forces acting on them.

The idea is to compare any slight deviations in the orientation of the gyroscopes with Einstein's predictions of spacetime warping and drag due to Earth. Leonard Schiff, who came up with the idea for the experiment in 1959, calculated that the curvature of spacetime would change the orientation of a gyroscope at an altitude of 400 miles, by 6.6 seconds of arc a year. Frame dragging would change it by 42 milliseconds of arc a year. (One arcsecond is equivalent to 1000 milliseconds of arc.) Since there are 1,296,000 seconds of arc in a complete circle, a second of arc is a minuscule angle.

How will such tiny changes be detected? The gyroscopes are coated with niobium, a superconductor that allows current to flow without any resistance. This generates a magnetic field with its north and south poles along the axis of rotation of the gyroscope. By measuring the magnetic fields, scientists can detect slight changes in the orientation of the gyroscopes. Niobium becomes a superconductor only at very low temperatures, so the gyroscopes have to be enclosed in liquid helium.

Scientists plan to analyze the data and announce the results of Gravity Probe B in January 2007.   — PH

The Gravity Probe B experiment is designed to measure slight fluctuations in the gravitational field around the earth as it orbits the sun.

into the basic principle of the theory—the principle that momenergy governs moment of curvature—the least unnatural change he could, a so-called cosmological term whose whole point and purpose was to provide a static universe. Then came 1929 and the discovery by Edwin Hubble at the Mt. Wilson Observatory that the universe is not static; it is expanding. Galaxies twice as far away as galaxies closer at hand move away from us at twice the speed. Such an expansion had already been predicted by Alexander Friedmann in 1922 on the foundation of Einstein's original simple theory. Each galaxy thinks it is the center of expansion, just as each star on the enlarging balloon, with equal justice—and equal self-centeredness—thinks of itself as the center of expansion. After Hubble's discovery, Einstein rated the idea of a cosmological constant as "the biggest blunder of my life." Nevertheless, all credit goes to the original 1915 theory and to the man who gave it to us. For it to have predicted correctly and against all expectations a phenomenon so fantastic as the expansion of the universe—when was mankind ever granted a more wonderful token that we will some day understand existence itself!

Many times in the half-century and more since Hubble's discovery, others besides the founding father himself lost faith in the simple straightforward predictions of Einstein's standard 1915 geometrodynamics. Many a radical cosmological concept has been put forward to explain observational findings that later proved to be mistaken or misinterpreted or both. Among such crises, none has endured longer or led to more numerous sound astrophysical observations and careful theoretical analyses than what was termed already in 1958 "the mystery of the 'missing' mass."

The Mystery of the Missing Mass

The mystery of the missing mass takes four forms: (1) Where is the mass needed to hold our Sun and other stars in their multimillion-year orbits about the center of the Milky Way? We don't see it! (2) Where is the mass that holds the other galaxies together as well? (3) Where do the pulls or pushes come from that—on the cosmological scale—assemble galaxies into great filaments separated by great voids? And, (4) where is the mass that is needed to curve up the universe into closure? We don't see it, either. The answers to these four questions are still being worked out in that marvelous world-wide collaboration that is science, in day-by-day consultations between colleagues, week-by-week journal publications, and season-by-season conferences. Already at this writing, however, it looks more and more as if the answer to all four questions may be the same: almost all of the missing mass sits out there right now as plain ordinary matter, not yet aggregated into assemblies large enough to have become luminous.

Let's look at the evidence on individual galaxies and assemblies of millions of galaxies before we turn back to the issue on which they all bear—the mass and fate of the universe.

Galileo, once described our galaxy, the Milky Way, as "nothing else than a mass of luminous stars…set thick together in a wonderful way." The galaxy has the flattened form of a fried egg. Our solar system lies about at what would be the division line between yolk and white, some 30,000 light years from the center. At the center of the yolk is a group of stars, visible by telescope, in rapid circulation about a central concentration of mass, believed to be a black hole of about 3.5 million times the mass of the Sun. It takes the far larger mass of the galaxy itself to hold the stars together—by gravitational attraction—in slow revolution about this center. First fully to appreciate this point and to present it in a book was Immanuel Kant (1724-1801). However, after his publisher went bankrupt, the chief creditor locked up all copies of Kant's book in a warehouse for years and years: thus his findings did not greatly influence the course of science. Otherwise his fame as an astrophysicist might have been comparable to his fame as the great philosopher of Königsberg.

Kant did not have the tools available to work out the details. Today we do. The mass of the stars observed to lie inside the orbit of the Sun adds up—within 10, 20, or 30 percent—to what is needed to hold the Sun in its orbit (with its observed orbital radius and observed speed, relative to the speed of light), as figured from the Kepler 1-2-3 law in the form:

$$\text{mass in meters} = \text{orbital radius in meters} \times (\text{speed})^2$$

For stars further from the center of the galaxy than the Sun, however, the deficiency of mass observed compared to mass required to hold those stars in their observed orbits grows greater, amounting to a factor of two or more.

It is easier in some ways to apply the Kepler 1-2-3 law to other galaxies than to our own. Stars on one side of an elliptic galaxy, seen edgewise, move toward us; on the other side, away. To plot out this velocity of selected stars and its dependence on their distance from the center is to come up with what is generally termed a "velocity curve." If the only source of gravitational attraction were the objects visible in the telescope, the pulling power of the galaxy, far out, could only hold a star in orbit at that distance if that star went around at something like half the speed it is observed to have. Conclusion? There is far more mass than what we see helping to hold that star in orbit!

One valuable clue to the mass of a distant galaxy or cluster of galaxies—called *gravitational lensing*—is beginning to be exploited. No one will feel baffled by this phenomenon who has looked at a distant streetlight through a pane of glass afflicted by a lenslike imperfection and seen a double or triple image. Likewise, a double or triple telescopic image is occasionally observed of an extremely distant, powerful, and concentrated source of light known as a quasistellar source, or *quasar*. This happens when the quasar lies in line with a closer, but still very distant, galaxy or cluster of galaxies, which provides gravitational pull. This pull takes rays of light that are diverging from the original source and causes them to converge—to arrive at the same telescope from different directions. The photographic plate registers the quasar as double, triple, or multiple. This gravitational lensing action provides a tool (not yet honed to precision!) to get at the mass of the intervening galaxy or cluster.

The mystery of the missing mass grows greater (and therefore closer to solution!) when we turn from the motions within one galaxy to the message of millions of galaxies. Take a telescope that has a 6 degree opening, swing it through the sky through an arc of 135 degrees, and chart the speed of recession (or, equivalently, the distance from us to it, according to Hubble's law connecting distance with speed) for every galaxy in that pie-shaped slice of space that (1) has a visual magnitude of 15.5 or brighter and (2) is receding from us at a speed of 15,000 kilometers a second or less. That's the task Valerie de Lapparent, Margaret Geller, and John P. Huchra set themselves...[T]heir spectacular 1986 finding [shows] great voids and great filaments in the distribution of the galaxies. The first reaction of many was astonishment: what strange pushes and pulls there must be at work in the universe to produce these inexplicable zones of clustering and zones of avoidance. Others, however, suspected there was nothing more needed to explain the broad features of this distribution of galaxies than plain ordinary gravity acting between plain ordinary masses. The most decisive evidence for this conservative interpretation reached me before the final revision of this chapter from Changbom Park, a graduate student at Princeton. Park made a printout of his computer findings... It shows an amazing similarity in character with the observations. Again clustering, again voids. Yet into those calculations went nothing but gravity. To look at what went into Park's analysis is to appreciate something of what the expanding universe is and means and does.

The calculations traced out the motion of two million mass points, each changing its location and speed from instant to instant in response to the Newtonian gravity of all the others. The number of points thus considered, unsurpassed in any such analysis ever made to date, is nevertheless sufficiently small compared to the scale of the universe as a whole that it is legitimate to neglect in the analysis all large-scale curvature of space. The points are followed from the very earliest time when, it is generally believed, today's whole visible universe occupied a space far smaller in dimensions than an elementary particle. A natural line of reasoning, based on modern quantum theory, indicates that the initial distribution of mass, though fantastically uniform, must nevertheless have presented percentagewise exceedingly small but predictable fluctuations from this ideal uniformity. Starting with this initial condition, the calculations reveal that two million "tracer points," each of galactic mass, gradually move, under gravity alone, into remarkable associations and avoidances. This effect is exaggerated, both in nature and in the calculations, by a built-in bias: those places where the tracer mass points come closer together symbolize locations where visible galaxies come into being; places where the tracer masses are further apart are locations where matter does not assemble itself into stars, and therefore these places will not be visible to any telescope... In brief, gravity, plus the physical requirements for the turning of matter into visible stars, bias telescopes against seeing any matter at all in great regions of space where matter has a density a little lower than elsewhere. Only when we recognize and come to terms with this bias can we hope to understand even the first thing about the gravitational dynamics of the biggest system of all, the universe itself.

The Universe: From Big Bang to What?

Somewhere between ten and twenty billion years ago, today's galactic recession velocities indicate, the universe began with a big bang. Small in the early days, the universe had high space curvature and needed a high density of energy to bend it up into that tight ball. Large today, it has a far smaller space curvature and needs much less density of mass to bend it up into closure—yet a density five to ten times greater than any visual observations directly evidence. The bias principle tells us, however, that we no longer have to worry about the mystery of the missing mass. Optical telescopes see only luminosity, and luminosity gives a quite unfairly low count of mass in regions of a little lower than average mass density.

As relics of the big bang we see around us not only today's rapidly receding galaxies but also today's abundance of the chemical elements—some among them still radioactive, the "still warm ashes of creation"(Victor F. Weisskopf)—and today's greatly cooled but still all pervasive "primordial cosmic fireball radiation." We now believe that in the first instants of its life, the entire universe filled an infinitesimally small space of enormous density and temperature where matter and energy fused in a homogeneous soup. Immediately the universe began expanding. After about 10^{-6} seconds it had cooled enough that subatomic particles condensed from the matter-energy soup. In the first three minutes after the big bang, neutrons and protons combined to make heavier elements. Eons later stars and galaxies formed. Never since has the universe paused in its continual spread outward.

Will the universe continue expanding forever? Or will its expansion slow, halt, and turn to contraction and crunch, a crunch similar in character but on a far larger scale than what happens in the formation of a back hole? For insight on this question, consider a Moon floating in the emptiness of space. Plant explosives within it and explode it. If we plant too much explosive, supply too much energy, start off the expansion with too much speed, the Moon is gone for keeps. If we implant less than the critical explosive charge, then contraction and final crunch will inevitably follow the initial bang. A great deal less charge, then a relatively short time between bang and crunch. A charge only moderately less than the critical amount, then a longer time before the outcome declares itself. But to tailor the explosive charge so delicately that the eventual outcome—expansion forever or contraction and final crunch—declares itself only after many billions of years seems to require such fine-tuning as to be incredible.

Right reasoning about the Moon has occasionally led to wrong reasoning about the universe. Some argue that the idea of a closed universe still expanding, though at a slowed rate, many billions of years after the initial big bang is preposterous. No one starting the whole show at time zero—it is sometimes said—could possibly adjust the initial propulsive energy with sufficient finesse to make things come out just right so far in the future. That mistaken argument is sometimes put forward to justify belief in, search for, and proclamation of one or another new idea about how things started, some new principle that will automatically guarantee the desired delicacy of adjustment of the energy. But Einstein's geometric theory of gravity tells us that no adjustment of energy needs to be made or even can be made.

One mass? One radius at maximum expansion, one time from bang to crunch. Twice that mass? Then twice the radius at maximum expansion, twice the time from bang to crunch. No tailoring of the "explosive charge." No fine-tuning of the initial expansion rate. No worry for fear that the closed model universe will go on expanding forever. The total mass of those clusters of galaxies, those "rocks," that make up the closed universe totally fixes the time scale of the expansion-plus-contraction, the size of the universe at the phase of maximum expansion, and the size and rate of expansion at any specified time after the big bang...

There are several reasons for caution about [the predicted features of a 3-sphere closed universe compatible with astronomical observations.] First, even today astronomers of the greatest skill and integrity have difficulty with the determination of galactic distances. Different groups may arrive at a distance for a particular faraway galaxy that differ from each other by a factor of two. The resulting uncertainty in the distance scale carries with it corresponding uncertainties in the size of the Friedmann model with which it should be compared and in the total mass. Second, close to the times of the big bang and the big crunch, radiation becomes so hot and energetic that it contributes more than matter to the density of mass in the universe... Finally, we have no absolute guarantee that the universe is indeed closed, or whether a deeper, quantum-based understanding of existence will still leave meaningful the concept of "space closure."

Einstein's battle-tested and still-standard geometric theory of gravitation leaves no room for ifs, ands, or buts about the dynamics of a 3-sphere model universe. It is a one-cycle only dynamics. It totally rules out any cycle after cycle after cycle history, world without end. No place whatsoever does it provide for any "before" before the big bang, any "after" after the big crunch. Never has there been a human without a navel. No other way offers itself for his creation. It is conceivable that the universe likewise could never have come into existence were it not endowed in its history-to-be with a double belly button: big bang and big crunch. What goes on at those two gates of time is still, however, beyond our ken.

We have come to the end of our journey. We have seen gravity turn to float, space and time meld into spacetime, and spacetime transformed from stage to actor. The grip by which spacetime acts on mass and mass on spacetime turns out not to be some elaborate gadgetry built of gears and pinions, but a mathematical principle of incredible simplicity—the principle that the boundary of a boundary is zero. Of all the indications that existence at bottom has a simplicity beyond anything we imagine today, there is none more inspiring than the unsurpassed simplicity of gravity, as we now see it. Someday, surely, we will see the principle underlying existence as so simple, so beautiful, so obvious that we will all say to each other, "Oh, how could we all have been so blind, so long!"

An Old Man's Toy

○

by A. Zee

His Last Birthday

On Einstein's seventy-sixth and last birthday, March 14, 1955, his neighbor Eric Rogers presented him with a toy constructed of a heavy ball, a spring, a broomstick, and other commonly found objects. The contraption is shown in Figures 1.1a, 1.1b, 1.1c. A brass ball attached to a string hangs outside a metal cup into which the ball can fit snugly. The string passes through a hole in the cup and down through a pipe, where it is tied to a spring. This entire assembly is mounted on a curtain rod so that one can hold on to the whole contraption easily. Finally, the cup and ball assembly is enclosed in a transparent glass sphere to give it a finished look.

Were the spring strong enough, it could pull the ball into the cup. By design, however, the spring is too weak to counteract the force of gravity. And so the ball hangs limply outside the cup.

By shaking the curtain rod, it's possible to pop the ball into the cup. But with a flabby spring and a small enough cup, this may be made frustratingly difficult. The challenge is to find a surefire way to pop the ball into the cup every time.

Einstein was delighted. He recognized immediately that the necessary trick hinged on a physical principle he himself had thought up half a century earlier. He was pleased that his friend Professor Rogers took the trouble to remind him in this fun way of what he had described as one of the happiest moments of his life, the moment when he understood gravity. In his delight, the old man's mind momentarily flashed back.

A Happy Thought

The scene of the birthday gathering dissolves, and we find ourselves in the summer of 1900. Albert Einstein, together with three other students, had just finished their physics studies at the technical university ETH (Eidgenössische Technische

---

Hochschule) in Zurich. While the authorities immediately appointed the other three students as assistants, they regarded the dreamy Einstein as unsuitable for a university position. Dejected, Einstein wrote letters all over, trying to get a job, but without success. He wrote later that he felt like "a pariah, discounted and little loved suddenly… abandoned by everyone, standing at a loss on the threshold of life."

Einstein spent the next two years in frustration, holding temporary jobs as a tutor and as a high-school substitute teacher in various small towns in Switzerland. Finally, in early 1902, he heard of an opening for a technical expert second class at the federal patent office in Bern, and as luck would have it, Einstein had a close friend whose father knew the director of the patent office. With a certain amount of string-pulling, Einstein was called for an interview. After a two-hour grilling, the official examiner noted for the record that the examination "unfortunately disclosed [Einstein's] obvious lack of technical training." Nevertheless, the much-maligned old boys' network triumphed, and in June 1902 Einstein was hired, but only as a technical expert third class.

Einstein was overjoyed. Now finally employed, he was able to marry Mileva Maric, a school friend from his Zurich days, and to start a family. He liked his job. The work was interesting and also left him time to think about his beloved physics. Bern was a university town, and Einstein soon befriended several young men with similar interests with whom he met regularly for intellectual discussions.

In 1905, Einstein shook physics to its foundation with his papers on the special theory of relativity. A year later, he was promoted to technical expert second class. Then one day in November 1907, inspiration struck: He had what he later called the happiest thought of his life.

"I was sitting in a chair in the patent office at Bern when all of a sudden a thought occurred to me: 'If a person falls freely he will not feel his own weight.' I was startled. This simple thought made a deep impression on me. It impelled me toward a theory of gravitation."

From this apparently nonsensical idea that a falling person feels no gravity emerged the secrets of gravity and the universe.

An April Fool's Prank

What was the patent clerk thinking about? Let us try to understand.

A sky diver jumps out of a plane and feels an exhilarating floating sensation. Her mind tells her that eventually she will reach solid ground; her parachute had better open or she will be turned to mud. But for now, she can close her eyes and enjoy

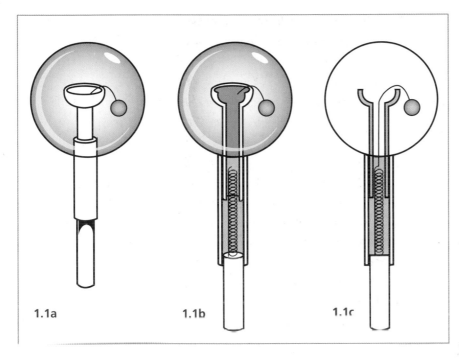

1.1a   1.1b   1.1c

Three views of "an old man's toy."

1.1a. A three-dimensional view.

1.1b. A three-dimensional cutaway drawing.

1.1c. A two-dimensional cross sectional drawing. The tension in the spring is counterbalanced by the weight of the ball, provided the toy is stationary. Once the toy is in free fall, the ball becomes "weightless" and only the spring acts on the ball, pulling it in.

the floating sensation. She knows that she is falling because of the air rushing up past her. Imagine removing the air. Were it feasible for a sky diver to fall through a pure vacuum, she would not know she is falling. Hearing nothing but the pulse of her blood, she would not feel the force of gravity at all.

To remove the effect of air rushing up past the sky diver, put her inside a box. In fact, let us go all out and play an elaborate April Fools' Day prank on one of our friends. Drug the poor victim in her sleep and put her in a spacious box elaborately furnished inside to look exactly like her living room. We then drop the box from a high-flying airplane.

When our friend wakes up, she will think that she is in her living room. Curiously, though, she feels that she is floating. To an observer on the ground (assume that the box is such that one can see in but not see out), our friend and her living room are hurtling toward a crunching rendezvous with the ground. Our friend, however, is blissfully unaware of the impending disaster. Since she is accelerating downward at the same rate as the box and all the objects contained inside, she feels that she is not moving downward at all relative to her surroundings. A slight spring in her step and she finds herself drifting toward the ceiling. She feels that she is floating—but this action is interpreted by the ground observer quite dif-

225

ferently: Our friend, by stepping on the floor, has at the same time decreased slightly her downward velocity and increased slightly the box's downward velocity. Our friend thinks she is floating upward but in reality her downward plunge is accelerating at the same rate as before.

Skeptical, you say that all this sounds like a physicist or a patent clerk playing with words. A falling person doesn't know she is falling? Tell me a better one. Surely, if I fell out of an airplane, even if I were inside a box, I would feel the force of gravity dragging me down!

I can easily overcome your skepticism: The awfully unethical April Fools' joke has already been tried. Surely, in the comfort and security of your own home, you must have seen on television the antics of astronauts floating in their spaceships orbiting the earth.

But, you object, the space shuttle is safely orbiting the earth, not falling freely.

Well, get ready, I am going to hit you with yet another apparently paradoxical statement: The orbiting shuttle is staying up there precisely because it is freely falling. Let me explain.

Staying Up by Falling

An enormous but common misconception is involved here. By now, everyone knows that the astronauts float because there is no gravity in space. Right?

Wrong! Surely, you don't believe everything television announcers say. Since the gravitational attraction between two bodies decreases as the distance between them increases, then deep in space, far away from any massive object, there is indeed no gravity. But the shuttle is orbiting the earth a mere one hundred miles or so above ground. Using the scales of everyday life, we naturally think of one hundred miles up as very high up indeed. After all, Mount Everest is only about five miles high. However, in this context one hundred miles is an insignificant height. The gravitational pull on the shuttle and on everything inside it is just about as strong as it would be were the shuttle parked on the ground.

The point is that in calculating the gravitational force on the shuttle, what counts is not the distance from the shuttle to the surface of the earth, but the distance from the shuttle to the center of the earth.

The radius of the earth is about four thousand miles. Thus, as the shuttle goes from its ground position to its orbital position one hundred miles above our heads, the distance from the shuttle to the center of earth increases from four thousand miles to forty-one hundred miles, a mere 2.5 percent increase. Recall that the gravitational force between two bodies decreases inversely as the square of the distance between them. As the shuttle goes into orbit, the earth's gravitational pull on it decreases by a factor of $(4000/4100)^2$, about a 5 percent decrease. The gravitational pull on the shuttle and on the astronauts inside it is almost as strong as would have been the case had they just been sitting on the ground.

The astronauts are definitely falling (although at a rate only 95 percent of the rate they would experience falling out of a tree). But yet they do not feel they are falling, because the shuttle around them and everything else inside it is also falling. In the same way, our friend does not feel that she is falling because her furniture and everything else around her is also falling.

A Clever Fellow

To see how it makes sense to say that the shuttle is constantly falling while it stays up there in orbit, we have to go back to Isaac Newton.

Newton, clever fellow he, figured out, using the following ingenious argument, how a satellite can be put into orbit. Imagine placing a cannon on the highest mountain peak you can find. Fire, and note the distance the cannonball travels before it strikes the ground. Increase the firepower and the cannonball will travel farther. (See Figure 1.2a) What would happen, Newton asked himself, if I keep on increasing the firepower? Obviously, the cannonball goes farther and farther.

But wait—Newton realized there is a catch. The picture we just drew is not quite right. The earth is round. As the cannonball flies farther and farther out, the curvature of the earth becomes more and more important. We have a curious situation: The cannonball is trying to fall to the ground, but the ground is trying its darndest to get away from the cannonball. (See Figure 1.2b.)

Newton realized that if the ground is "dropping away" as fast as the cannonball is falling, the cannonball is in orbit.

Satellites and shuttles stay "up there" in orbit not because they are *not* falling. They are falling all right, day and night, seven days a week. Their trick is to move forward fast enough so that by the time they have fallen by a foot the ground has also dropped away by a foot.

(It is perhaps worth emphasizing a well-known distinction: Ordinary aircraft and orbiting spacecraft rely on different physics to stay up. An aircraft stays up because of the pressure generated by the rapid air flow around its specially shaped wings. The presence of air is essential. In contrast, any air present at the orbiting altitudes would cause the spacecraft to slow down and eventually plunge back into the atmosphere.)

When I first came across Newton's argument, I was impressed. It is one thing to talk about centrifugal force, vectors, kinetic energy, and so on, but to say that a satellite is staying up there because the ground is dropping away represents an extra measure of inspiration and understanding.

Thus, quite literally, we put astronauts inside a box called a spaceship and drop it out of the sky. To be humane, we give the box a forward motion so that as soon as the box drops, the ground would have the good sense of curving away by just the right amount.

Next time you turn on the TV and see an astronaut floating in space, with the announcer commenting in the background that the astronaut is in the zero-g environment of space one hundred miles above our heads, you will know better. The astronaut is in a 0.95-g environment. He is subject to only 5 percent less gravity

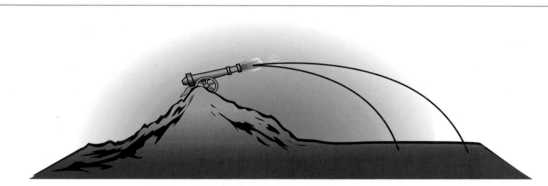

1.2a If the earth were flat, a cannonball fired from a high altitude would eventually crash to earth. The faster the cannonball leaves the cannon, the farther it travels, but it always lands somewhere.

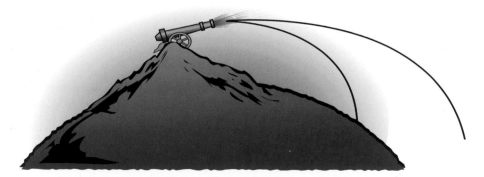

1.2b. On a round earth. the cannonball falls towards earth, but the earth recedes from the cannonball. If the cannonball is fast enough, the earth recedes faster than the ball falls and it ends up in orbit.

than we are. *He is floating because he is falling*, and because he is falling he does not feel that he is falling, just as the young technical expert second class thought.

A Big Grin

We can now see how the old man's toy illustrates this fundamental truth about gravity. The eminent historian of science I. Bernard Cohen visited Einstein not long after his birthday and wrote about his visit:

> At last I was taking my leave. Suddenly [Einstein] turned and called "Wait. Wait. I must show you my birthday present."
>
> Back in the study I saw Einstein take from the corner of the room what looked like a curtain rod five feet tall, at the top of which was a plastic sphere about four inches in diameter. "You see," said Einstein, "this is designed as a model to illustrate the equivalence principle..."
>
> A big grin spread across his face and his eyes twinkled with delight as he said, "And now the equivalence principle." Grasping the gadget in the middle of the long brass curtain rod, he thrust it upwards until the sphere touched the ceiling. "Now I will let it drop," he said, "and according to the equivalence principle there will be no gravitational force. So the spring will now be strong enough to bring the little ball into the plastic tube." With that he suddenly let the gadget fall freely and vertically, guiding it with his hand, until the bottom reached the floor. The plastic sphere at the top was now at eye level. Sure enough, the ball rested in the tube.

(The idea that a falling object does not feel any gravity is part of the equivalence principle, about which more later.)

The little brass ball is just as easily fooled as an astronaut. When Einstein let his toy fall, the little brass ball, precisely because it was falling, did not feel any gravity; the ball was the stand-in for the falling person in the patent clerk's daydream. The spring, normally too weak to pull the ball up against gravity, now seized the chance to yank the ball into the bowl. Get it? It is really quite simple. As you are falling, you are not aware of gravity. Of course, as an animate being, you are concerned about how the eventual impact will hurt. But that is psychology, not physics.

The Falling Candle

Einstein loved to pop playful little puzzles on his visitors. He was equally delighted whether or not they knew the answers. If they didn't, he would get a big kick out of explaining it. Here is one that he asked on more than one occasion.

Suppose you have just lighted a candle in an elevator when, unfortunately, the cable breaks. The elevator falls freely. What happens to the candle flame?

First of all, we have to understand how a burning candle works normally. The hot gas produced by the burning candle, being less dense than air, rushes upward. The upward rush of the glowing gas is what we see as the flame. The candle is thus assured of a steady supply of oxygen from the ambient air as the gas rushes out of the way. The second point is that the upward rush of the gas can be better interpreted as due to gravity pulling the denser air down. By moving downward, the ambient air is actually displacing the gas upward.

Good. We can now answer the grinning old man looking at us with a twinkle in his eyes. The falling candle feels no gravity, and neither does the air around it. The hot gas expands outward rather than rushing upward out of the way. For a moment, the candle is deprived of air supply and goes out.

Universality

To understand gravity in more detail, let us consider our April Fools' prank again. For the prank to work, it is crucial that all objects fall at exactly the same rate. Suppose to the contrary that the box falls faster than our friend. Then our friend would find herself pinned to the ceiling, which she would interpret as being due to the presence of a force pushing her up. Conversely, if the box were to fall slower, our friend would feel a force pulling her to the floor. The extreme case in which the box is not falling at all is of course the normal situation, with the box resting on the house foundation.

That objects all fall at the same rate regardless of their composition is contrary to everyday intuition, but as Galileo suspected, our everyday experiences are distorted by air resistance. In a vacuum, a feather and a cannonball fall at the same rate. Hard to believe, but true. Remember how Galileo is supposed to have demonstrated this by dropping several objects simultaneously from the Leaning Tower of Pisa? (Incidentally, the temptation to drop things off the Leaning Tower is so overpowering—I was barely to able to restrain myself—that I could well believe that Galileo actually did it, even though historians now say the whole story is apocryphal.) In the three hundred-odd years since Galileo, physicists have performed ever more sophisticated versions of his experiment with ever increasing accuracy. That all objects fall at the same rate is extremely well verified.

Physicists refer to the fact that different objects all respond in exactly the same way to gravity as the universality of gravity.

The gravitational force acts on all objects equally; it is universal. In sharp contrast, the electromagnetic force is not universal. Indeed, the electric force does not act on objects with zero charge, while it pushes positively charged objects and negatively charged objects in opposite directions. Also, when acted upon by the electric force, an object with twice the electric charge as another (but with the same mass) will accelerate twice as much.

Any deep theory must be able to account for the universality of the gravitational force. In Newtonian physics, universality follows from the law of motion and the supposition that the gravitational force on any object is proportional to its mass.

But let us go back and try to imagine the patent clerk's train of thought. A falling person does not know she is falling, because everything around her is falling at the same rate; in other words, because of universality. Hey, wait a minute,

An elderly Einstein in Princeton, photographed in April 1955, shortly before his death. Isolated from mainstream quantum physics, Einstein continued to work—unsuccessfully—on a unified field theory up to the very day he died.

can I turn this around? Gravity must be universal because a falling person does not know she is falling. In a way, falling cancels out gravity. Hmmm, suppose I somehow reverse falling by thrusting upward. Can I then produce gravity? Aha!

To understand what Einstein had in mind, let us inflict an even more elaborate April Fools' joke on our friend. This time, after drugging her and putting her inside the box, we fly her deep into intergalactic space, far away from any gravitational field of force. Now rev up the engine and accelerate the whole contraption at a constant rate. When she wakes up, she notices nothing unusual at all. No floating sensation this time. She drops an earring, and it promptly falls to the floor

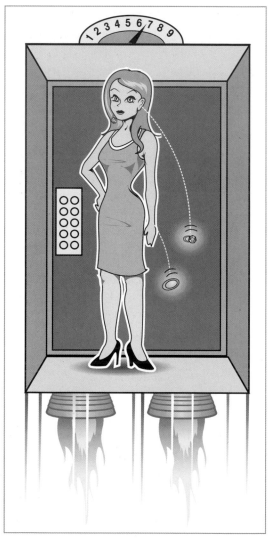

1.3. April Fools' Prank: a dropped earring and a bracelet "fall" at the same rate. The passenger cannot distinguish between being in an upward-accelerating spaceship and being near a planet.

(Figure 1.3). But to an outside observer, floating in space and watching the spaceship go zinging by, the dropped earring is actually floating in space in happy ignorance of the fact that the floor is rushing at it with ever-increasing speed. If we accelerate the rocket at precisely the right rate, our friend would see her earring falling to the floor exactly as always. By accelerating the rocket—in effect, reversing free-fall—we can "produce" gravity.

Clearly, if our friend had dropped her bracelet as well as her earring, releasing them both simultaneously from the same height, she would observe them hitting the floor at precisely the same instant. But what is to her a mysterious universality is laughably obvious to the observer floating about outside: The floor is moving up to meet the floating earring and bracelet and so obviously arrives at the two objects at the same time.

A Ball of Whiskey

That a person inside an accelerating rocket ship feels an effect equivalent to gravity, an effect that disappears as soon as the ship stops accelerating, is illustrated wonderfully by the Belgian children's-book writer Hergé in his *Explorers on the Moon*, published in 1954. The story is part of a marvelous Belgian-French series of comic books describing the adventures of a redheaded boy named Tintin.

On a trip to the moon, Captain Haddock, an alcoholic sea captain and Tintin's faithful friend, had sneaked some whiskey on board the spaceship, hiding the bottles inside a hollowed-out astronomy book. The first panel shows the captain about to enjoy his drink. When the spaceship was much more than four thousand miles away from the center of the earth, the effect of the earth's gravity became negligible, and any effect of gravity was due to the spaceship's acceleration. Since Hergé showed the characters going about their business as in a normal gravitational field, we can infer that the spaceship was accelerating at precisely the rate needed to reproduce the earth's gravitational field.

Just as the captain was about to set lip to glass, Thomson, a fumbling detective along for the trip, turned off the rocket engine, and the spaceship stopped accelerating. The whiskey, suddenly feeling no gravity, had no further compunction to stay inside the glass. The so-called surface tension, a force due to the attraction felt by the whiskey molecules for each other, curled the whiskey up into a ball, as Hergé correctly depicted. Eventually, Tintin managed to turned the engine back on, at which point the captain and the whiskey both fell crashing to the floor (or, as an outside observer would say, the accelerating floor rushed up to the captain and the whiskey). Incidentally, had the captain taken the old man's toy along, he would have noticed the brass ball popping neatly into the cup just as his whiskey floated out of his glass.

That acceleration can produce effects identical to gravity has become quite familiar in this age of rocketry and fast elevators. Riding an elevator, we feel our internal organs sag momentarily in response to the additional gravitational pull produced as the elevator accelerates upward. For that moment, our stomachs have not yet realized that the elevator has taken off. Remarkably, this deceptively simple thought of the patent clerk, when suitably formulated into a principle of physics, will turn out to have far-reaching implications.

Before going on, let us summarize: (1) A falling person does not know she is falling because everything around her is falling at the same rate. (2) This apparently paradoxical thought has actually been verified by the experience of astronauts. To understand this, we have to understand in turn Newton's insight that orbiting space stations are actually in free-fall. They stay up because of their forward movement. (3) A person in an accelerating box is fooled into thinking that she is in a gravitational field, while in fact she could be light-years away from any gravitational field. To an outside observer floating by, however, a dropped bracelet is not falling to the floor. Rather, the floor is rushing up to meet the bracelet.

Desire to Fall

While with her husband on the top floor of a skyscraper, [the patient] had an image of falling out of the window. The fantasy was so vivid that she shouted for help. When her husband questioned her, she realized that the fall was purely in her imagination… Numerous clinical observations suggest that a person visualizing a scene may react as though it were actually occurring… Even though the daydream may be temporarily experienced as reality (as a sleeping dreamer experiences a dream), anxious patients are able to regain their objectivity and label the phenomenon a fantasy.

–A. T. BECK et al., *Anxiety
Disorders and Phobias*

Standing on the edge of a cliff facing the breaking waves, or leaning over the balcony of an upper floor of one of those towerlike contemporary hotels, looking down into the lobby, I have often mused about how a falling person feels no gravity. Surely, the patent clerk's daydream surfaced from the subconscious of an acrophobe.

Of all the phobias, acrophobia is perhaps the most common. At a conscious level, the acrophobe is afraid of falling. At the subconscious, he is afraid of his desire to go over the edge and fall. At least, that is how I feel—the urge to go into free-fall!

Falling—we have all had the experience of awaking suddenly from a dream in which we are falling into a dark void. Millions of years of experience instilled in our arboreal ancestors the fear of falling. Yet we cherish the experience: Children love the playground slide—it is the primordial fear momentarily mastered.

In the dream of falling, curiously enough, it is not entirely discomforting to be falling into a dark void. As long as we don't think there is solid ground at the bottom, we don't mind. We are terrified not by the fall itself but by the end of falling, by the sickening crunch of bone on ground. When we are no longer allowed by the ground to fall, that's when we feel gravity. But when we are falling, we are free from the grasp of gravity. It is the beckoning temptation of this freedom that terrifies the acrophobe.

By Einstein's own admission, he was daydreaming when what he called the happiest thought of his life occurred to him. Perhaps he was having the dream of falling. The subconscious can work in strange ways.

Making Einstein's Birthday Toy

You can make a simplified version of Einstein's 76th birthday toy from household items.

What you'll need:

1 large (wide) unbreakable plastic cup, about 5 inches tall;

1 thick nail (or a drill will do);

2 paper clips;

2 rubber bands, each shorter than the height of the cup, but able to be stretched easily to a length larger than the height of the cup;

2 rubber balls, each an inch in diameter (or at least able to both fit easily into the cup);

2 screws (or anything that can securely attach the rubber bands to the balls).

How to make the toy:

First, make a hole in the middle of the bottom of the cup (with the nail or drill).

Second, straighten out one of the paper clips, but leave a hook at one end.

Now stick the hook through the hole from the bottom into the cup and hook the two rubber bands, and pull them slightly through the hole. Then use the second paper clip to anchor the rubber bands to the bottom of the cup (see the illustration).

Finally, attach the other ends of the rubber bands to the balls. The balls need to be heavy enough so that they can stretch the rubber bands and hang over the lip of the cup when the cup is stationery.

The demonstration:

Hold the cup up with the balls hanging over the edge of the cup (as

Left: The pull of the rubber bands and the weight of the balls are equal and opposite, so the balls stay outside the cup. **Right:** while falling, only the rubber bands exert force on the balls, so they are pulled into the cup.

shown). Now let go of the cup and let it fall a few feet—but catch it or have someone at the ready to catch it. What you will see is that the balls are pulled into the cup by the rubber bands as the cup falls.

The explanation:

When you hold the cup up and at rest, the force of gravity pulling down on the balls is counterbalanced by the tension in the rubber bands pulling up on the balls. When you let go of the cup, however, it is in free fall; the cup and the balls accelerate downward at the same rate, eliminating any visible effect of gravity. This allows the force of the elasticity of the rubber bands to be the only effective force on the balls. The toy is a (kind of) demonstration of the equivalence principle of general relativity—the equivalence of gravity and the effect of accelerating.

For the more ambitious and adroit, one can attempt to construct the (working) model of Einstein's Birthday toy, created from household items by Canadian artist Tim Westbury, pictured at left.   —HR

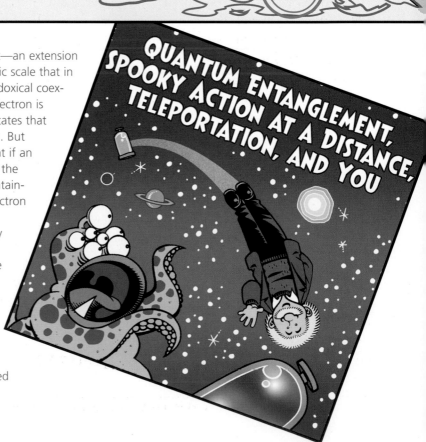

Above: the paradox of Schrödinger's Cat—an extension of quantum mechanics to the macroscopic scale that in this thought experiment implies the paradoxical coexistence of live and dead cats. When an electron is aimed at a double slit, quantum theory states that it is uncertain which slit it passes through. But suppose we create an apparatus such that if an electron goes through one slit, it triggers the hammer to fall, which breaks the vial containing poison—and kitty's a goner. If the electron goes through the other slit, the hammer stays up, the vial is not smashed and kitty lives to purr another day. So, Einstein asked, in our experiment, is the kitty alive or dead?

Right: the cover of Jim Ottaviani and Roger Langridge's little comic book on the application of Einstein's views of quantum theory to the possibility of teleportation. The comic book is continued on page 239.

Quantum Trickery: Testing Einstein's Strangest Theory

by Dennis Overbye

Einstein said there would be days like this.

This fall scientists announced that they had put a half dozen beryllium atoms into a "cat state."

No, they were not sprawled along a sunny windowsill. To a physicist, a "cat state" is the condition of being two diametrically opposed conditions at once, like black and white, up and down, or dead and alive.

These atoms were each spinning clockwise and counterclockwise at the same time. Moreover, like miniature Rockettes they were all doing whatever it was they were doing together, in perfect synchrony. Should one of them realize, like the cartoon character who runs off a cliff and doesn't fall until he looks down, that it is in a metaphysically untenable situation and decide to spin only one way, the rest would instantly fall in line, whether they were across a test tube or across the galaxy.

The idea that measuring the properties of one particle could instantaneously change the properties of another one (or a whole bunch) far away is strange to say the least—almost as strange as the notion of particles spinning in two directions at once. The team that pulled off the beryllium feat, led by Dietrich Leibfried at the National Institute of Standards and Technology, in Boulder, Colo., hailed it as another step toward computers that would use quantum magic to perform calculations.

But it also served as another demonstration of how weird the world really is according to the rules, known as quantum mechanics.

The joke is on Albert Einstein, who, back in 1935, dreamed up this trick of synchronized atoms—"spooky action at a distance," as he called it—as an example of the absurdity of quantum mechanics.

"No reasonable definition of reality could be expected to permit this," he, Boris Podolsky and Nathan Rosen wrote in a paper in 1935.

Today that paper, written when Einstein was a relatively ancient 56 years old, is the most cited of Einstein's papers. But far from demolishing quantum theory, that paper wound up as the cornerstone for the new field of quantum information.

Nary a week goes by that does not bring news of another feat of quantum trickery once only dreamed of in thought experiments: particles (or at least all their properties) being teleported across the room in a microscopic version of *Star Trek* beaming; electrical "cat" currents that circle a loop in opposite directions at the same time; more and more particles farther and farther apart bound together in Einstein's spooky embrace now known as "entanglement." At the University of California, Santa Barbara, researchers are planning an experiment in which a small mirror will be in two places at once.

Niels Bohr, the Danish philosopher king of quantum theory, dismissed any attempts to lift the quantum veil as meaningless, saying that science was about the results of experiments, not ultimate reality. But now that quantum weirdness is not confined to thought experiments, physicists have begun arguing again about what this weirdness means, whether the theory needs changing, and whether in fact there is any problem.

This fall two Nobel laureates, Anthony Leggett of the University of Illinois and Norman Ramsey of Harvard argued in front of several hundred scientists at a conference in Berkeley about whether, in effect, physicists were justified in trying to change quantum theory, the most successful theory in the history of science. Dr. Leggett said yes; Dr. Ramsey said no.

It has been, as Max Tegmark, a cosmologist at the Massachusetts Institute of Technology, noted, "a 75-year war." It is typical in reporting on this subject to bounce from one expert to another, each one shaking his or her head about how the other one just doesn't get it. "It's a kind of funny situation," N. David Mermin of Cornell, who has called Einstein's spooky action "the closest thing we have to magic," said, referring to the recent results. "These are extremely difficult experiments that confirm elementary features of quantum mechanics." It would be more spectacular news, he said, if they had come out wrong.

Anton Zeilinger of the University of Vienna said that he thought, "The world is not as real as we think."

"My personal opinion is that the world is even weirder than what quantum physics tells us," he added.

The discussion is bringing renewed attention to Einstein's role as a founder and critic of quantum theory, an "underground history," that has largely been overlooked amid the celebrations of relativity in the past Einstein year, according to David Z. Albert, a professor of philosophy and physics at Columbia. Regarding the 1935 paper, Dr. Albert said, "We know something about Einstein's genius we didn't know before."

The Silly Theory

From the day 100 years ago that he breathed life into quantum theory by deducing that light behaved like a particle as well as like a wave, Einstein never stopped warning that it was dangerous to the age-old dream of an orderly universe.

Comic continued from page 236; continues on page 241.

If light was a particle, how did it know which way to go when it was issued from an atom?

"The more success the quantum theory has, the sillier it seems," Einstein once wrote to friend.

The full extent of its silliness came in the 1920's when quantum theory became quantum mechanics.

In this new view of the world, as encapsulated in a famous equation by the Austrian Erwin Schrödinger, objects are represented by waves that extend throughout space, containing all the possible outcomes of an observation—here, there, up or down, dead or alive. The amplitude of this wave is a measure of the probability that the object will actually be found to be in one state or another, a suggestion that led Einstein to grumble famously that God doesn't throw dice.

Worst of all from Einstein's point of view was the uncertainty principle, enunciated by Werner Heisenberg in 1927.

Certain types of knowledge, of a particle's position and velocity, for example, are

Einstein and Bohr

Einstein argued with Bohr over quantum mechanics for almost 30 years. Nevertheless both men greatly admired each other and refined their understanding through these intellectual duels. "Not often in life has a man given me so much happiness by his mere presence as you have done," Einstein wrote to Bohr. "I have learned much from you, mainly from your sensitive approach to scientific problems."

The debates took place during the Solvay Conferences, held every three years, and sponsored by Belgium industrialist Ernest Solvay. Einstein was the youngest attendee of the first 1911 conference in Brussels. The twenty-third conference on "The Quantum Structure of Space and Time" occurred in 2005.

Einstein first challenged Bohr with a series of thought experiments at the 1927 conference, which was attended by 29 physicists, including 17 current or future Nobel laureates. At breakfast, according to Werner Heisenberg, Einstein would pose a *gedanken* experiment—a thought experiment—to Bohr, which challenged the uncertainty principle. The proposition would be discussed throughout the day, but by evening Bohr would answer Einstein. "Einstein would look a bit worried, but

Bohr and Einstein in Paul Ehrenfest's home in Brussels, 1930. Their intense discussions, coupled with their profound respect for one another, made these discussions landmark events in the history of science.

by the next morning he was ready with a new imaginary experiment," recalled Heisenberg.

Einstein then moved on to question the completeness of quantum mechanics. In 1935, Einstein and two young assistants, Boris Podolsky and Nathan Rosen, raised the now famous issue of entanglement, from which the phrase "spooky action at a distance" was born. The EPR paper, as it was called (after its authors) questioned the idea that measuring the attributes of one particle could determine the attributes of another distant particle.

In 1979, Nathan Rosen wrote in a critical appraisal of Bohr's original response to the EPR paper, "What it seems to amount to is that [according to Bohr] the description of reality by quantum mechanics is complete because reality is whatever quantum mechanics is capable of describing." Recent experiments have, however, proved Bohr correct.

Bohr and Einstein aso shared a profound commitment to limiting the development of atomic weapons—a commitment that only strengthened their friendship.   — PH

incompatible: the more precisely you measure one property, the blurrier and more uncertain the other becomes.

In the 1935 paper, Einstein and his colleagues, usually referred to as E.P.R., argued that the uncertainty principle could not be the final word about nature. There must be a deeper theory that looked behind the quantum veil.

Imagine that a pair of electrons are shot out from the disintegration of some other particle, like fragments from an explosion. By law certain properties of these two fragments should be correlated. If one goes left, the other goes right; if one spins clockwise, the other spins counterclockwise.

That means, Einstein said, that by measuring the velocity of, say, the left hand electron, we would know the velocity of the right hand electron without ever touching it.

Conversely, by measuring the position of the left electron, we would know the position of the right hand one.

Since neither of these operations would have involved touching or disturbing the right hand electron in any

Comic continued from page 239; continues on page 243.

way, Einstein, Podolsky and Rosen argued that the right hand electron must have had those properties of both velocity and position all along. That left only two possibilities, they concluded. Either quantum mechanics was "incomplete," or measuring the left hand particle somehow disturbed the right hand one.

But the latter alternative violated common sense. Such an influence, or disturbance, would have to travel faster than the speed of light. "My physical instincts bristle at that suggestion," Einstein later wrote.

Bohr responded with a six-page essay in *Physical Review* that contained but one simple equation, Heisenberg's uncertainty relation. In essence, he said, it all depends on what you mean by "reality."

Enjoy the Magic

Most physicists agreed with Bohr, and they went off to use quantum mechanics to build atomic bombs and reinvent the world.

241

The consensus was that Einstein was a stubborn old man who "didn't get" quantum physics. All this began to change in 1964 when John S. Bell, a particle physicist at the European Center for Nuclear Research near Geneva, who had his own doubts about quantum theory, took up the 1935 E.P.R. argument. Somewhat to his dismay, Bell, who died in 1990, wound up proving that no deeper theory *that contained the presumption of locality (namely that distant particles cannot affect each other instantaneously)* could reproduce the predictions of quantum mechanics. Bell went on to outline a simple set of experiments that could settle the argument and decide who was right, Einstein or Bohr.

When the experiments were finally performed—first by John Clauser and Stuart Freedman at the University of California, Berkeley, in the 1970s, and more definitively by Alain Aspect and his colleagues at the University of Orsay in France in 1982—the results agreed with quantum mechanics and not reality as Einstein had always presumed it should be. Apparently a particle in one place could be affected by what you do somewhere else.

"That's really weird," Dr. Albert said, calling it "a profoundly deep violation of an intuition that we've been walking with since caveman days."

Physicists and philosophers are still fighting about what this means. Many of those who care to think about these issues (and many prefer not to), concluded that Einstein's presumption of locality—the idea that physically separated objects are really separate—is wrong.

Dr. Albert said, "The experiments show locality is false, end of story." But for others, it is the notion of realism, that things exist independent of being perceived, that must be scuttled. In fact, physicists don't even seem to agree on the definitions of things like "locality" and "realism."

"I would say we have to be careful saying what's real," Dr. Mermin said. "Properties cannot be said to be there until they are revealed by an actual experiment."

What everybody does seem to agree on is that the use of this effect is limited. You can't use it to send a message, for example.

Leonard Susskind, a Stanford theoretical physicist, who called these entanglement experiments "beautiful and surprising," said the term "spooky action at a distance," was misleading because it implied the instantaneous sending of signals. "No competent physicist thinks that entanglement allows this kind of nonlocality."

Indeed the effects of spooky action, or "entanglement," as Schrödinger called it, only show up in retrospect when the two participants in a Bell-type experiment compare notes. Beforehand, neither has seen any violation of business as usual; each sees the results of his measurements of, say, whether a spinning particle is pointing up or down, as random.

In short, as Brian Greene, the Columbia theorist wrote in *The Fabric of the Cosmos*, Einstein's special relativity, which sets the speed of light as the cosmic speed limit, "survives by the skin of its teeth."

In an essay in 1985, Dr. Mermin said that "if there is spooky action at a distance, then, like other spooks, it is absolutely useless except for its effect, benign or otherwise, on our state of mind."

He added, "The E.P.R. experiment is as close to magic as any physical phe-

Comic continued from page 241; continues on page 245.

nomenon I know of, and magic should be enjoyed." In a recent interview, he said he still stood by the latter part of that statement. But while spooky action remained useless for sending a direct message, it had turned out to have potential uses, he admitted, in cryptography and quantum computing.

Nine Ways of Killing a Cat

Another debate, closely related to the issues of entanglement and reality, concerns what happens at the magic moment when a particle is measured or observed.

Before a measurement is made, so the traditional story goes, the electron exists in a superposition of all possible answers, which can combine, adding and interfering with one another.

Then, upon measurement, the wave function "collapses" to one particular value. Schrödinger himself thought this was so absurd that he dreamed up a counter example. What is true for electrons, he said, should be true as well for cats.

In his famous thought experiment, a cat is locked in a box where the decay of a radioactive particle will cause the release of poison that will kill it. If the particle has a 50-50 chance of decaying, then according to quantum mechanics the cat is both alive and dead before we look in the box, something the cat itself, not to mention cat lovers, might take issue with.

But cats are always dead or alive, as Dr. Leggett of Illinois said in his Berkeley talk. "The problem with quantum mechanics," he said in an interview, "is how it explains definite outcomes to experiments."

If quantum mechanics is only about information and a way of predicting the results of measurements, these questions don't matter, most quantum physicists say.

"But," Dr. Leggett said, "if you take the view that the formalism is reflecting something out there in real world, it matters immensely." As a result, theorists have come up with a menu of alternative interpretations and explanations. According to one popular notion, known as decoherence, quantum waves are very fragile and collapse from bumping into the environment. Another theory, by the late David Bohm, restores determinism by postulating a "pilot wave" that acts behind the scenes to guide particles.

In yet another theory, called "many worlds," the universe continually branches so that every possibility is realized: the Red Sox win and lose and it rains; Schrödinger's cat lives, dies, has kittens and scratches her master when he tries to put her into the box.

Recently, as Dr. Leggett pointed out, some physicists have tinkered with Schrödinger's equation, the source of much of the misery, itself.

A modification proposed by the Italian physicists Giancarlo Ghirardi and Tullio Weber, both of the University of Trieste, and Alberto Rimini of the University of Pavia, makes the wave function unstable so that it will collapse in a time depending on how big a system it represents.

In his standoff with Dr. Ramsey of Harvard last fall, Dr. Leggett suggested that his colleagues should consider the merits of the latter theory. "Why should we think of an electron as being in two states at once but not a cat, when the theory is ostensibly the same in both cases?" Dr. Leggett asked.

Dr. Ramsey said that Dr. Leggett had missed the point. How the wave function mutates is not what you calculate. "What you calculate is the prediction of a measurement," he said.

"If it's a cat, I can guarantee you will get that it's alive or dead," Dr. Ramsey said. David Gross, a recent Nobel winner and director of the Kavli Institute for

Theoretical Physics in Santa Barbara, leapt into the free-for-all, saying that 80 years had not been enough time for the new concepts to sink in. "We're just too young. We should wait until 2200 when quantum mechanics is taught in kindergarten."

The Joy of Randomness

One of the most extreme points of view belongs to Dr. Zeilinger of Vienna, a bearded, avuncular physicist whose laboratory regularly hosts every sort of quantum weirdness.

In an essay recently in Nature, Dr. Zeilinger sought to find meaning in the very randomness that plagued Einstein.

Comic continued from page 243; continues on page 249.

"The discovery that individual events are irreducibly random is probably one of the most significant findings of the 20th century," Dr. Zeilinger wrote.

Dr. Zeilinger suggested that reality and information are, in a deep sense, indistinguishable, a concept that Dr. Wheeler, the Princeton physicist, called "it from bit."

In information, the basic unit is the bit, but one bit, he says, is not enough to specify both the spin and the trajectory of a particle. So one quality remains unknown, irreducibly random.

As a result of the finiteness of information, he explained, the universe is fundamentally unpredictable.

"I suggest that this randomness of the individual event is the strongest indication we have of a reality 'out there' existing independently of us," Dr. Zeilinger wrote in *Nature*.

He added, "Maybe Einstein would have liked this idea after all."

Einstein and Gödel were both on the faculty of the Institute for Advanced Study in Princeton for many years. The two men—one a founder of modern physics and the other a founder of modern mathematical logic—were seen walking through the streets of the quiet New Jersey town and discussing matters, both trivial and ponderous, virtually daily.

Time Bandits: What Were Einstein and Gödel Talking About?

by Jim Holt

In 1933, with his great scientific discoveries behind him, Albert Einstein came to America. He spent the last twenty-two years of his life in Princeton, New Jersey, where he had been recruited as the star member of the Institute for Advanced Study. Einstein was reasonably content with his new milieu, taking its pretensions in stride. "Princeton is a wonderful piece of earth, and at the same time an exceedingly amusing ceremonial backwater of tiny spindle-shanked demigods," he observed. His daily routine began with a leisurely walk from his house, at 112 Mercer Street, to his office at the institute. He was by then one of the most famous and, with his distinctive appearance—the whirl of pillow-combed hair, the baggy pants held up by suspenders—most recognizable people in the world.

A decade after arriving in Princeton, Einstein acquired a walking companion, a much younger man who, next to the rumpled Einstein, cut a dapper figure in a white linen suit and matching fedora. The two would talk animatedly in German on their morning amble to the institute and again, later in the day, on their way homeward. The man in the suit may not have been recognized by many towns-people, but Einstein addressed him as a peer, someone who, like him, had single-handedly launched a conceptual revolution. If Einstein had upended our everyday notions about the physical world with his theory of relativity, the younger man, Kurt Gödel, had had a similarly subversive effect on our understanding of the abstract world of mathematics.

Gödel, who has often been called the greatest logician since Aristotle, was a strange and ultimately tragic man. Whereas Einstein was gregarious and full of laughter, Gödel was solemn, solitary, and pessimistic. Einstein, a passionate amateur violinist, loved Beethoven and Mozart. Gödel's taste ran in another direction: his favorite movie was Walt Disney's "Snow White and the Seven Dwarfs," and

---

Originally published in *The New Yorker* February 28, 2005, pp. 80–85 Reprinted by permission of the author.

when his wife put a pink flamingo in their front yard he pronounced it *furchtbar herzig*—"awfully charming." Einstein freely indulged his appetite for heavy German cooking; Gödel subsisted on a valetudinarian's diet of butter, baby food, and laxatives. Although Einstein's private life was not without its complications, outwardly he was jolly and at home in the world. Gödel, by contrast, had a tendency toward paranoia. He believed in ghosts; he had a morbid dread of being poisoned by refrigerator gases; he refused to go out when certain distinguished mathematicians were in town, apparently out of concern that they might try to kill him. "Every chaos is a wrong appearance," he insisted—the paranoiac's first axiom.

Although other members of the institute found the gloomy logician baffling and unapproachable, Einstein told people that he went to his office "just to have the privilege of walking home with Kurt Gödel." Part of the reason, it seems, was that Gödel was undaunted by Einstein's reputation and did not hesitate to challenge his ideas. As another member of the institute, the physicist Freeman Dyson, observed, "Gödel was the only one of our colleagues who walked and talked on equal terms with Einstein." But if Einstein and Gödel seemed to exist on a higher plane than the rest of humanity, it was also true that they had become, in Einstein's words, "museum pieces." Einstein never accepted the quantum theory of Niels Bohr and Werner Heisenberg. Gödel believed that mathematical abstractions were every bit as real as tables and chairs, a view that philosophers had come to regard as laughably naïve. Both Gödel and Einstein insisted that the world is independent of our minds, yet rationally organized and open to human understanding. United by a shared sense of intellectual isolation, they found solace in their companionship. "They didn't want to speak to anybody else," another member of the institute said. "They only wanted to speak to each other."

People wondered what they spoke about. Politics was presumably one theme. (Einstein, who supported Adlai Stevenson, was exasperated when Gödel chose to vote for Dwight Eisenhower in 1952.) Physics was no doubt another. Gödel was well versed in the subject; he shared Einstein's mistrust of the quantum theory, but he was also skeptical of the older physicist's ambition to supersede it with a "unified field theory" that would encompass all known forces in a deterministic framework. Both were attracted to problems that were, in Einstein's words, of "genuine importance," problems pertaining to the most basic elements of reality. Gödel was especially preoccupied by the nature of time, which, he told a friend, was the philosophical question. How could such a "mysterious and seemingly self-contradictory" thing, he wondered, "form the basis of the world's and our own existence"? That was a matter in which

Einstein had shown some expertise.

A century ago, in 1905, Einstein proved that time, as it had been understood by scientist and layman alike, was a fiction. And this was scarcely his only achievement that year, which John S. Rigden skillfully chronicles, month by month, in *Einstein 1905: The Standard of Greatness*. As it began, Einstein, twenty-five years old, was employed as an inspector in a patent office in Bern, Switzerland. Having earlier failed to get his doctorate in physics, he had temporarily given upon the idea of an academic career, telling a friend that "the whole comedy has become boring." He

Comic continued from page 245; continues on page 251.

had recently read a book by Henri Poincaré, a French mathematician of enormous reputation, which identified three fundamental unsolved problems in science. The first concerned the "photoelectric effect": how did ultraviolet light knock electrons off the surface of a piece of metal? The second concerned "Brownian motion": why did pollen particles suspended in water move about in a random zigzag pattern? The third concerned the "luminiferous ether" that was supposed to fill all of space and serve as the medium through which light waves moved, the way sound waves move through air, or ocean waves through water: why had experiments failed to detect the earth's motion through this ether?

Each of these problems had the potential to reveal what Einstein held to be the underlying simplicity of nature. Working alone, apart from the scientific community, the unknown junior clerk rapidly managed to dispatch all three. His solutions were presented in four papers, written in the months of March, April, May, and June of 1905. In his March paper, on the photoelectric effect, he deduced that light came in discrete particles, which were later dubbed "photons." In his April and May papers, he established once and for all the reality of atoms, giving a theoretical estimate of their size and showing how their bumping around caused Brownian motion. In his June paper, on the ether problem, he unveiled his theory, of relativity. Then, as a sort of encore, he published a three-page note in September containing the most famous equation of all time: $E = mc^2$.

All of these papers had a touch of magic about them, and upset deeply held convictions in the physics community. Yet, for scope and audacity, Einstein's June paper stood out. In thirty succinct pages, he completely rewrote the laws of physics, beginning with two stark principles. First, the laws of physics are absolute: the same laws must be valid for all observers. Second, the speed of light is absolute; it, too, is the same for all observers. The second principle, though less obvious, had the same sort of logic to recommend it. Since light is an electromagnetic wave (this had been known since the nineteenth century), its speed is fixed by the laws of electromagnetism; those laws ought to be the same for all observers; and therefore everyone should see light moving at the same speed, regardless of the frame of reference. Still, it was bold of Einstein to embrace the light principle, for its consequences seemed downright absurd.

Suppose—to make things vivid—that the speed of light is a hundred miles an hour. Now suppose I am standing by the side of the road and I see a light beam pass by at this speed. Then I see you chasing after it in a car at sixty miles an hour. To me, it appears that the light beam is outpacing you by forty miles an hour. But you, from inside your car, must see the beam escaping you at a hundred miles an hour, just as you would if you were standing still: that is what the light principle demands. What if you gun your engine and speed up to ninety-nine miles an hour? Now I see the beam of light outpacing you by just one mile an hour. Yet to you, inside the car, the beam is still racing ahead at a hundred miles an hour, despite your increased speed. How can this be? Speed, of course, equals distance divided by time. Evidently, the faster you go in your car, the shorter your ruler must become and the slower your clock must tick relative to mine; that is the only way we can continue to agree on the speed of light. (If I were to pull out a pair of binoculars and look at your speeding car, I would actually see its length contracted and you moving in slow motion inside.) So Einstein set about recasting the laws of physics accordingly. To make these laws absolute, he made distance and time relative.

It was the sacrifice of absolute time that was most stunning. Isaac Newton believed that time was regulated by a sort of cosmic grandfather clock. "Absolute, true, mathematical time, of itself, and from its own nature, flows equably without relation to anything external," he declared at the beginning of his *Principia*. Einstein, however, realized that our idea of time is something we abstract from our experience with rhythmic phenomena: heartbeats, planetary rotations and revolutions, the ticking of clocks. Time judgments always come down to judgments of simultaneity. "If, for instance, I say, 'That train arrives here at 7 o'clock,' I mean something like this: 'The pointing of the small hand of my watch to 7 and the arrival of the train are simultaneous events.'"

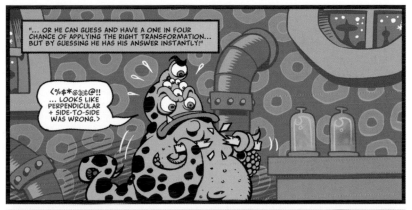

Comic continued from page 249; continues on page 253.

Einstein wrote in the June paper. If the events in question are at some distance from one another, judgments of simultaneity can be made only by sending light signals back and forth. Working from his two basic principles, Einstein proved that whether an observer deems two events to be happening "at the same time" depends on his state of motion. In other words, there is no universal *now*. With different observers slicing up the timescape into "past," "present," and "future" in different ways, it seems to follow that all moments coexist with equal reality.

Einstein's conclusions were the product of pure thought, proceeding from the most austere assumptions about nature. In the century since he derived them, they have been precisely confirmed by experiment after experiment. Yet his June, 1905, paper on relativity was rejected when he submitted it as a dissertation. (He then submitted his April paper, on the size of atoms, which he thought would be less likely to startle the examiners; they accepted it only after he added one sentence to meet the length threshold.) When Einstein was awarded the 1921 Nobel Prize in Physics, it was for his work on the photoelectric effect. The Swedish Academy forbade him to make any mention of relativity in his acceptance speech. As it happened, Einstein was unable to attend the ceremony in Stockholm. He gave his Nobel lecture in Gothenburg, with King Gustav V seated in the front row. The King wanted to learn about relativity, and Einstein obliged him.

In 1906, the year after Einstein's annus mirabilis, Kurt Gödel was born in the city of Brno (now in the Czech Republic). As Rebecca Goldstein recounts in her enthralling intellectual biography *Incompleteness: The Proof and Paradox of Kurt Gödel*, Kurt was both an inquisitive child—his parents and brother gave him the nickname *der Herr Warum*, "Mr. Why?"—and a nervous one. At the age of five, he seems to have suffered a mild anxiety neurosis. At eight, he had a terrifying bout of rheumatic fever, which left him with the lifelong conviction that his heart had been fatally damaged.

Gödel entered the University of Vienna in 1924. He had intended to study physics, but he was soon seduced by the beauties of mathematics, and especially by the notion that abstractions like numbers and circles had a perfect, timeless existence independent of the human mind. This doctrine, which is called Platonism, because it descends from Plato's theory of ideas, has always been popular among mathematicians. In the philosophical world of nineteen-twenties Vienna, however, it was considered distinctly old-fashioned. Among the many intellectual movements that flourished in the city's rich café culture, one of the most prominent was the Vienna Circle, a group of thinkers united in their belief that philosophy must he cleansed of metaphysics and made over in the image of science. Under the influence of Ludwig Wittgenstein, their reluctant guru, the members of the Vienna Circle regarded mathematics as a game played with symbols, a more intricate version of chess. What made a proposition like "2 + 2 = 4" true, they held, was not that it correctly described some abstract world of numbers but that it could be derived in a logical system according to certain rules.

Gödel was introduced into the Vienna Circle by one of his professors, but he kept quiet about his Platonist views. Being both rigorous and averse to controversy, he did not like to argue his convictions unless he had an airtight way of demonstrating that they were valid. But how could one demonstrate that mathematics could not be reduced to the artifices of logic? Gödel's strategy—one of "heart-stopping beauty," as Goldstein justly observes—was to use logic against itself. Beginning with a logical system for mathematics, one presumed to be free of contradictions, he invented an ingenious scheme that allowed the formulas in it to engage in a sort of double speak.

A formula that said something about numbers could also, in this scheme, be interpreted as saying something about other formulas and how they were logically related to one another. In fact, as Gödel showed, a numerical formula could even be made to say something about itself. (Goldstein compares this to a play in which the characters are also actors in a play within the play; if the playwright is sufficiently clever, the lines the actors speak in the play within the play can be interpreted as having a "real life" meaning in the play proper.) Having painstakingly built this apparatus of mathematical self-reference, Gödel came up with an astonishing twist:

Comic continued from page 251; continues on page 255.

he produced a formula that, while ostensibly saying something about numbers, also says, "I am not provable." At first, this looks like a paradox, recalling as it does the proverbial Cretan who announces, "All Cretans are liars." But Gödel's self-referential formula comments on its provability, not on its truthfulness. Could it be lying? No, because if it were, that would mean it could be proved, which would make it true. So, in asserting that it cannot be proved, it has to be telling the truth. But the truth of this proposition can be seen only from outside the logical system. Inside the system, it is neither provable nor disprovable. The system, then, is incomplete. The conclusion—that no logical system can capture all the truths of mathematics—is known as the first incompleteness theorem. Gödel also proved that no logical system for mathematics could, by its own devices, be shown to be free from inconsistency, a result known as the second incompleteness theorem.

Wittgenstein once averred that "there can *never* be surprises in logic." But

Gödel's incompleteness theorems did come as a surprise. In fact, when the fledgling logician presented them at a conference in the German city of Königsberg in 1930, almost no one was able to make any sense of them. What could it mean to say that a mathematical proposition was true if there was no possibility of proving it? The very idea seemed absurd. Even the once great logician Bertrand Russell was baffled; he seems to have been under the misapprehension that Gödel had detected an inconsistency in mathematics. "Are we to think that 2 + 2 is not 4, but 4.001?" Russell asked decades later in dismay, adding that he was "glad [he] was no longer working at mathematical logic." As the significance of Gödel's theorems began to sink in, words like "debacle," "catastrophe," and "nightmare" were bandied about. It had been an article of faith that, armed with logic, mathematicians could in principle resolve any conundrum at all—that in mathematics, as it had been famously declared, there was no *ignorabimus*. Gödel's theorems seemed to have shattered this ideal of complete knowledge.

That was not the way Gödel saw it. He believed he had shown that mathematics has a robust reality that transcends any system of logic. But logic, he was convinced, is not the only route to knowledge of this reality; we also have something like an extrasensory perception of it, which he called "mathematical intuition." It is this faculty of intuition that allows us to see, for example, that the formula saying "I am not provable" must be true, even though it defies proof within the system where it lives. Some thinkers (like the physicist Roger Penrose) have taken this theme further, maintaining that Gödel's incompleteness theorems have profound implications for the nature of the human mind. Our mental powers, it is argued, must outstrip those of any computer, since a computer is just a logical system running on hardware, and our minds can arrive at truths that are beyond the reach of a logical system.

Gödel was twenty-four when he proved his incompleteness theorems (a bit younger than Einstein was when he created relativity theory). At the time, much to the disapproval of his strict Lutheran parents, he was courting an older Catholic divorcée by the name of Adele, who, to top things off, was employed as a dancer in a Viennese night club called Der Nachtfalter (the Moth). The political situation in Austria was becoming ever more chaotic with Hitler's rise to power in Germany, although Gödel seems scarcely to have noticed. In 1936, the Vienna Circle dissolved, after its founder was assassinated by a deranged student. Two years later came the Anschluss. The perilousness of the times was finally borne in upon Gödel when a band of Nazi youths roughed him up and knocked off his glasses, before retreating under the umbrella blows of Adele. He resolved to leave for Princeton, where he had been offered a position by the Institute for Advanced Study. But, the war having broken out, he judged it too risky to cross the Atlantic. So the now married couple took the long way around, traversing Russia, the Pacific, and the United States, and finally arriving in Princeton in early 1940. At the institute, Gödel was given an office almost directly above Einstein's. For the rest of his life he rarely left Princeton, which he came to find "ten times more congenial" than his once beloved Vienna.

"There it was, inconceivably, K. Goedel, listed just like any other name in the bright orange Princeton community phone book,"

Comic continued from page 253; continues on page 257.

writes Goldstein, who came to Princeton University as a graduate student of philosophy in the early nineteen-seventies. (It's the setting of her novel *The Mind-Body Problem*.) "It was like opening up the local phone book and finding *B. Spinoza* or *I. Newton*." Although Gödel was still little known in the world at large, he had a godlike status among the cognoscenti. "I once found the philosopher Richard Rorty standing in a bit of a daze in Davidson's food market," Goldstein writes. "He told me in hushed tones that he'd just seen Gödel in the frozen food aisle."

So naïve and otherworldly was the great logician that Einstein felt obliged to help look after the practical aspects of his life. One much retailed story concerns Gödel's decision after the war to become an American citizen. The character witnesses at his hearing were to be Einstein and Oskar Morgenstern, one of the founders of game theory. Gödel took the matter of citizenship with great solemni-

255

ty, preparing for the exam by making a close study of the United States Constitution. On the eve of the hearing, he called Morgenstern in an agitated state, saying he had found an "inconsistency" in the Constitution, one that could allow a dictatorship to arise. Morgenstern was amused, but he realized that Gödel was serious and urged him not to mention it to the judge, fearing that it would jeopardize Gödel's citizenship bid. On the short drive to Trenton the next day, with Morgenstern serving as chauffeur, Einstein tried to distract Gödel with jokes. When they arrived at the courthouse, the judge was impressed by Gödel's eminent witnesses, and he invited the trio into his chambers. After some small talk, he said to Gödel, "Up to now you have held German citizenship."

No, Gödel corrected, Austrian.

"In any case, it was under an evil dictatorship," the judge continued. "Fortunately that's not possible in America."

"On the contrary, I can prove it is possible!" Gödel exclaimed, and he began describing the constitutional loophole he had descried. But the judge told the examinee that "he needn't go into that," and Einstein and Morgenstern succeeded in quieting him down. A few months later, Gödel took his oath of citizenship.

Around the same time that Gödel was studying the Constitution, he was also taking a close look at Einstein's relativity theory. The key principle of relativity is that the laws of physics should be the same for all observers. When Einstein first formulated the principle in his revolutionary 1905 paper, he restricted "all observers" to those who were moving uniformly relative to one another—that is, in a straight line and at a constant speed. But he soon realized that this restriction was arbitrary. If the laws of physics were to provide a truly objective description of nature, they ought to be valid for observers moving in any way relative to one another—spinning, accelerating, spiralling, whatever. It was thus that Einstein made the transition from his "special" theory of relativity of 1905 to his "general" theory, whose equations he worked out over the next decade and published in 1916. What made those equations so powerful was that they explained gravity, the force that governs the overall shape of the cosmos.

Decades later, Gödel, walking with Einstein, had the privilege of picking up the subtleties of relativity theory from the master himself. Einstein had shown that the flow of time depended on motion and gravity, and that the division of events into "past" and "future" was relative. Gödel took a more radical view: he believed that time, as it was intuitively understood, did not exist at all. As usual, he was not content with a mere verbal argument. Philosophers ranging from Parmenides, in ancient times, to

Immanuel Kant, in the eighteenth century, and on to J. M. E. McTaggart, at the beginning of the twentieth century, had produced such arguments, inconclusively. Gödel wanted a proof that had the rigor and certainty of mathematics. And he saw just what he wanted lurking within relativity theory. He presented his argument to Einstein for his seventieth birthday, in 1949, along with an etching. (Gödel's wife had knitted Einstein a sweater, but she decided not to send it.)

What Gödel found was the possibility of a hitherto unimaginable kind of universe. The equations of general relativity can be solved in

End of the Comic.

a variety of ways. Each solution is, in effect, a model of how the universe might be. Einstein, who believed on philosophical grounds that the universe was eternal and unchanging, had tinkered with his equations so that they would yield such a model—a move he later called "my greatest blunder." Another physicist (a Jesuit priest, as it happens) found a solution corresponding to an expanding universe born at some moment in the finite past. Since this solution, which has come to be known as the Big Bang model, was consistent with what astronomers observed, it seemed to be the one that described the actual cosmos. But Gödel came up with a third kind of solution to Einstein's equations, one in which the universe was not expanding but rotating. (The centrifugal force arising from the rotation was what kept everything from collapsing under the force of gravity.) An observer in this universe would see all the galaxies slowly spinning around him; he would know it was the universe doing the spinning, and not himself because he would feel no dizziness. What makes this rotating universe truly weird, Gödel showed, is the way its geometry mixes up space and time. By completing a sufficiently long round trip in a rocket ship, a resident of Gödel's universe could travel back to any point in his own past.

Einstein was not entirely pleased with the news that his equations permitted something as Alice in Wonderland-like as spatial paths that looped backward in time; in fact, he confessed to being "disturbed" by Gödel's universe. Other physicists marvelled that time travel, previously the stuff of science fiction, was apparently consistent with the laws of physics. (Then they started worrying about what would happen if you went back to a time before you were born and killed your own grandfather.) Gödel himself drew a different moral. If time travel is possible, he submitted, then time itself is impossible. A past that can be revisited has not really passed. And the fact that the actual universe is expanding, rather than rotating, is irrelevant. Time, like God, is either necessary or nothing; if it disappears in one possible universe, it is undermined in every possible universe, including our own.

Gödel's conclusion went almost entirely unnoticed at the time, but it has since found a passionate champion in Palle Yourgrau, a professor of philosophy at Brandeis. In *A World Without Time: The Forgotten Legacy of Gödel and Einstein,* Yourgrau does his best to redress his fellow philosophers neglect of the case that Gödel made against time. The "deafening silence," he submits, can be blamed on the philosophical prejudices of the era. Behind all the esoteric mathematics, Gödel's reasoning looked suspiciously metaphysical. To this day, Yourgrau complains, Gödel is treated with condescension by philosophers, who regard him, in the words of one, as "a logician par excellence but a philosophical fool." After ably tracing Gödel's life, his logical achievements, and his friendship with Einstein, Yourgrau elaborately defends his importance as a philosopher of time. "In a deep sense," he concludes, "we all do live in Gödel's universe."

Gödel's strange cosmological gift was received by Einstein at a bleak time in his life. His quest for a unified theory of physics was proving fruitless, and his opposition to quantum theory alienated him from the mainstream of physics. Family life provided little consolation. His two marriages had been failures; a daughter born out of wedlock seems to have disappeared from history; of his two sons one was schizophrenic, the other estranged. Einstein's circle of friends had shrunk to Gödel and a few others. One of them was Queen Elisabeth of Belgium, to whom he confided, in March, 1955, that "the exaggerated esteem in which my lifework is held makes me very ill at ease. I feel compelled to think of myself as an involuntary swindler." He died a month later, at the age of seventy-six. When Gödel and another colleague went to his office at the institute to deal with his papers,

Wormholes

Time travel has become a popular subject in modern science fiction. *The Time Machine* by H.G. Wells has twice been adapted to the screen and the subject has fascinated writers, film makers and scientists for decades. Speculation about time travel has long been hampered by the paradoxes that arise from such a prospect—usually labeled "grandfather paradoxes," these take the form of, "what if you were to go back to the past and kill your own grandfather before your father was born? Wouldn't that mean *you* were never born?"

The entire time travel business received a shot in the arm in 1916, just a few months after Einstein published his seminal work on general relativity. Within a few months of that publication, solutions were found that allowed for the possibility of something called a "wormhole"—a tear in the fabric of spacetime that connects great distances by providing a "trap door" from one point in the universe (or, one might say, "from one place-moment in spacetime") to another.

The mathematics of wormholes was worked out in detail in the 1950s by a team headed by John Archibald Wheeler (though, as the current essay by Jim Holt shows, Gödel had some interesting things to say about this as early as the 1940s). According to the "Wheelerites," a wormhole is created when a large mass creates a singularity, or "pinch" in the fabric of spacetime, something made possible by general relativity. When the singularity of one mass meets the singularity of another in "hyperspace"—the multi-dimen-

sional universe in which our paltry universe exists—they can fuse and create a passageway through which…something—matter, light, radiation—might pass relatively quickly in spite of the

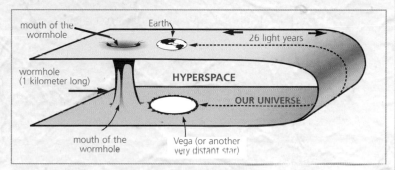

A simple schematic of a wormhole between Earth and the star Vega 26 light years away. Even a photon travelling at the speed of light would require 26 years to make the trip. But if a wormhole opened up between Earth and Vega, the trip would take only a small fraction of a second.

great distance between the masses.

One recent depiction of this theory was in the 1985 Carl Sagan novel (made into a movie in 1997), *Contact*, in which an astronaut (named Eleanore Arroway, played in the movie by Jodie Foster) travels to a star many light years away by being propelled through such a wormhole by a device created by a *very* advanced civilization. Dr. Arroway takes this incredible journey of many hours in what seems on Earth to be just a few seconds.

Sagan worked with astrophysicist Kip Thorne to concoct the wormhole when it became clear that a black hole (which might also serve as a shortcut to places very distant in space and time) was not a congenial conveyance for a human astronaut (certainly not for a star like Jodie Foster).

Three problems remain, however. First, the theory of wormholes gives us no clue as to how the two singularities of the two masses happen to meet each other in hyperspace to form this

tunnel through spacetime. Second, under ordinary conditions (meaning, without any intervention from some outside agent), wormholes are fleeting things, available for small fractions of a second before they disappear and the tunnel is closed. Sagan and Thorne proposed that this was why one needs the civilzation that created this spacetime travel system to be a highly advanced one that would have figured out how to keep the wormhole open longer.

But the biggest difficulty has to be that, while black holes are a natural, inevitable consequence of the evolution of stars, and we now have direct evidence that such things actually exist out there in space, there is no evidence as yet that wormholes exist anywhere in the universe.   — HR they found the blackboard covered with dead-end equations.

After Einstein's death, Gödel became ever more withdrawn. He preferred to conduct all conversations by telephone, even if his interlocutor was a few feet distant. When he especially wanted to avoid someone, he would schedule a rendezvous at a precise time and place, and then make sure he was somewhere far away. The honors the world wished to bestow upon him made him chary. He did show up to collect an honorary doctorate in 1953 from Harvard, where his incompleteness theorems were hailed as the most important mathematical discovery of the previous hundred years; but he later complained of being "thrust quite undeservedly into the most highly bellicose company" of John Foster Dulles, a co-honoree. When he was awarded the National Medal of Science, in 1975, he refused to go to Washington to meet Gerald Ford at the White House, despite the offer of a chauffeur for him and his wife. He had hallucinatory episodes and talked darkly of certain forces at work in the world "directly submerging the good." Fearing that there was a plot to poison him, he persistently refused to eat. Finally, looking like (in the words of a friend) "a living corpse," he was taken to the Princeton Hospital. There, two weeks later, on January 14, 1978, he succumbed to self-starvation. According to his death certificate, the cause of death was "malnutrition and inanition" brought on by "personality disturbance."

A certain futility marked the last years of both Gödel and Einstein. What may have been most futile, however, was their willed belief in the unreality of time. The temptation was understandable. If time is merely in our minds, perhaps we can hope to escape it into a timeless eternity. Then we could say, like William Blake, "I see the Past, Present and Future, existing all at once / Before me." In Gödel's case, Rebecca Goldstein speculates, it may have been his childhood terror of a fatally damaged heart that attracted him to the idea of a timeless universe. Toward the end of his life, he told one confidant that he had long awaited an epiphany that would enable him to see the world in a new light, but that it never came. Einstein, too, was unable to make a clean break with time. "To those of us who believe in physics," he wrote to the widow of a friend who had recently died, "this separation between past, present, and future is only an illusion, if a stubborn one." When his own turn came, a couple of weeks later, he said, "It is time to go."

PART FOUR

EINSTEIN'S ENDURING INFLUENCE

Einstein on the cover of *Wired* Magazine (top, left). Clockwise, Einstein has also been depicted as an action doll, and portrayed by Walter Matthau in the romantic comedy *IQ*, released in 1994.

Einstein and Charlie Chaplin at the opening of *City Lights* in 1931, two of the most recognizable faces in the world at that time.

Einstein and the Shaping of Our Imagination

by Gerald Holton

Einstein reached out not only to scholars in science and outside but also to all others who are touched by the fruits of the scientific imagination. He wrote, "All religions, arts, and sciences are branches of the same tree. All these aspirations are directed towards ennobling man's life, lifting it from the sphere of mere physical existence and leading the individual toward freedom."

One's first thought must of course go to science itself. And there it is significant to note how long it took, by present standards, for his seminal, early work to be understood even by his fellow physicists. Six years elapsed after the first publication of the special theory of relativity before it established itself sufficiently to merit a textbook (Max von Laue's *Das Relativitatsprinzip*), and for some years after that the theory continued to be confused by most scientists with the electrodynamics of H. A. Lorentz. Einstein's ideas on quantum physics, published from 1905 on, were also generally neglected or discounted for years. R. A. Millikan, on accepting his Nobel Prize for 1923, confessed that the validity of Einstein's "bold, not to say reckless" explanation of the photoelectric effect forced itself on him slowly, "after ten years of testing [and] contrary to my own expectation." The transcripts of the questions asked in scientific meetings in the decade after 1905 contain many passages that demonstrated to the historian of science the large intellectual effort required at the time to enter fully into the meaning of the new physics.

Today, virtually every student who wishes can learn at least the elements of relativity or quantum physics before leaving high school, and the imprint of Einstein's work on the different areas of physical science is so large and varied that a scientist who tries to trace it would be hard put to know where to start. A modern dictionary of scientific terms contains thirty-five entries bearing his name, from "einstein: A unit of light energy used in photochemistry" and "Einstein-Bose statistics" to "Einstein

Top: President Harding and the National Academy of Sciences at the White House, Washington, D.C., April 1921. Einstein and the president are standing in the center clearing.
Bottom: Einstein characteristically bemused by an Einstein puppet.

tensor" and "Einstein viscosity equation."[3] It is ironic that now, a quarter of a century after his death, there is in many branches of the physical sciences more awareness of his generative role than would have been credited during the last decade or two of his life. His ideas became essential for laying out conceptual paths for contemporary work in astronomy or cosmology, for unifying gravitation with the quantum field theory of gauge fields, or even for understanding new observations that were not possible in his time but were predicted by him (as in his 1936 paper on the optical lens formed by gravitational fields).

Apart from changing science itself, Einstein has reached into the daily life of virtually every person on the globe in direct or indirect ways through the incorporation of his ideas on physics into a vast range of technical devices and processes. I need cite only some of the most obvious ones. Every photoelectric cell can be considered one of his intellectual grandchildren. Hence, we are in his debt whenever photo emission or absorption is used, in the home or on the job, to capture an image by means of a television camera, or to project the optical soundtrack of a motion picture, or to set the page of a book or newspaper by photocomposition, or to make a telephone call over a modem fiber cable, or (eventually) to replace the oil-fired heater by an array of photovoltaic cells. In each case, if a law required a label on the appliance giving its intellectual content or pedigree, such a display would list prominently: "Einstein, *Annalen der Physik* 17 (1905), pp. 132–148; 20 (1906), pp. 199–206," etc.

One would find an entry of this sort also on the laser, whose beam was probably used to lay out the highway on which one travels to the office or to site the office building itself ("Einstein, *Physikalische Zeitschrift* 18 (1917), pp. 121–128," etc.). Or again, the same kind of answer comes if one lists key ideas that helped to make possible modern electric machinery, such as power generators, or precision clocks that allow the course of planes and ships to be charted. Einstein appears also, if one looks for the ancestry of the ideas, in quantum and statistical physics, by which solid-state devices operate, from calculators and computers to the transistor radio and the ignition system; and once more, even when one takes one's vitamin pill or other pharmaceutical drug, for it is likely that its commercial production involved diffusion processes, first explained in Einstein's papers on Brownian movement and statistical mechanics ("Einstein, *Annalen der Physik* 17 (1905), pp. 549–560," etc.).

As Edward M. Purcell remarked in his lecture at the Centennial Symposium at Princeton a few days before the Jerusalem Symposium, since the magnetism set up by electric currents is a strictly relativistic effect, derivable from Coulomb's law of electrostatics and the kinematics of relativity, and nothing more, it requires no elaboration to discuss "special relativity in engineering": "This is the way the world is. And it does not really take gigavolts or nanoseconds to demonstrate it; stepping on the starter will do it!" It is not too much to say that even in our most common experiences, that unworldly theoretician's publications help to explain what happens to us all day—indeed, from the moment we open our eyes on the light of the morning, since the act of seeing is initiated by a photochemical reaction ("Einstein, *Annalen der Physik* 37 (1912), pp. 832-838; 38 (1912), pp. 881–884," etc.).

The proverbial man in the street is quite blissfully ignorant of all that, and has preferred to remain so, even while expecting fully that, mysteriously yet automatically, a stream of practical, benign "spin-offs" continues from the pursuit of pure science. But the philosopher, the writer, the artist, and many others outside the scientific laboratories could not help but be caught up to some extent by the wave that spread beyond science and technology, at first slowly, then with astonishing intensity. As the best scientists were coming to understand what Einstein had done, the trumpets began to sound. Even Max Planck, a person conservative in thought and expression, enthused by 1912: "This new way of thinking ... well surpasses in daring everything that has been achieved in speculative scientific research, even in the theory of knowledge... This revolution in the physical *Weltanschauung*, brought about by the relativity principle, is to be compared in scope and depth only with that caused by the introduction of the Copernican system of the world."[4] At the

"People slowly accustomed themselves to the idea that the physical states of space itself were the final physical reality."
- from *The Evolution of Physics* by Albert Einstein and Leopold Infeld

Drawing by Rea Irvin from *The New Yorker*, 1929

same time, on the other end of the philosophical spectrum, the followers and heirs of Ernst Mach rushed to embrace relativity as a model triumph of positivistic philosophy. In the inaugural session of the *Gesellschaft für positivistische Philosophie* in Berlin (11 November 1912), relativity was interpreted as an antimetaphysical and instrumentalist tract and was hailed as "a mighty impulse for the development of the philosophical point of view of our time." When in London on 6 November 1919 the result of the British eclipse expedition was revealed to bear out one of the predictions of general relativity theory, the discussion of implications rose to fever pitch among scholars and laymen, beginning with declarations such as that in *The Times* of London (8 November 1919): the theory had served "to overthrow the certainty of ages, and to require a new philosophy, a philosophy that will sweep away nearly all that has hitherto been accepted as the axiomatic basis of physical thought." It became evident that, as Newton had "demanded the muse" alter the *Principia*, now it would be Einstein's turn.

In fact, Einstein did his best to defuse the euphoria and excess of attention that engulfed and puzzled him from that time on. When asked to explain the mass enthusiasm, his answer in 1921 was that "it seemed psychopathological!" The essence of the theory was chiefly "the logical simplicity with which it explained apparently conflicting facts in the operation of natural law," freeing science of the burden of "many general assumptions of a complicated nature."[5] That was all. As for being labeled a great revolutionary, as his friends and opponents did equally, Einstein took every opportunity to disavow it. He saw himself essentially as a continuist, had specific ideas on the way scientific theory developed by evolution,[6] and attempted to keep the discussion limited to work done and yet to be done in science. He did not get much help, however. Thus, the physicist J. J. Thomson reported that the Archbishop of Canterbury, Randall Davidson, had been told by Lord Haldane "that relativity was going to have a great effect upon theology, and that it was his duty as Head of the English Church to make himself acquainted with it. . . . The Archbishop, who is the most conscientious of men, has procured several books on the subject and has been trying to read them, and they have driven him to what it is not too much to say is a state of *intellectual desperation*." On Einstein's first visit to England in June 1921, the Archbishop of Canterbury therefore sought him out to ask what effect relativity would have on religion. Einstein replied briefly and to the point: "None. Relativity is a purely scientific matter and has nothing to do with religion."[7] But of course this did not dispose of the question. Later that year, even the scientific journal *Nature* felt it necessary to print opposing articles on whether "Einstein's space-time is the death-knell of materialism."[8]

Although the crest of the flood, and the worst excesses, have now passed, debates of this sort continue. More constructively, since modern philosophy is concerned in good part with the nature of space and time, causality, and other conceptions to which relativity and quantum physics have contributed, Einstein has had to be dealt with in the pages of philosophers, from Henri Bergson and A. N. Whitehead to the latest issues of the professional journals. As John Passmore observed correctly, it appeared in this century that "physics fell heir to the responsibility of metaphysics."

Some philosophers and philosopher-scientists have confessed that Einstein's work started them off on their speculations in the first place, thus giving some direction to their very careers. One example is P. W. Bridgman, who disclosed that the effort to clarifying his mind the issues in relativistic electrodynamics, when first asked to teach that course, drew him to the task of writing the influential book *The Logic of Modern Physics* (1927). Another case is Karl Popper, who in his recent autobiography reveals that his falsification criterion originated in his interpretation of a passage in Einstein's popular exposition of relativity, which Popper says he read with profound effect when he was still in his teens.

Philosophy was no doubt destined to be the most obvious and often the earliest and most appropriate field, outside science itself, that the radiation from Einstein's work would reach. But soon there were others, even though the connections made or asserted were not always valid. From Einstein's wide-ranging output, relativity was invoked most frequently. Cultural anthropology, in Claude Levi-Strauss's phrase, had evolved the doctrine of cultural relativism "out of a deep feeling of respect toward other cultures than our own" but this doctrine often invited confusion with physical relativity. Much that has been written on "ethical relativity" and on "relativism" is based on a seductive play with words. And painters and art critics have helped to keep alive the rumor of a supposed genetic connection of visual arts with Einstein's 1905 publication.

Here again, Einstein protested when he could and, as so often, without effect. One art historian submitted to him a draft of an essay entitled "Cubism and the Theory of Relativity," which argued for such a connection—for example, that in both fields "attention was paid to relationships, and allowance was made for the simultaneity of several views." Politely but firmly, Einstein tried to put him straight, and he explained the difference between physical relativity and vulgar relativism so succinctly as to invite an extensive quotation:

> The essence of the theory of relativity has been incorrectly understood in it [your paper], granted that this error is suggested by the attempts at popularization of the theory. For the description of a given state of facts one uses almost always only one system of coordinates. The theory says only that the general laws are such that their form does not depend on the choice of the system of coordinates. This logical demand, however, has nothing to do with how the single, specific case is represented. A multiplicity of systems of coordinates is not needed for its representation. It is completely sufficient to describe the whole mathematically in relation to one system of coordinates.[10]

This is quite different in the case of Picasso's painting, as I do not have to elaborate any further. Whether, in this case, the representation is felt as artistic unity depends, of course, upon the artistic antecedents of the viewer. This new artistic "language" has nothing in common with the Theory of Relativity.[11]

Einstein might well have added here, as he did elsewhere, that the existence of a multiplicity of frames, each one as good as the next for solving some problems in mechanics, went back to the seventeenth century (Galilean relativity). As to the superposition of different aspects of an object on a canvas, that had been done for a long time; thus Canaletto drew various parts of a set of buildings from different places and merged them in a combined view on the painting (for example, *Campo S. S. Giovanni e Paolo*), the view becoming thereby an impossible *veduta*. It was therefore doubly wrong to invoke Einstein as authority in support of the widespread misunderstanding that physical relativity meant that all frameworks, points of view, narrators, fragments of plot or thematic elements are created equal, that each of the polyphonic reports and contrasting perceptions is as valid or expedient as any other, and that all of these, when piled together or juxtaposed, *Rashomon*-like, somehow constitute the real truth. If anything, twentieth-century relativistic physics has taught the contrary: that under certain conditions we can extract from different reports, or even from the report originating in one frame properly identified, all the laws of physics, each applicable in any framework, each having therefore an invariant meaning, one that does not depend on the accident of which frame one inhabits. It is for this reason that, by comparison with classical physics, modern relativity is simple, universal, and, one may even say, "absolute." The cliché became, erroneously, that "everything is relative," whereas the whole point is that out of the vast flux one can distill the very opposite: "some things are invariant."

The cost of the terminological confusion has been so great that a brief elaboration on this point will be relevant. Partly because he

Yusef Bulos and Polly Draper as "The Professor" (Albert Einstein) and "The Actress" (Marilyn Monroe) in the play "Insignificance," made into a film in 1985. They are later joined by "The Senator" (Joe McCarthy) and "The Ball-player," (Monroe's former husband, Joe DiMaggio).

saw himself as a continuist rather than as an iconoclast, Einstein was reluctant to present this new work as a new theory. The term "relativity theory," which made the confusions in the long run more likely, was provided by Max Planck in 1907. For a time Einstein referred to it in print as the "so-called relativity theory," and until 1911 he avoided using the term altogether in the titles of his papers on the subject. In informal correspondence Einstein seemed happier with the term *Invariantentheorie*, which is of course much more true to its method and aim. How much nonsense we might have been spared if Einstein had adopted that term, even with all its shortcomings! To a correspondent who suggested just such a change, Einstein replied (letter to E. Zschimmer, 30 September 1921): "Now to the name relativity theory. I admit that it is unfortunate, and has given occasion to philosophical misunderstandings ... The description you proposed would perhaps be better; but I believe it would cause confusion to change the generally accepted name after

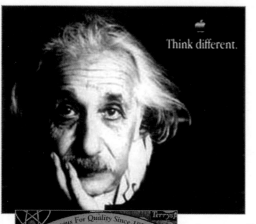

Einstein's face appears in many advertisements to this day—selling such diverse items as computers and knitting yarn.

all this time."

To come back to Einstein's careful disavowal of a substantive genetic link between modern art and relativity: far from abandoning the quest for it, his correspondent forged onward enthusiastically and published three such essays instead of one. Newton did not always fare better at the hands of eighteenth-century literati and divines who thought they were following in his foot steps. Poets rush in where scientists no longer fear to tread. And why not, if the apparent promises are so great? In April 1921, at the height of what Einstein on his first journey to the United States all too easily diagnosed as a pathological mass reac-tion, William Carlos Williams published a poem entitled "St. Francis Einstein of the Daffodils,"[12] containing such lines as "April Einstein /... has come among the daffodils / shouting / that flowers and men were created / relatively equal   Declaring simply that "relativity applies to everything"[13] and that "Relativity gives us the clue. So, again, mathematics comes to the rescue of the arts." Williams felt encouraged to adopt a new variable measure for his poems—calling it "a *relatively* stable foot, not a rigid one"—that proved of considerable influence on other poets.[14]

Williams was of course not alone. Robert Frost, Archibald MacLeish, E. E. Cummings, Ezra Pound, T. S. Eliot, and some of their disciples (and outside the English-speaking world, others such as Thomas Mann and Hermann Broch) referred directly to Einstein or to his work. Some were repelled by the vision thought to be opened by the new science, but there were at least as many who seemed to be in sympathy with Jean-Paul Sartre's remark that "the theory of relativity applies in full to the universe of fiction."[15] Perhaps the most cheerful of the attempts to harness science and literature to common purpose is Lawrence Durrell's entertaining set of novels, *The Alexandria Quartet*, of which its author says by way of preface: "Modern literature offers us no Unities, so I have turned to science and am trying to complete a four-decker novel whose form is based on the relativity proposition. Three sides of space and one of time constitute the soup-mix recipe of a continuum."[16] The intention is to use the properties of space and time as determining models for the structure of the book.

Durrell says "the first three parts…are to be deployed spatially… and are not linked in a serial form…The fourth part alone will represent time and be a true sequel."

For that alone one would not have had to wait for Einstein. But more seems to be hoped for; that, and the level of understanding, is indicated by the sayings of Pursewarden recorded in the novel. Pursewarden—meant to be one of the foremost writers in the English language, his death mask destined to be placed near those of Keats and Blake—is quoted as saying, "In the Space and Time marriage we have the greatest Boy meets Girl story of the age. To our great-grandchildren this will be as poetical a union as the ancient Greek marriage of Cupid and Psyche seems to us." More over, "the Relativity proposition was directly responsible for abstract painting, atonal music, and formless…literature."[17]

Throughout the novel it is evident that Durrell has taken the trouble to read up on relativity, but chiefly out of impressionistic popularizations such as *The Mysterious Universe* by James Jeans, even though Durrell readily confessed that "none of these attempts has been very successful."[18] There is something touching and, from the point of view of an intellectual historian, even a bit tragic about the attempt. In his study, *A Key to Modern British Poetry*, Durrell revealed his valid concern to show that as a result of "the far-reaching changes in man's ideas" about the outer and inner universe, "language has undergone a change in order to keep in line with cosmological inquiry (of which it forms a part)."[19] Yet on page after page the author demonstrates that he has been misled by the simplifications of H. V. Routh and Jeans; he believes that Rutherford and Soddy suggested that the "ultimate laws of nature were not simply causal at all," that "Einstein's theory joined up subject and object," that "so far as phenomena are concerned ... the uniformity of nature disappears," and so forth.[20] The terrible but clarifying remark of Wolfgang Pauli comes to mind, who said about a physical theory that seemed to him doomed: "It is not even wrong."

If I have spelled out some of the misunderstandings by which Einstein's work, for better or worse, has been thought to have found its way into contemporary culture, I had to do so in justice to any occasion associated with his name. But the examples of incorrect interpretation prepare us to appreciate that much more the correct ones. I should confess that my own favorite example of the successful transmutation of scientifically based conceptions in the writer's imagination is a novel, and a controversial one. William Faulkner's *The Sound and the Fury* is more like an earthquake than a book. Immediately on publication in 1929 it caused universal scandal; for example, not until judge Curtis Bok's decision in 1949 was this, among Faulkner's other novels, allowed to be sold in Philadelphia. On the surface it seems unlikely that this book—even a friendly reviewer characterized it as "designedly a silo of compressed sin"—has any resonance with the ideas of modern physics, by intent or otherwise. At the time he poured himself into the book, Faulkner was still almost unknown, largely self-taught, eking out a meager living as a carpenter, hunter, and coal carrier on the night shift of a power station, his desk the upturned wheelbarrow on which he would write while kneeling on the floor. Yet, even there, he was not isolated if he read even a small part of the flood of articles in newspapers, periodicals, and popular books in the 1920s dealing with the heady concepts of relativity theory—such as the time dilation experienced by a clock traveling through space, the necessity to recognize the meaninglessness of absolute time and space and the recent quantum physics, with its denial of the comforts of classical causality. Particularly in America, Einstein was quoted down to the level of local evening papers and *Popular Mechanics*, resulting in wide circulation of such haunting epigrams as his remark, made in exasperation to Max Born (1926), that "God does not throw dice." Could any of this have reached Faulkner?

In the second of the four chapters of *The Sound and the Fury* we follow Quentin Compton of Jefferson, Mississippi, as he lives through a day in June 1910. It is the end of his freshman year at college and the culmination of a short life wrenched by the degeneration and guilt, the fixations and tribal racism of his whole haunted family—from his father Jason, drinking himself to death, to his idiot brother Benjamin, whose forty-acre pasture has been sold to send Quentin to college. The only resource of human affection he has known came from black laborers and servants, although they have been kept in the centuries-old state of terror, ignorance, and obeisance. But the Comptons are doomed. As the day unfolds, Quentin moves toward the suicide he knows he will commit at midnight.

It is all too easy to discover theological and Freudian motives woven into the text, and one must not without provocation drag an author for cross-examination into the physics laboratory when he has already suffered through interrogations at the altar and on the couch. But Faulkner asked for it. Let me select here from a much more extensive body of evidence in the novel itself.

Quentin's last day on earth is a struggle against the flow of time. He attempts to stem the flow, first by deliberately breaking the cover glass of the pocket watch passed down to him from grandfather and father, then twisting off the hands of the watch, and then launching on seemingly random travel, by streetcar and on foot, across the whole city. His odyssey brings him to the shop of an ominous, cyclopean watch repairer. Quentin forbids him to tell him the time but asks if any of the watches in the shop window "are right." The answer he gets is "No." But wherever he then turns, all day and into the night, he encounters chimes, bells ringing the quarter hours, a factory whistle, a clock in the Unitarian steeple, the long, mournful sound of the train tracing its trajectory in space and time, "dying away, as though it were running through another month." Even his stomach is a kind of space-time metronome. "The business of eating inside of you space too space and time confused stomach says noon brain says eat o'clock All right I wonder what time it is what of it." Throughout, Quentin carries the blinded watch with him, the watch that never knew how to tell real time and cannot even tell relative time. But it is not dead: "I took out my watch and listened to it clicking away, not knowing it could not even lie."[21] And in the streetcar, the clicking away of time is audible to him only while the car has come to a stop.

Quentin has taken a physics course that freshman year and uses it to calculate how heavy the weights must be that he buys to help drown himself. It is, he says wryly to himself, "the only application of Harvard," and as he reflects on it: "The displacement of water is equal to the something of something. Reducto absurdum of all human experience, and two six-pound flat-irons weigh more than one tailor's goose. What a sinful waste, Dilsey would say. Benjy knew it when Damuddy died. He cried."

As midnight approaches, before he is ready to put his "hand on the light switch" for the last time,[22] Quentin is overcome by torment, caused by the shamed memory of his incestuous love for his sister Candace, by her loss, and by his own sense of loss even of the meaningfulness of that double betrayal. In anguish he remembers his father's terrible prediction after he had made his confession:

You cannot bear to think that some day it will no

The famous photo of Einstein sticking out his tongue was taken in 1951 and has been reproduced on everything from postage stamps (top) to key chains. At the time, Einstein was sitting between two people in the back seat of a car and had already endured a full day of picture-taking by UPI photographer Arthur Sasse, for a story on a day in the life of Einstein. He was probably quite annoyed by them and may have simply been expressing his desire for a little privacy.

273

longer hurt you like this now were getting at it. . . . you wont do it under these conditions it will be a gamble and the strange thing is that man who is conceived by accident and whose every breath is a fresh cast with dice already loaded against him will not face that final main which he knows beforehand he has assuredly to face without essaying expedients. . . .that would not deceive a child until some day in very disgust he risks everything on a single blind turn of a card no man ever does that under the first fury of despair or remorse or bereavement he does it only when he has realized that even the despair or remorse or bereavement is not particularly important to the dark dice man. . . . it is hard believing to think that a love or a sorrow is a bond purchased without design and which matures willynilly and is recalled without warning to be replaced by whatever issue the gods happen to be floating at the time.

This was not the God Newton had given to his time—Newton, of whom, just two centuries before Faulkner's soaring outcry, the poet James Thomson had sung in 1729 that "the heavens are all his own, from the wide role of whirling vortices, and circling spheres, to their great simplicity restored." Nor, of course, was it Einstein's God, a God whose laws of nature are both the testimony of His presence in the universe and the proof of its saving rationality. But this, it seems to me, defines the dilemma precisely. If the poet neither settles for the relief of half-understood analogies nor can advance to an honest understanding of the rational structure of that modern world picture, and if he is sufficiently sensitive to this impotency, he must rage against what is left him: time and space are then without meaning; so is the journey through them; so is grief itself, when the very gods are playing games of chance, and all the sound and the fury signify nothing. And this leads to recognizing the way out of the dilemma, at least for a few. At best, as in the case of Faulkner, this rage itself creates the energy needed for a grand fusion of the literary imagination with perhaps only dimly perceived scientific ideas. There are writers and artists of such inherent power that the ideas of science they may be using are dissolved, like all other externals, and rearranged in their own glowing alchemical cauldron.

It should not, after all, surprise us; it has always happened this way. Dante and Milton did not use the cosmological ideas of their time as tools to demarcate the allowed outline or content of their imaginative constructs. Those students of ours who, year after year, write us dutifully more or less the same essay, explaining the

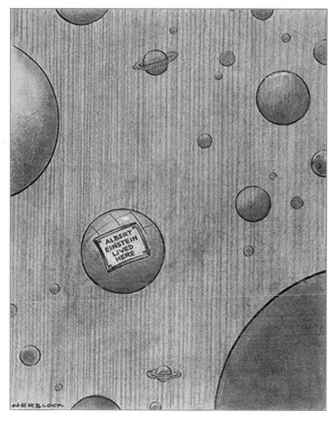

Cartoon by Herblock, published in the *Washington Post* some days after Einstein's death on April 18, 1955.

structure of the *Divine Comedy* or *Paradise Lost* by means of astronomy, geography, and the theory of optical phenomena—they may get the small points right, but they miss the big one, which is that the good poet is a poet surely because he can transcend rather than triangulate. In Faulkner, in Eliot's *The Waste Land*, in Woolf's *The Waves*, in Mann's *Magic Mountain*, it is futile to judge whether the traces of modern physics are good physics or bad, for these trace elements have been used in the making of a new alloy. It is one way of understanding Faulkner's remark on accepting his Nobel Prize in 1950: the task was "to make out of the material of the human spirit something which way not there before."[23] And insofar as an author *fails* to produce the feat of recrystallization, I suspect this lack would not be cured by more lessons on Minkowski's space-time, or Heisenberg's indeterminacy principle, or even thermodynamics, although these lessons could occasionally have a prophylactic effect that might not be without value.

Here we suddenly remember that, of course, the very same thing is true for scientists themselves. The most creative ones, almost by definition, do not build their constructs patiently by assembling blocks that have been precast by others and certified as sound. On the contrary, they too melt down the ready-made materials of science and recast them in a way that their contemporaries tend to think is outrageous. That is why Einstein's own work took so long to be appreciated even by his best fellow physicists, as I noted earlier. His physics looked to them like alchemy, not because they did not understand it at all, but because, in one sense, they understood it all too well. From their thematic perspective, Einstein's was anathema. Declaring, by simple postulation rather than by proof, Galilean relativity to be extended from mechanics to optics and all other branches of physics; dismissing the ether, the playground of most nineteenth-century physicists, in a peremptory half-sentence; depriving time intervals of inherent meaning; and other such outrages, all delivered in a casual, confident way in the first, short paper on relativity—those were violent and "illegitimate" distortions of science to almost every physicist. As for Einstein's new ideas on the quantum physics of light emission, Max Planck felt so embarrassed by it when he had to write Einstein a letter of recommendation seven years later that he asked that this work be overlooked in judging the otherwise promising young man.

Moreover, the process of transformation characterizes not only science itself and the flow of ideas from high science to high literature. It also works across the boundaries in other ways. The most obvious example is Einstein's importation into his early physics of an epistemology that he himself thought, with some enthusiasm, to be based on Ernst Mach's kind of positivism. Mach had begun to find him out on this point even while Einstein was still signing his letters to Mach as "Your devoted Student."

It seems clear to me that without this process of transformation, willing or unwilling, of ideas from science and from philosophy, physics would not have come into its twentieth-century form. (A similar statement may well be made for the cases of Copernicus,

The statue sculpted by Robert Berks in 1974 sits casually on the grounds of the National Academy of Sciences in Washington, DC. On the tablet are three equations that are central to Einstein's science.

Kepler, Galileo, and Newton.) The case of Einstein suggests, therefore, that the accomplishments of the major innovator—and not only in science—depend on his ability to persevere in four ways: by giving his loyalty primarily to his own belief system rather than to the current faith; by perceiving and exploiting the man-made nature and plasticity of human conceptions; by demonstrating eventually that the new unity he has promised does become lucid and convincing to lesser mortals active in his field—that he has it all "wrong" in the right way; and, in those rare cases, by even issuing ideas that lend themselves, quite apart from misuse and oversimplification, to further adaptation and transformation in the imagination of similarly exalted spirits who live on the other side of disciplinary boundaries.

It remains to deal with one more, somewhat different mechanism by which Einstein's imprint came to be felt far beyond his own field of primary attention: the power of his personal intervention on behalf of causes ranging from the establishment of a homeland for a persecuted people to his untiring efforts, over four decades, for peace and international security. In retrospect we can see that he had the skill, at strategic periods of history, to lend his ideas and prestige to the necessary work of a Chaim Weizmann or a Bertrand Russell. Even the most famous of these personal interventions, the call on President Roosevelt in 1939 to initiate a study whether the laws of nature allow anyone to produce an atomic weapon, was of that sort, although it has perhaps been misunderstood more widely than anything else Einstein did. He was, after all, correct in his perception that the Germans, who were pushing the world into a war, had all the skill and intention needed to start production of such a weapon if it was feasible. In fact, they had a head start, and but for some remarkable blunders, they might have fulfilled the justified fears,with incalculable consequences on the course of civilization.

To highlight these personal interactions, I select one as more or less paradigmatic of the considerable effect Einstein had even in some brief or seemingly casual discussion with the right person. The illustration is particularly apt because, in response to our invitation, the Swiss psychologist Jean Piaget had expressed the hope that he could come to speak about it at the Symposium. This proved not possible after all; but the record allows a certain amount of reconstruction. Piaget's work entered its most important phase with the publication in 1946 of *The Child's Conception of Time*. The book begins with a plain acknowledgment: "This work was promoted by a number of questions kindly suggested by Albert Einstein more than fifteen years ago [1928, at a meeting in Davos]... Is our intuitive grasp of time primitive or derived? Is it identical with our intuitive grasp of velocity? What if any bearing do these questions have on the genesis and development of the child's conception of time? Every year since then we have made a point of looking into these questions... The results [concerning time] are presented in this volume; those bearing on the child's conception of motion and speed are reserved for "a later work."[24] Throughout his later writings, Piaget remarks on this debt: "It was the author of the theory of relativity who suggested to us our work,"[25] or "Einstein once suggested we study the question from the psychological viewpoint and try to discover if there existed an intuition of speed independent of time."[26] In addition, Piaget refers explicitly to notions of relativity and other aspects of Einstein's work.[27]

Einstein came to have an immense correspondence with leaders in virtually every type of endeavor. Much of that has been preserved, thanks largely to the devoted labors and care of his longtime secretary, Miss Helen Dukas. These documents have now been catalogued for scholarly research, and there is hope that the task of actual publication of the letters, draft manuscripts, and other materials, which Princeton University Press has accepted, will begin soon. Working in this archive has made it already clear that it will not be difficult to reconstruct instances of the same sort as Einstein's brief but seminal interaction with Piaget.[28]

Looking back at the variety of ways in which Einstein came to impress the imagination of his time and ours, we can discern some rough categories, spread out, as it were, in a spectrum from left to right. At the center portion, corresponding to the largest intensity, one finds the widespread but unfocused and mostly uninformed fascination, manifested in a variety of ways, from enthusiastic mass gatherings to glimpse the man, to the outpouring of popularizations with good intentions, to responses that betray the vague discomfort aroused by the ideas. A good example of the last is an editorial entitled "A Mystic Universe" in the *New York Times* of 28 January 1928 (p. 14): "The new physics comes perilously close to proving what most of us cannot believe... Not even the old and much simpler Newtonian physics was comprehensible to the man on the street. To understand the new physics is apparently given only to the highest flight of mathematicians. We cannot grasp it by sequential thinking. We can only hope for dim enlightenment." The editorial writer then notes that the ever-changing scene in physics does offer some "comfort":

Earnest people who have considered it their duty to keep abreast of science by readapting their lives to the new physics may now safely wait until the results of the new discoveries have been fully tested out by time, harmonized and sifted down to a formula that will hold for a fair term of years. It would be a pity to develop an electronic marriage morality and find that the universe is after all ether, or to develop a wave code for fathers and children only to have it turn out that the family is determined not by waves but by particles. Arduous enough is the task of trying to understand the new physics, but there is no harm in trying. Reshaping life in accordance with the new physics is no use at all. Much better to wait for the new physics to reshape our lives for us as the Newtonian science did.

Similarly, in Tom Stoppard's play *Jumpers* a philosopher is heard to ask: "If one can no longer believe that a twelve-inch ruler is always a foot long, how can one be sure of relatively less certain propositions?"

Near this position, as we said, are the enthusiastic misapplications, usually achieved by an illicit shortcut of meaning from, say, the true statement that the operational definition of length is "framework" dependent, to the invalid deduction that mental phenomena in a human observer have thereby been introduced into the very definitions of physical science. (To be sure, the layman has not always been served by the explanation on this point given by the scientists themselves; for example, in such pronouncements as "the object of research is no longer nature in itself but rather nature exposed to man's questioning, and to this extent man here also meets himself.") The irony here is that the first lessons we learned from relativity physics were that short circuits in signification must be avoided, for they were just what burdened down classical physics, and that attention must be paid as never before to the meaning of the terms we use.

When we now glance further toward the left, or blue, end of the spectrum, the expressions of resignation or futility become more explicit. Indeed, among some of the most serious intellectuals there seems, on this point, a sense of despair. By the very nature of their deep motivation they must feel most alienated from a universe whose scientific description they can hardly hope to understand except in a superficial way. The much-admired humanistic scholar Lionel Trilling spoke for many when he stated the dilemma frankly and eloquently:

The operative conceptions [of science] are alien to the mass of educated persons. They generate no cosmic speculation, they do not engage emotion or challenge imagination. Our poets are indifferent to them…

This exclusion of most of us from the mode of thought which is habitually said to be the characteristic achievement of the modern age is bound to be experienced as a wound given to our intellectual self-esteem. About this humiliation we all agree to be silent, but can we doubt that it has its consequences, that it introduced into the life of mind a significant element of dubiety and alienation which must be taken into account in any estimate that is made of the present fortunes of mind?

Einstein, who had intended originally to become a science teacher, came to understand this syndrome, and the obligation it put on him. He devoted a good deal of time to popularization of his own. His avowed aim was to simplify short of distortion. In addition to a large number of essays and lectures, he wrote, and repeatedly updated, a short book on relativity that he promised in the very title to be *gemeinverständlich*.[31] It is, however, overly condensed for most nonscientific beginners. Later, Einstein collaborated with Leopold Infeld in a second attempt to reach out to the population at large by means of a book-length treatment of modern physics. As the preface acknowledged, the authors no longer attempted "a systematic course in elementary facts and theories." Rather, they aimed at a historical account of how the ideas of relativity and quanta entered science, "to give some idea of the eternal struggle of the inventive human mind for a fuller understanding of the laws governing physical phenomena."[32] In fact, there is to this day no generally agreed source, the reading of which by itself will bring a large fraction of its nonscientific audiences to a sound enough understanding of these ideas, even for those who truly want to attain it and are willing to pay close attention. I believe it is a fact of great consequence that it takes a much larger effort, and one starting earlier than most people undertake. To make matters worse, so little has been found out about how scientific literacy is achieved or resisted that little blame can be spun off on the would-be students, young or old.

Going now further along the spectrum in the same direction, we encounter outright hostility and opposition to Einstein's work, either on scientific or on ideological grounds. Almost all scientists, even those initially quite reluctant, became eventually at least reconciled, save (to this day) for Einstein's famous refusal to regard the statistical interpretation as fundamental. Even on that point, the last word may not have been said. On the other hand, the opposition to Einstein's work on grounds other than scientific has had a longer history. Thus, a number of studies now exist that show the lengths to which various totalitarian groups, for various reasons, felt compelled to go in their attacks.

Turning now to the other, more "positive" half of the spectrum, we see there the gradual acceptance and elaboration of Einstein's work within the corpus of physical science; its penetration into technology (largely unmarked) and into the more thoughtful philosophies of science; Einstein's effect through his personal intervention, causing some historic redirections of research; and its passage into the scientific world picture of our time, as it tries to achieve a unification that eluded Einstein. And beyond that, at the end of the spectrum, where the number of cases is small but the color deep and vibrant, we perceive the examples of creative transformation beyond science. Those are the works of the few who found that scientific ideas, or rather *metaphors* embodying such ideas, released in them a fruitful response with an authenticity of its own, far removed from textbook physics.

This last is the oldest and surely still the most puzzling interplay between the science and the rest of culture. Evidently, the mediation occurs through a sharing of an analogy or metaphor—irresistible, despite the dangers inherent in the obvious differences or discontinuities. We know that such a process exists, because any major work of science itself, in its nascent phase, is connected analogically rather than fully logically, both with the historic past in that science and with its supporting data. The scientist's proposal may fit the facts of nature as a glove fits a hand, but the glove does not uniquely imply the hand, nor the hand the glove.

Einstein spoke insistently over the decades about the need to recognize the existence of such a discontinuity, one that in his early scientific papers asserted itself first in his audacious method of postulation. In essay after essay, he tried to make the same point, even though it had little effect on the then reigning positivism. Typical are the phrases in his Herbert Spencer Lecture of 1933.[33] The rational and empirical components of human knowledge stand in "eternal antithesis," for "propositions arrived at by purely logical means are completely empty as regards reality." In this sense, the "fundamentals of scientific theory," being initially free inventions of the human mind, are of "purely fictional character." The phenomenic-analytic dichotomy makes it inherently impossible to claim that the principles of a theory are "deduced from experience" by "abstraction," that is, by logically complete claims of argument. As he put it soon afterwards, the relation between sense experience and concept "is analogous not to that of soup to beef, but rather that of check number to overcoat."[34]

If this holds for the creative act in science itself, we should hardly be surprised to find the claim to be extended to more humanistic enterprises. The test, in both instances, is of course whether the freely invented check token yields the intended use of a suitable overcoat. The existence of both splendid scientific theories and splendid products of the humanistic imagination shows that despite all their other differences, they share the ability to build on fundamentals of a "purely fictional character." And even the respective fundamentals, despite all their differences, can share a common origin. That is to say, at a given time the cultural pool contains a variety of themata and metaphors, and some of these have a plasticity and applicability far beyond their initial provenance. The innovator, whether a scientist or not, necessarily dips into this pool for his fundamental notions and in turn may occasionally deposit into it new or modified themata and metaphors of general power.

Examples of such science-shaped metaphors, each of these by no means a "fact" of the external world, yet revealing immense explanatory energy, are easy to find: Newton's "innate force of matter (*vis insita*)" and the Newtonian clockwork universe; Faraday's space-filling electric and magnetic lines of force; Niels Bohr's examples of complementarity in physics and in daily life, Mendeleev's neat tableau setting for the families of elements and Rutherford's long parent-daughter-granddaughter chains of decaying atoms, Minkowski's space-time "World," of which our perceptible space and time are like shadows playing on the wall of Plato's cave; and of course the imaginative scenes Einstein referred to—the traveler along the light beam, the calm experimenter in the freely falling elevator, the dark, dice-playing God, the closed but unbounded cosmos, the Holy Grail of complete unification of all forces of nature. So it continues in science.

The allegorical use of such conceptions may, as we have noted, help to shape works of authenticity outside the sciences. And the process works both ways. Thus Niels Bohr acknowledged that his reading in Kierkegaard and William James helped him to the imaginative leap embodied in his physics, Einstein stressed the influence on his early scientific thinking of the philosophical tracts of that period, and Heisenberg noted the stimulus of Plato's *Timeaus* read in his school years. No matter if such "extraneous" elements are eventually suppressed or forgotten, or even have to be overcome; at an early point they can encourage the mind's struggle.

We conclude, then—and, I trust, in the spirit of Einstein—that in pursuing the evident and documentable cases of "impact" of one person or field on another, we have been led to a more mysterious fact, namely, the mutual adaptation and resonance of the innovative mind with portions of the total set of metaphors current at a given time. The philosopher José Ortega y Gasset was one of those who struggled with this idea. In 1921–1922, evidently caught up by the rise of the new physics, he began an essay on "The Historical Significance of the Theory of Einstein."[35] There he remarked quite correctly that the most relevant issue was not that the triumph of the theory will influence the spirit of mankind by imposing on it the adoption of a definite route. What is really interesting is the inverse proposition: the spirit of man has set out, of its own accord, upon a definite route, and it has therefore been possible for the theory of relativity to be born and to triumph. The more subtle and technical ideas are, the more remote they seem from the ordinary preoccupations of men, the more authentically they denote the profound variations produced in the historical mind of humanity.[36]

But to this day, attempts to go much beyond that point, it seems to me, have not been very successful. Perhaps the tantalizing task will have borne fruit by the time scholars meet for discussion at Einstein's Second Centennial Symposium.

Notes

[Editor's note: Footnotes 1 and 2 do not appear because material relevant only to *Albert Einstein, Historical and Cultural Perspectives*, from which this introduction was taken, have been edited from this selection.]

3. Daniel N. Lapedes, ed., *McGraw-Hill Dictionary of Scientific and Technical Terms*, 2d ed., New York, McGraw-Hill, 1978, pp. 512–513. As another measure in the continuing, albeit sometimes only ritualistic, reference made in the ongoing research literature to Einstein's publications, Eugene Garfield has found that during the period 1961-1975 the serious scientific journals in total carried no less than 40 million citations to previously published articles. Of these, 58 cited articles stand out by virtue of having been published before 1930 and cited over 100 times each; and among these 58 enduring classics, ranging from astronomy and physics to biomedicine and psychology, 4 are Einstein's. See E. Garfield, *Current Contents* 21 (1976), p. 59.

4. Quoted in Ernst Lechner, *Physikalische Weltbilder*, Leipzig, Theodore Thomas Verlag, [1912], p. 84.

5. Quoted in the *New York Times*, 3, 4, and 5 April 1921.

6. For discussion and documentation, see G. Holton, "Einstein's Search for the 'Weltbild,' *Proceedings of the American Philosophical Society* 125, no. 1 (1981), pp. 1–15.

7. Quoted from J. J. Thomson, *Reflections and Recollections*, London, G. Bell and Sons, Ltd., 1936, p. 431 italics in original). See also Philipp Frank, *Einstein: His Life and Times*, New York, Alfred A. Knopf, 1947, p. 190. Frank's book is one of the good sources for documentation on the reception and rejection of Einstein's theories by various religions and philosophic and political systems, ranging from Cardinal O'Connell's assessment that "those theories (Einstein's as well as Darwin's) became

outmoded because they were mainly materialistic and therefore unable to stand the test of time" (p. 262), to the attempt of a Nazi scientist to overcome his aversion sufficiently to "recommend Einstein's theory of relativity to National Socialists" as a weapon in the fight against "materialistic philosophy" (p. 351).

8. These articles, and excerpts from some other publications dealing with the influence of Einstein's work, have been gathered in L. Pearce Williams, ed., *Relativity Theory: Its Origins and Impact on Modern Thought*, New York, John Wiley and Sons, 1968. It is a useful compendium, and I am indebted to it for a number of illustrations to be referred to below.

9. John Passmore, *A Hundred Years of Philosophy*, rev. ed., New York, Basic Books, 1966, p. 332.

10. As reported by Paul M. LaPorte, "Cubism and Relativity, with a Letter of Albert Einstein," *Art Journal* 25, no. 3 (1966), p. 246.

11. Ibid. See also C. H. Waddington, *Behind Appearances: A Study of the Relations between Painting and the Natural Sciences in this Century*, Edinburgh, Edinburgh University Press, 1969, Cambridge, Mass., MIT Press, 1970, pp. 9-39. At the Jerusalem Symposium, Professor Meyer Schapiro presented an extensive and devastating critique of the frequently proposed relation between modern physics and modern art.

12. *Contact* 4 (1923), p. 3. I am indebted to Carol Donley's draft paper, "Einstein, Too, Demands the Muse" for this lead and others in the following paragraphs.

13. Selected Essays of William Carlos Williams, New York, Random House, 1954, p. 283.

14. *Ibid.*, p. 340.

15. J.-P. Sartre, "François Mauriac and Freedom," in *Literary and Philosophical Essays*, New York, Criterion Books, 1955, p. 23.

16. Lawrence Durrell, *Balthazar*, New York, E. P. Dutton, 1958, Author's Note, p. 9.

17. Ibid., p. 142.

18. For a good review of details, to which l am indebted, see Alfred M. Bork, "Durrell and Relativity," *Centennial Review* 7 (1963), pp. 191–203.

19. L. Durrell, *A Key to Modern British Poetry*, Norman, University of Oklahoma Press, 1952, p. 48

20. Ibid., pp. 25, 26, 29.

21. All quotations are from William Faulkner, *The Sound and the Fury*, London, Chatto and Windus, 1961, pp. 81-177. I thank Dr. J. M. Johnson for a draft copy of her interesting essay, "Albert Einstein and William Faulkner," and have profited from some passages even while differing with others.

22. The chapter is shot through with references to light, light rays, even to travel "down the long and lonely light rays."

23. In *Les Prix Nobel* in 1950, Stockholm, Imprimierie Royale, 1951, p. 71

24. Jean Piaget, *The Child's Conception of Time*, New York, Ballantine Books, 1971, p. vii.

25. Jean Piaget, *Genetic Epistemology*, New York, Columbia University Press, 1970, p. 69, see also p. 7.

26. Jean Piaget, *Psychology and Epistemology*, New York, Grossman Publishers, 1971, p. 82 see also pp. 10, 110. A similar statement is to be found in Piaget's *Six Psychological Studies*, New York, Vintage Books, 1968, p. 85.

27. For example, Jean Piaget, with Barbel Inhelder, *The Child's Conception of Space*, New York, W. W. Norton, 1967, pp. 232–233 *The Child's Conception of Time*, London, Routledge and Kegan Paul, 1969, pp. 305–306, *Biology and Knowledge*, Chicago, University of Chicago Press, 1971, pp. 308, 337, 341–342. I wish to express my thanks to Dr. Katherine Sopka for help in tracing these references.

28. All this is quite apart from the role of model or personal culture hero that Einstein played in the lives of a great many individuals whom he never met. Indeed, several participants at the Symposium reminisced that they had tried to correspond with Einstein at a point in their lives when they were deciding to become scientists. To this day, his picture can be found in wide circulation and in the most unlikely places, from the tee shirt of a student in high school to the workbench of a cobbler in Shanghai.

29. Werner Heisenberg, "The Representation of Nature in Contemporary Physics," *Daedalus*, Summer 1958, pp. 103–105.

30. Lionel Trilling, *Mind in the Modern World: The 1972 Jefferson Lecture in the Humanities*, New York, Viking Press, 1972, pp. 13-14.

31. Albert Einstein, *Uber die spezielle und die allgemeine Relativitätstheorie gemeinver-ständlich*, Braunschweig, Vieweg, 1917. It was often translated and to this day is perhaps his most widely known work.

32. Albert Einstein and Leopold Infeld, *The Evolution of Physics*, New York, Simon and Schuster, 1938.

33. Albert Einstein, *On the Method of Theoretical Physics*, Oxford, Clarendon Press, 1933.

34. Albert Einstein, "Physics and Reality," *Journal of the Franklin Institute* 221, 1936, pp. 349–382.

35. First published in English in 1933, in his *The Modern Theme*, New York, W. W. Norton.

36. Ibid., pp. 135-136.

In a popular poster featuring Einstein, the association with the famous equation certifies the identification.

That Famous Equation and You

by Brian Greene

During the summer of 1905, while fulfilling his duties in the patent office in Bern, Switzerland, Albert Einstein was fiddling with a tantalizing outcome of the special theory of relativity he'd published in June. His new insight, at once simple and startling, led him to wonder whether "the Lord might be laughing and leading me around by the nose."

But by September, confident in the result, Einstein wrote a three-page supplement to the June paper, publishing perhaps the most profound afterthought in the history of science. A hundred years ago this month, the final equation of his short article gave the world $E = mc^2$.

In the century since, $E = mc^2$ has become the most recognized icon of the modern scientific era. Yet for all its symbolic worth, the equation's intimate presence in everyday life goes largely unnoticed. There is nothing you can do, not a move you can make, not a thought you can have, that doesn't tap directly into $E = mc^2$. Einstein's equation is constantly at work, providing an unseen hand that shapes the world into its familiar form. It's an equation that tells of matter, energy and a remarkable bridge between them.

Before $E = mc^2$, scientists described matter using two distinct attributes: how much the matter weighed (its mass) and how much change the mater could exert on its environment (its energy). A 19th-century physicist would say that a baseball resting on the ground has the same mass as a baseball speeding along at 100 miles per hour. The key difference between the two balls, the physicist would emphasize, is that the fast-moving baseball has more energy: if sent ricocheting through a china shop, for example, it would surely break more dishes than the ball at rest. And once the moving ball has done its damage and stopped, the 19th-century physicist would say that it has exhausted its capacity for exerting change and hence contains no energy.

After $E = mc^2$, scientists realized that this reasoning, however sensible it once seemed, was deeply flawed. Mass and energy are not distinct. They are the same basic stuff packaged in forms that make them appear different. Just as solid ice can melt into liquid water, Einstein showed, mass is a frozen form of energy that can be converted into the more familiar energy of motion. The amount of energy (E) produced by the conversion is given by his formula: multiply the amount of mass converted (m) by the speed of light squared (c^2). Since the speed of light is a few hundred million meters per second (fast enough to travel around the earth seven times in a single second), c^2, in these familiar units, is quite a huge number, about 100,000,000,000,000,000.

A little bit of mass can thus yield enormous energy. The destruction of Hiroshima and Nagasaki was fueled by converting less than an ounce of matter into energy; the energy consumed by New York City in a month is less than that contained in the book that you're holding. Far from having no energy, the baseball that has come to rest on the china shops floor contains enough energy to keep an average car running continuously at 65 m.p.h. for about 5,000 years.

Before 1905, the common view of energy and matter thus resembled a man carrying around his money in a box of solid gold. After the man spends his last dollar, he thinks he's broke. But then someone alerts him to his miscalculation; a substantial part of his wealth is not what's in the box, but the box itself. Similarly, until Einstein's insight, everyone was aware that matter, by virtue of its motion or position, could possess energy. What everyone missed is the enormous energetic wealth contained in mass itself.

The standard illustrations of Einstein's equation—bombs and power stations—have perpetuated a belief that $E = mc^2$ has a special association with nuclear reactions and is thus removed from ordinary activity.

This isn't true. When you drive your car, $E = mc^2$ is at work. As the engine burns gasoline to produce energy in the form of motion, it does so by converting some of the gasoline's mass into energy, in accord with Einstein's formula. When you use your MP3 player, $E = mc^2$ is at work. As the player drains the battery to produce energy in the form of sound waves, it does so by converting some of the battery's mass into energy, as dictated by Einstein's formula. As you read this text, $E = mc^2$ is at work. The processes in the eye and brain, underlying perception and thought, rely on chemical reactions that interchange mass and energy, once again at accord with Einstein's formula.

The point is that although $E = mc^2$ expresses the interchangeability of mass and energy, it doesn't single out any particular reaction for executing the conversion.

The distinguishing feature of nuclear reactions, compared with the chemical reactions involved in burning gasoline or running a battery, is that they generate less waste and thus produce more energy—by a factor of roughly a million. And when it comes to energy, a factor of a million justifiably commands attention. But don't let the spectacle of $E = mc^2$ in nuclear reactions inure you to its calmer but thoroughly pervasive incarnations in everyday life.

That's the content of Einstein's discovery. Why is it true?

Einstein's derivation of $E = mc^2$ was wholly mathematical. I know his derivation, as does just about anyone who has taken a course in modern physics. Nevertheless, I consider my understanding of a result incomplete if I rely solely on the math. Instead, I've found that thorough understanding requires a mental image an analogy or a story—that may sacrifice some precision but captures the essence of the result.

Here's a story for $E = mc^2$. Two equally strong and skilled jousters, riding identical horses and gripping identical (blunt) lances, head toward each other at an identical speed. As they pass, each thrusts his lance across his breastplate toward his opponent, slamming blunt end into blunt end. Because they're equally matched, neither lance pushes farther than the other, and so the referee calls it a draw.

This story contains the essence of Einstein's discovery. Let me explain.

Einstein's first relativity paper, the one in June 1905, shattered the idea that time elapses identically for everyone. Instead, Einstein showed that if from your perspective someone is moving, you will see time elapsing slower for him than it does for you. Everything he does—sipping his coffee, turning his head, blinking his eyes—will appear in slow motion.

This is hard to grasp because at everyday speeds the slowing is less than one part in a trillion and is thus imperceptibly small. Even so, using extraordinarily precise atomic clocks, scientists have repeatedly confirmed that it happens just as Einstein predicted. If we lived in a world where things routinely traveled near the speed of light, the slowing of time would be obvious.

Let's see what the slowing of time means for the joust. To do so, think about the story not from the perspective of the referee, but instead imagine you are one of the jousters. From your perspective, it is your opponent—getting ever closer-who is moving. Imagine that he is approaching at nearly the speed of light so the slowing of all his movements—readying his joust, tightening his face—is obvious. When he shoves his lance toward you in slow motion, you naturally think he's no match for your swifter thrust; you expect to win. Yet we already know the outcome. The referee calls it a draw and no matter how strange relativity is, it can't change a draw into a win.

After the match, you naturally wonder how your opponent's slowly thrusted-lance hit with the same force as your own. There's only one answer. The force with which something hits depends not only on its speed but also on its mass. That's why you don't fear getting hit by a fast-moving Ping-Pong ball (tiny mass) but you do fear getting bit by a fast-moving Mack truck (big mass). Thus, the only explanation for how the slowly thrust lance hit with the same force as your own is that its more massive.

This is astonishing. The lances are identically constructed. Yet you conclude that one of them—the one that from your point of view is in motion, being carried toward you by your opponent on his galloping horse—is more massive than the other. That's the essence of Einstein's discovery. Energy of motion contributes to an object's mass.

As with the slowing of time, this is unfamiliar because at everyday speeds the effect is imperceptibly tiny. But if, from your viewpoint, your opponent were to approach at 99.99999999 percent of the speed of light, his lance would be about 70,000 times more massive than yours. Luckily, his thrusting speed would be 70,000 times slower than yours, and so the resulting force would equal your own.

Once Einstein realized that mass and energy were convertible, getting the exact formula relating them—$E = mc^2$—was a fairly basic exercise, requiring nothing-more than high school algebra. His genius was not in the math; it was in his ability to see beyond centuries of misunderstanding and recognize that there was a connection between mass and energy at all.

A little known fact about Einstein's September 1905 paper is that he didn't actually write $E = mc^2$; he wrote the mathematically equivalent (though less euphonious) $m = E/c^2$, placing greater emphasis on creating mass from energy (as in the joust) than on creating energy from mass (as in nuclear weapons and power stations).

Over the last couple of decades, this less familiar reading of Einstein's equation has helped physicists explain why everything ever encountered has the mass that it does. Experiments have shown that the subatomic particles making up matter-have almost no mass of their own. But because of their motions and interactions inside of atoms, these particles contain substantial energy—and it's this energy that gives matter its heft. Take away Einstein's equation, and matter loses its mass. You can't get much more pervasive than that.

Einstein and the Bomb

In January 1939, Lise Meitner and her nephew Otto Frisch published a paper explaining Otto Hahn's recent experiments in Berlin as nuclear fission reactions. Nuclear fission became the hot topic in physics research, but one man, Leo Szilard, seemed to understand the devastating potential of this discovery.

Unlike most physicists at that time Szilard believed in the possibility of the atomic bomb. In July 1939, he and fellow Hungarian Eugene Wigner visited Einstein who was vacationing on Long Island. They conveyed to Einstein their concern that Germany might be developing an atomic bomb. Einstein, originally a pacifist, now felt that the force emanating from Germany could

Einstein with Leo Szilard, re-enacting the signing of the 1939 "A-bomb" letter to Franklin Delano Roosevelt.

only be met with force. Szilard asked him to write a letter to his friend the Belgian Queen Mother, asking her to prevent the Nazis from gaining control of the vast quantities of uranium in the Belgian Congo. Einstein agreed but proposed a letter should first be sent to the Belgian ambassador.

Szilard mentioned his plan to Alexander Sachs, unofficial advisor to President Franklin D. Roosevelt. Sachs recommended that Einstein write a letter of warning to the president. The letter, dated August 2, 1939, was eventually delivered by Sachs on October 11, just after World War II began. The letter included the disturbing news that "Germany has actually stopped the sale of uranium from the Czechoslovakian mines which she has taken over."

The president appointed the Briggs Committee to study uranium chain reactions but the committee moved slowly. After six months, Einstein signed another letter to the president with an implied threat that Szilard would publish the details of how to set

up a chain reaction in uranium, unless more action was taken by the American government. Within two months the Briggs committee was reorganized as part of the National Defense Research Committee.

In December 1941, Vannevar Bush asked Einstein to work on the theoretical problem of separating isotopes by gaseous diffusion. By this time, it was clear that only the rare uranium 235 nuclei could be split by fission, but uranium ore consists of mostly U-238. Einstein worked on the problem briefly, probably without knowing the proposed application, but Bush did not trust Einstein enough to involve him further.

A large-scale U.S. atomic bomb program finally began on December 6, 1941, the day before the Pearl Harbor attack. After the nuclear explosions in Japan, Szilard was devastated and turned to molecular biology, never to study physics again. Einstein spent the rest of his life promoting peaceful uses of atomic energy. — PH

Einstein's August 2, 1939 letter to Franklin D. Roosevelt, a copy of which was on FDR's desk when the president died.

Its singular fame notwithstanding, $E = mc^2$ fits into the pattern of work and discovery that Einstein pursued with relentless passion throughout his entire life. Einstein believed that deep truths about the workings of the universe would always be "as simple as possible, but no simpler." And in his view, simplicity was epitomized by unifying concepts—like matter and energy—previously deemed separate. In 1916, Einstein simplified our understanding even further by combining gravity with space, time, matter and energy in his General Theory of Relativity. For my money, this is the most beautiful scientific synthesis ever-achieved.

With these successes. Einstein's belief in unification grew ever stronger. But the sword of his success was double-edged. It allowed him to dream of a single theory encompassing all of nature's laws, but led him to expect that the methods that had worked so well for him in the past would continue to work for him in the future.

It wasn't to be. For the better part of his last 30 years, Einstein pursued the "unified theory," but it stubbornly remained beyond his grasp. As the years passed, he became increasingly isolated; mainstream physics was concerned with prying apart the atom and paid little attention to Einstein's grandiose quest. In a 1942 letter, Einstein described himself as having become a "a lonely old man who is displayed now and then as a curiosity because he doesn't wear socks."

Today, Einstein's quest for unification is no curiosity—it is the driving force for-many physicists of my generation. No one knows how close we've gotten. Maybe the unified theory will elude us just as it dodged Einstein last century. Or maybe the new approaches being developed by contemporary physics will finally prevail, giving us the ultimate explanation of the cosmos. Without a unified theory it's hard to imagine we will ever resolve the deepest of all mysteries—how the universe began—so the stakes are high and the motivation strong.

But even if our science proves unable to determine the origin of the universe, recent progress has already established beyond any doubt that a fraction of a second after creation (however that happened), the universe was filled with tremendous energy in the form of wildly moving exotic particles and radiation. Within a few minutes, this energy employed $E = mc^2$ to transform itself into more familiar matter—the simplest atoms—which, in the course of about a billion years, clumped into planets and stars.

During the 13 billion years that have followed, stars have used $E = mc^2$ to transform their mass back into energy in the form of heat and light; about five billion years ago, our closest star—the sun— began to shine, and the heat and light generated was essential to theformation of life on our planet. If prevailing theory and observations are correct, the conversion of matter to energy throughout the cosmos, mediated by stars, black holes and various forms of radioactive decay, will continue unabated.

In the far, far future, essentially all matter will have returned to energy. But because of the enormous expansion of space, this energy will be spread so thinly that it will hardly ever convert back to even the lightest particles of matter. Instead, a faint mist of light will fall for eternity through an ever colder and quieter cosmos.

The guiding hand of Einstein's $E = mc^2$ will have finally come to rest.

Albert Einstein and David Ben-Gurion in Princeton, New Jersey, 1951.
Einstein took seriously Ben-Gurion's offer to become President of Israel,
but he declined because of a realistic assessment of his own political talents.

Einstein and the Jews

—————————◦—————————

by Abraham Pais

In his early years Einstein had already been aware of being a Jew and of the existence of anti-Semitism, but had not paid particular attention to his descent... [H]is first exposure to Zionism dated from the years just after the First World War... I shall now follow his later public comments until, literally, the end of his life.[1]

1926. From 1921 onwards, Arabs in Palestine frequently engaged in anti-Jewish riots, some of them bloody. This was one of the reasons for occasional proposals to colonize Jews elsewhere. In 1926 Einstein commented as follows on the idea to do so in Russia:

> Although I believe that it is only in Palestine that work of lasting value can be achieved and that everything that is done in the Diaspora countries is only a palliative, I nevertheless hold that the efforts which are being made to colonize Jews in Russia must not be opposed because they aim at assisting thousands of Jews whom Palestine cannot immediately absorb. On this ground, these efforts seem to me worthy of support. I do not, therefore, believe that the money which is expended in Russia on Jewish colonization is being wasted. Whether the necessary guarantees exist for the success of this colonization work, I cannot say without first having been on the spot. But if the colonization is successful, it will ultimately be of benefit also to us because it will mean a strengthening of the Jewish people and every effort, every factor, which strengthens our people, even if only morally or indirectly, is justified.

1929. In 1929 there occurred a series of extremely violent attacks by Arabs against Jewish settlers. To add insult to injury, certain British circles used the occasion for making anti-Zionist statements. Einstein reacted with an indignant letter to a British newspaper which included these lines:

> Arab mobs, organised and fanaticised by political intriguers working on the religious fury of the ignorant, attacked scattered Jewish settlements and murdered and plundered wherever no resistance was offered. In Hebron, the inmates of a rabbinical college, innocent youths who had never handled weapons in their lives, were butchered in cold blood; in Safed the same fate befell aged rabbis and their wives and children. Recently some Arabs raided a Jewish settlement where the pathetic remnants of the great Russian pogroms had found a haven of refuge. Is it not then amazing that an orgy of such primitive brutality upon a peaceful population has been utilised by a certain section of the British press for a campaign of propaganda directed, not against the authors and instigators of these brutalities, but against their victims?

1930. In January Einstein addressed the Arabs directly:

> One who, like myself, has cherished for many years the conviction that the humanity of the future must be built up on an intimate community of the nations, and that aggressive nationalism must be conquered, can see a future for Palestine only on the basis of peaceful co-operation between the two peoples who are at home in the country. For this reason I should have expected that the great Arab people will show a truer appreciation of the need which the Jews feel to rebuild their national home in the ancient seat of Judaism; I should have expected that by common effort ways and means would be found to render possible an extensive Jewish settlement in the country. I am convinced that the devotion of the Jewish people to Palestine will benefit all the inhabitants of the country, not only materially, but also culturally and nationally. I believe that the Arab renaissance in the vast expanse of territory now occupied by the Arabs stands only to gain from Jewish sympathy. I should welcome the creation of an opportunity for absolutely free and frank discussion of these possibilities, for I believe that the two great Semitic peoples, each of which has in its way contributed something of lasting value to the civilisation of the

West, may have a great future in common, and that instead of facing each other with barren enmity and mutual distrust, they should support each other's national and cultural endeavors, and should seek the possibility of sympathetic cooperation. I think that those who are not actively engaged in politics should above all contribute to the creation of this atmosphere of confidence.

I deplore the tragic events of last August not only because they revealed human nature in its lowest aspects, but also because they have estranged the two peoples and have made it temporarily more difficult for them to approach one another. But come together they must, in spite of all.

September. In an address to the first international congress of Palestine workers, held in Berlin: "It does not matter how many Jews are in Palestine but it does matter what they produce there. That should be something the Jews of the whole world can point to as ideal creative work and with which they can identify themselves."

October. In a speech delivered on 29 October at the Savoy Hotel in London, Einstein said in part:

The position of our scattered Jewish community is a moral barometer for the political world. For what surer index of political morality and respect for justice can there be than the attitude of the nations toward a defenceless minority, whose peculiarity lies in their preservation of an ancient cultural tradition?

This barometer is low at the present moment, as we are painfully aware from the way we are treated. But it is this very lowness that confirms me in the conviction that it is our duty to preserve and consolidate our community. Embedded in the tradition of the Jewish people there is a love of justice and reason which must continue to work for the good of all nations now and in the future...

Remember that difficulties and obstacles are a valuable source of health and strength to any society. We should not have survived for thousands of years as a community if our bed had been of roses; of that I am quite sure.

1934. April. Einstein gave his opinion on revisionism, a right-wing Jewish movement that advocated and executed terrorist actions aiming at the realization of a

Jewish state in Palestine:

> Revisionism is the modern embodiment of those harmful forces which Moses with foresight sought to banish when he formulated his model codes of social law. The secret of our apparently inexhaustible vitality lies in our strong traditions of social justice and of modest service to our immediate community and society as a whole. The Jews must beware of viewing Palestine merely as a place of refuge.

Also in 1934 Einstein issued a statement entitled "Let's not forget":

> If we as Jews can learn anything from these politically sad times, it is the fact that destiny has bound us together, a fact which, in times of quiet and security, we often so easily and gladly forget. We are accustomed to lay too much emphasis on the differences that divide the Jews of different lands and different religious views. And we forget often that it is the concern of every Jew, when anywhere the Jew is hated and treated unjustly, when politicians with flexible consciences set into motion against us the old prejudices, originally religious, in order to concoct political schemes at our expense. It concerns every one of us because such diseases and psychotic disturbances of the folk-soul are not stopped by oceans and national borders, but act precisely like economic crises and epidemics.

1935. On 24 March Einstein spoke at a Purim dinner of the German-Jewish club in New York: "There are no German Jews, there are no Russian Jews, there are no American Jews. Their only difference is their daily language. There are in fact only Jews."

April. At a Passover celebration in New York, Einstein again condemned revisionism and urged Jewish-Arab unity. He reminded the audience that the founders of the Zionist movement worked for the traditional ideals of justice and the selfless love of mankind.

June.

> The intellectual decline brought on by shallow materialism is a far greater menace to the survival of the Jew than the numerous external foes who threaten its existence with violence. We must never forget

that through all the severe afflictions for twenty centuries our ancestors found consolation, refuge and strength in the fostering of our spiritual traditions.

1938. January. Einstein sent greetings to the National Council for Jewish Women, in session in Pittsburgh:

Mutual assistance is, God be thanked, our one weapon in our bitter struggle for existence. Weakened through dispersion in countless factions, we, nonetheless, remain united through the fairest of all duties—the duty of unselfish mutual aid. Never has Jewry denied itself to the demands of this duty.

April. Einstein spoke in German at the Seder by the National Labor Committee for Palestine at the Hotel Astor in New York.

To be a Jew... means first of all, to acknowledge and follow in practice those fundamentals in humaneness laid down in the Bible—fundamentals without which no sound and happy community of men can exist.

We meet today because of our concern for the development of Palestine. In this hour one thing, above all, must be emphasized: Judaism owes a great debt of gratitude to Zionism. The Zionist movement has revived among Jews the sense of community...

Now the fateful disease of our time—exaggerated nationalism, borne up by blind hatred—has brought our work in Palestine to a most difficult stage. Fields cultivated by day must have armed protection at night against fanatical Arab outlaws...

Just one more personal word on the question of partition. I should much rather see reasonable agreement with the Arabs on the basis of living together in peace than the creation of a Jewish state... We are no longer the Jews of the Maccabee period. A return to a nation in the political sense of the word would be equivalent to turning away from the spiritualization of our community which we owe to the genius of our prophets. If external necessity should after all compel us to assume this burden, let us bear it with tact and patience.

Also, a message by Einstein urging Great Britain not to yield to the pressure of terrorism:

Einstein and God

Except for a brief period at the age of 12 when Einstein earnestly practiced Jewish rituals and even adhered to the kosher laws—to his parents' surprise and, quite possibly, their chagrin—Einstein spent his life alienated from organized religion. Einstein describes this period as his "religious paradise of youth." He tells of composing songs in praise of God and singing them on his way to school.

This period was not to last long enough for him to participate in a Bar Mitzvah ceremony; shortly before his thirteenth birthday, he came to believe that belief in God was "naive" and that religion was part of the general deceit of the State, perpetrated on youth to maintain control of the masses.

Einstein's marriages were civil and he never attended services (though, contrary to popular belief, he was not averse to setting foot in a synagogue or to wearing a traditional yarmulke). He left instructions to be cremated after he died, yet he often participated in Jewish activities. He supported Israel, fought zealously against anti-Semitism, and showed little tolerance for anyone who disparaged religion. He would refer to God reverentially as "the Old One" and spoke often of "cosmic religion," in which the physical mysteries of the universe pointed to the work and existence of an Almighty—or at least to a level of existence beyond human experience and comprehension.

Einstein visited the question of the relationship between science and religion publicly many times in his lifetime, rejecting efforts to interpret relativity as an indication of a higher plane of existence, yet claiming, in an oft-quoted remark, that "science without religion is blind; religion without science is lame." In a symposium on the relationship between science and religion held in 1941, Einstein suggested that science could be a helpful ally to religion in fulfilling its goal "to liberate mankind as far as possible from the egocentric cravings, desires, and fears," to which mortals are prone.

Religion played a particularly poignant role in Einstein's lifelong debate with Bohr and in his critique of quantum physics. In rejecting the quantum interpretation of physical processes—in which strict causality was replaced with a probabilistic description of the atomic world—Einstein often couched his criticsms in religious terms: "The theory yields much," he would say, "but it hardly brings us closer to the Old One's secrets. I, in any case, am convinced that He does not play dice."

The most effective expression of Einstein on religion appears in his descriptions of the scientific personality, which he saw closely related to that of the deeply spiritual and religious temperament. "The scientist is possessed by the sense of universal causation," he wrote in 1934. "The future, to him, is every whit as necessary and determined as the past. There is nothing divine about morality; it is a purely human affair. His religious feeling takes the form of a rapturous amazement at the harmony of natural law, which reveals an intelligence of such superiority that, compared with it, all the systematic thinking and acting of human beings is an utterly insignifigant reflection. This feeling is the guiding principle of his life and work…It is beyond ques-

Einstein at a concert, in a skull cap (yarmulke), in a Berlin synagogue.

tion closely akin to that which has possessed the religious geniuses of all ages."

"Those individuals," he wrote in 1948, "to whom we owe the great creative achievements of science were all of them imbued with the truly religious conviction that this unverse of ours is something perfect and susceptible to the rational striving of knowledge."

A rabbi, having heard that Einstein was an atheist, once sent a telegram to him that read: DO YOU BELIEVE IN GOD STOP PREPAID REPLY 50 WORDS. Einstein replied, "I believe in Spinoza's God who reveals himself in the harmony of all being, not in a God who concerns himself with the fate and actions of men."   — HR

Our bitter appeal is addressed to the nations and to England, and our demand should have the support of unwritten law and justice. We ask England not to compel by the sword what she has promised, but not to permit a minority to impose its will through terror on the majority of Arabs and Jews.

November. "Why do they hate the Jews?", a long magazine article by Einstein, was published in a New York magazine. From its text:

Why did the Jews so often happen to draw the hatred of the masses? Primarily because there are Jews among almost all nations and because they are everywhere too thinly scattered to defend themselves against violent attack.

A few examples from the recent past will prove the point: Toward the end of the nineteenth century the Russian people were chafing under the tyranny of their government. Stupid blunders in foreign policy further strained their temper until it reached the breaking point. In this extremity the rulers of Russia sought to divert unrest by inciting the masses to hatred and violence toward the Jews. These tactics were repeated after the Russian government had drowned the dangerous revolution of 1905 in blood—and this maneuver may well have helped to keep the hated regime in power until near the end of the World War.

When the Germans had lost the World War hatched by their ruling class, immediate attempts were made to blame the Jews, first for instigating the war and then for losing it. In the course of time, success attended these efforts.

The crimes with which the Jews have been charged in the course of history—crimes which were to justify the atrocities perpetrated against them—have changed in rapid succession. They were supposed to have poisoned wells. They were said to have murdered children for ritual purposes. They were falsely charged with a systematic attempt at the economic domination and exploitation of all mankind. Pseudo-scientific books were written to brand them an inferior, dangerous race. They were reputed to foment wars and revolutions for their own selfish purposes. They were presented at once as dangerous innovators and as enemies of true progress. They were charged with falsifying the culture of nations by penetrating the national life under the guise of becoming assimilated. In the same breath they were accused of being so stubbornly inflexible that it was impossible for them to fit into any society.

[What is the basis for these allegations?]

The members of any group existing in a nation are more closely bound to one another than they are to the remaining population. Hence a nation will never be free of friction while such groups continue to be distinguishable. In my belief, uniformity in a population would not be desirable, even if it were attainable. Common convictions and aims, similar interests, will in every society produce groups that, in a certain sense, act as units. There will always be friction between such groups— the same sort of aversion and rivalry that exists between individuals

Were anyone to form a picture of the Jews solely from the utterances of their enemies, he would have to reach the conclusion that they represent a world power. At first sight that seems downright absurd; and yet, in my view, there is a certain meaning behind it. The Jews as a group may be powerless, but the sum of the achievements of their individual members is everywhere considerable and telling, even though these achievements were made in the face of obstacles. The forces dormant in the individual are mobilized, and the individual himself is stimulated to self-sacrificing effort, by the spirit that is alive in the group.

Hence the hatred of the Jews by those who have reason to shun popular enlightenment. More than anything else in the world, they fear the influence of men of intellectual independence…

1939. March. From a radio address for United Jewish Appeal, broadcast on the 22nd:

In the past we were persecuted despite the fact that we were the people of the Bible; today, however, it is just because we are the people of the Book that we are persecuted. The aim is to exterminate not only ourselves but to destroy, together with us, that spirit expressed in the Bible and in Christianity which made possible the rise of civilization in Central and Northern Europe. If this aim is achieved, Europe will become a barren waste. For human community life cannot long endure on a basis of crude force, brutality, terror, and hate…

One of the most tragic aspects of the oppression of Jews and other groups has been the creation of a refugee class. Many distinguished men in science, art, and literature have been driven from the lands which they enriched with their talents... As one of the former citizens of Germany who have been fortunate enough to leave that country, I know I can speak for my fellow refugees, both here and in other countries, when I give thanks to the democracies of the world for the splendid manner in which they have received us. We, all of us, owe a debt of gratitude to our new countries, and each and every one of us is doing the utmost to show our gratitude by the quality of our contributions to the economic, social, and cultural work of the countries in which we reside.

May. A radio address broadcast from Einstein's home to a meeting of the Jewish National Workers Alliance, held in Town Hall, New York:

"Remember in the midst of your justified embitterment that England's opponents are also our bitterest enemies and that, in spite of everything, the maintenance of England's position is of utmost importance to us."
(On 5 April 1920, at the San Remo Conference, Britain had been assigned by the League of Nations as mandatory power for Palestine, according to principles laid down in the Treaty of Versailles. In 1939 Britain was still in control over that area.)
On 3 September the Second World War broke out.

1940. At a testimonial dinner to Einstein, given by the friends of the Haifa Technion (an Institute of Technology), Einstein said:

I can remember very well the time when Jews in Germany laughed over Palestine. I remember, when I spoke with Rathenau about Palestine, he said: "Why go to this land that is only sand and worth nothing and which can never be developed?" This was his idea. But, if he had not been murdered, he probably would now be in Palestine. You can therefore see that the development of Palestine is of real tremendous importance for all of Jewry.[2]

1944. June. Message by Einstein to a dinner by the American Fund for Palestinian Institutions:

The spirit of the Jews in Palestine has remained fresh and resilient. I have no doubt that they will succeed in a good measure of cooperation with the Arab people if only both our people and the Arabs succeed in conquering that childhood complaint of a narrow-minded nationalism imported from Europe and aggravated by professional politicians. Both peoples, it is to be hoped, will soon recognize that no rigid legal formula but only a lively mutual understanding and faithful cooperation in the daily tasks can open the right way.

1946. January. Papers carry a detailed account of Einstein's testimony, given in Washington, D.C., before the Anglo-American Committee of Inquiry on Palestine.

"Smiling benignly, Professor Einstein launched into the most whole-hearted denunciation of British colonial policy that the committee has yet heard." He makes four points.

(1) He is convinced, though regretfully, that British colonial policy is such as to render Great Britain unfit for further administration of her mandate over Palestine. (2) A trusteeship should be set up by the United Nations to administer Palestine and it should not be confined to any single power, including the United States. (3) The great majority of Jewish refugees in Europe should be settled in Palestine. (4) He has never seen any necessity for a Jewish commonwealth such as is advocated by the Zionist Organization for Palestine. He repeated two or three times his belief that inquiry commissions such as the one before which he was appearing have been created only to give the impression of goodwill without any intention on the part of responsible authorities to pay any attention to their findings and recommendations. Richard Grossman, MP, got the witness to agree that it was not "a British imperialist fiction" that Arabs might shoot Jewish refugees if they came in large numbers. When asked what would happen if the Arabs resisted Jewish immigration, Einstein replied, "They won't if they are not instigated. If people work together, they won't worry who has the larger number." Also 'the State [of Israel] idea is not according to my heart. I cannot understand why it is needed. It is connected with narrow-mindedness and economic obstacles. I believe it is bad. I have always been against it… [It is] an imitation of Europe —the end of Europe was brought by nationalism."

February. In a letter to the Progressive Palestine Association, Einstein expressed the belief that "A Government in Palestine under the United Nation's direct control and a constitution assuring Jews' and Arabs' security against being outvoted by each other would solve the Jewish-Arab difficulties."

1947. Einstein sends a message to a dinner at the Waldorf Astoria Hotel in New York honoring Weizmann: "In these days of fateful decision you have presented our case before the world with a vision that no one among us could muster."

On 29 November 1947, four days after that dinner, the United Nations General Assembly passed a resolution to the effect that the British Mandate for Palestine should be terminated. This decision was the result of a recommendation by a special UN Committee for Palestine which had started deliberations the preceding May. Arab violence broke out immediately after the news of the resolution became known.

1948. The British Mandate ended on 15 May 1948. On 14 May the State of Israel was officially proclaimed. On the 15th, Israel was attacked by combined forces of Egypt, Transjordan, Syria, Lebanon, and Iraq. The resulting War of Independence ended formally in July 1949, in defeat of the Arab Alliance.

A month before the outbreak of that War, Einstein had sent the following letter to the *New York Times*:

> Both Arab and Jewish extremists are today recklessly pushing Palestine into a futile war... We feel it to be our duty to declare emphatically that we do not condone methods of terrorism and of fanatical nationalism any more if practiced by Jews than if practiced by Arabs... A decisive victory by either would yield a corroding bitterness... We appeal to the Jews in this country and in Palestine not to permit themselves to be driven into a mood of despair or false heroism which eventually results in suicidal measures.

1949. March. Upon receiving an honorary degree from the Hebrew University, Einstein declared: "The wisdom and moderation the leaders of the new State have shown gives me confidence that gradually relations will be established with the Arab people which are based on fruitful cooperation and mutual trust."

November. Einstein was re-elected Chairman of the National Council of Friends of the Hebrew University. On the 27th of that month, Einstein made a radio broadcast for the United Jewish Appeal in which he said in part:

> There is no problem of such overwhelming importance to us Jews as consolidating that which has been accomplished in Israel with amaz-

ing energy and an unequaled willingness for sacrifice. May the joy and admiration that fill us when we think of all that this small group of energetic and thoughtful people has achieved give us the strength to accept the great responsibility which the present situation has placed upon us. [They have created] a community which conforms as closely as possible to the ethical ideals of our people as they have been formed on the course of a long history.

One of these ideals is peace, based on understanding and self-restraint, and not on violence. If we are imbued with this ideal, our joy becomes somewhat mingled with sadness, because our relations with the Arabs are far from this ideal at the present time. It may well be that we would have reached this ideal had we been permitted to work out, undisturbed by others, our relations with our neighbors, for we want peace and we realize that our future development depends on peace.

It was much less our own fault or that of our neighbors than of the Mandatory Power that we did not achieve an undivided Palestine in which Jews and Arabs would live as equals, free, in peace. If one nation dominates the other nations, as was the case in the British Mandate over Palestine, she can hardly avoid following the notorious device of *Divide et Impera*. In plain language this means: create discord among the governed people so they will not unite in order to shake off the yoke imposed upon them. Well, the yoke has been removed, but the seed of dissension has borne fruit and may still do harm for some time to come—let us hope not for too long.

The economic means of the Jewish Community in Israel do not suffice to bring this tremendous enterprise to a successful end. For a hundred thousand out of more than three hundred thousand persons who immigrated to Israel since May 1948, no homes or work could be made available. They had to be concentrated in improvised camps under conditions which are a disgrace to all of us.

It must not happen that this magnificent work breaks down because the Jews of this country do not help sufficiently or quickly enough. Here, to my mind, is a precious gift with which all Jews have been presented: the opportunity to take an active part in this wonderful task.

1951. Einstein bought the 200,000th $500 State of Israel bond of a recent issue.

1953. In a plea of support for the Hebrew University, Einstein said: "For the young State to achieve real independence, and conserve it, there must be a group of intellectuals and experts produced in the country itself."

1954. Einstein spoke at a planning conference in Princeton of the American Friends of the Hebrew University.

> Israel is the only place on earth where Jews have the possibility to shape public life according to their own traditional ideals... In our tradition it is neither the ruler nor the politician, neither the soldier nor the merchant, who represents the ideal. The ideal is represented by the teacher who... enrich[es] the intellectual, moral, and artistic life of the people. This implies a definite repudiation of what is commonly called 'materialism'. Human beings can attain a worthy and harmonious life only if they are able to rid themselves, within the limits of human nature, of the striving for the wish fulfillments of material kinds. The goal is to raise the spiritual values of society.

1955. On the morning of Wednesday 13 April 1955, the Israeli consul called on Einstein at his home in order to discuss the draft of a statement Einstein intended to make on television and radio on the occasion of the forthcoming anniversary of Israel's independence. The incomplete draft ends as follows.

> No statesman in a position of responsibility has dared to take the only promising course [toward a stable peace] of supranational security, since this would surely mean his political death. For the political passions, aroused everywhere, demand their victims.

These may well be the last phrases Einstein committed to paper. Two hours after this visit, Einstein was fatally stricken. He died in the early morning hours of 18 April. In the afternoon of that day he was cremated. The ashes were scattered in an unknown place.

References

1.   See also the collection of Einstein's essays *On Zionism*.
2.   On 24 June 1922 Walther Rathenau, the German Foreign Minister, a Jew, and an acquaintance of Einstein, had been assassinated in Berlin.

Einstein peering into the eyepiece of the 100-inch Hooker telescope at the Mt. Wilson Observatory, California, 1931. Edwin Hubble (left) and Walter Adams, director of the observatory, look on. Adams was the discoverer of the first white dwarf star and confirmed Einstein's prediction of the relativistic effects of stars on their own light spectra.

Einstein's Mistakes

by Steven Weinberg

Albert Einstein was certainly the greatest physicist of the 20th century, and one of the greatest scientists of all time. It may seem presumptuous to talk of mistakes made by such a towering figure, especially in the centenary of his annus mirabilis. But the mistakes made by leading scientists often provide a better insight into the spirit and presuppositions of their times than do their successes. Also, for those of us who have made our share of scientific errors, it is mildly consoling to note that even Einstein made mistakes. Perhaps most important, by showing that we are aware of mistakes made by even the greatest scientists, we set a good example to those who follow other supposed paths to truth. We recognize that our most important scientific forerunners were not prophets whose writings must be studied as infallible guides—they were simply great men and women who prepared the ground for the better understandings we have now achieved.

The cosmological constant

In thinking of Einstein's mistakes, one immediately recalls what Einstein (in a conversation with George Gamow[1]) called the biggest blunder he had made in his life: the introduction of the cosmological constant. After Einstein had completed the formulation of his theory of space, time, and gravitation—the general theory of relativity—he turned in 1917 to a consideration of the spacetime structure of the whole universe. He then encountered a problem. Einstein was assuming that, when suitably averaged over many stars, the universe is uniform and essentially static, but the equations of general relativity did not seem to allow a time-independent solution for a universe with a uniform distribution of matter. So Einstein modified his equations, by including a new term involving a quantity that he called the cosmological constant. When it was discovered that the universe is not static, but expanding. Einstein came to regret that he had needlessly mutilated his original

Reprinted with permission from *Physics Today*, November 2005, pp. 31-35. Copyright © 2005, American Institute of Physics.

Einstein with (from left to right) Arthur Eddington, Paul Ehrenfest, Henrik Lorentz, and Willem de Sitter at the Leiden Observatory in Germany, September 1923.

theory. It may also have bothered him that he had missed predicting the expansion of the universe.

This story involves a tangle of mistakes, but not the one that Einstein thought he had made. First, I don't think that it can count against Einstein that he had assumed the universe is static. With rare exceptions, theorists have to take the world as it is presented to them by observers. The relatively low observed velocities of stars made it almost irresistible in 1917 to suppose that the universe is static. Thus when Willem de Sitter proposed an alternative solution to the Einstein equations in 1917, he took care to use coordinates for which the metric tensor is time-independent. However, the physical meaning of those coordinates is not transparent, and the realization that de Sitter's alternate cosmology was not static—that matter particles in his model would accelerate away from each other—was considered to be a drawback of the theory.

It is true that Vesto Melvin Slipher, while observing the spectra of spiral nebulae in the 1910s, had found a preponderance of redshifts, of the sort that would be produced in an expansion by the Doppler effect, but no one then knew what the spiral nebulae were; it was not until Edwin Hubble found faint Cepheid variables in the Andromeda Nebula in 1923 that it became clear that spiral nebulae were distant galaxies, clusters of stars far outside our own galaxy. I don't know if Einstein had heard of Slipher's redshifts by 1917, but in any case he knew very well about at least one other thing that could produce a redshift of spectral lines: a gravitational field. It should be acknowledged here that Arthur Eddington, who had learned about general relativity during World War I from de Sitter, did in 1923 interpret Slipher's redshifts as due to the expansion of the universe in the de Sitter model. (The two scientists are pictured with Einstein and others above.) Nevertheless, the expansion of the universe was not generally accepted until Hubble announced in 1929—and actually showed in 1931—that the redshifts of distant galaxies increase in proportion to their distance, as would be expected for a uniform expansion (see page 311). Only then was much attention given to the expanding-universe models introduced in 1922 by Alexander Friedmann, in which no cosmological constant is needed. In 1917 it was quite reasonable for Einstein to assume that the universe is static.

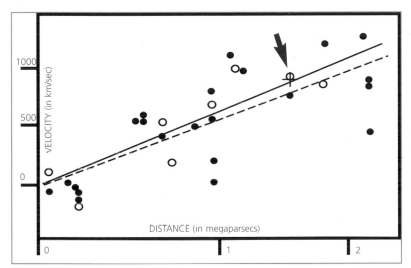

Recessional velocities of nearby galaxies vary linearly with distance, as Edwin Hubble demonstrated in these 1929 readings. Filled circles and the solid line approximation describe individual galaxies. Open circles and the broken line correspond to galaxies combined into groups. The cross (red arrow) represents the mean velocity and distance for 22 galaxies whose individual distances could not be estimated. The slope of the graph is about seven times the now accepted value. (A parsec is 3.26 light-years.)

Einstein did make a surprisingly trivial mistake in introducing the cosmological constant. Although that step made possible a time-independent solution of the Einstein field equations, the solution described a state of unstable equilibrium. The cosmological constant acts like a repulsive force that increases with distance, while the ordinary attractive force of gravitation decreases with distance. Although there is a critical mass density at which this repulsive force just balances the attractive force of gravitation, the balance is unstable; a slight expansion will increase the repulsive force and decrease the attractive force so that the expansion accelerates. It is hard to see how Einstein could have missed this elementary difficulty.

Einstein was also at first confused by an idea he had taken from the philosopher Ernst Mach: that the phenomenon of inertia is caused by distant masses. To keep inertia finite, Einstein in 1917 supposed that the universe must be finite, and so he assumed that its spatial geometry is that of a three-dimensional spherical surface. It was therefore a surprise to him that when test particles are introduced into the empty universe of de Sitter's model, they exhibit all the usual properties of inertia. In general relativity the masses of distant bodies are not the cause of inertia, though they do affect the choice of inertial frames. But that mistake was harmless. As Einstein pointed out in his 1917 paper, it was the assumption that the universe is static, not that it is finite, that had made a cosmological constant necessary.

Aesthetically motivated simplicity

Einstein made what from the perspective of today's theoretical physics is a deeper mistake in his dislike of the cosmological constant. In developing general relativity, he had relied not only on a simple physical principle—the principle of the equivalence of gravitation and inertia that he had developed from 1907 to 1911—but also on a sort of Occam's razor, that the equations of the theory should be not only consistent with this principle but also as simple as possible. In itself, the principle of equivalence would allow field equations of almost unlimited complexity. Einstein could have included terms in the equations involving four spacetime derivatives, or six spacetime derivatives, or any even number of spacetime derivatives, but he limited himself to second-order differential equations.

This could have been defended on practical grounds. Dimensional analysis shows that the terms in the field equations involving more than two spacetime derivatives would have to be accompanied by constant factors proportional to positive powers of some length. If this length was anything like the lengths encountered in elementary-particle physics, or even atomic physics, then the effects of these higher derivative terms would be quite negligible at the much larger scales at which all observations of gravitation are made. There is just one modification of Einstein's equations that could have observable effects: the introduction of a term involving no spacetime derivatives at all—that is, a cosmological constant.

But Einstein did not exclude terms with higher derivatives for this or for any other practical reason, but for an aesthetic reason: They were not needed, so why include them? And it was just this aesthetic judgment that led him to regret that he had ever introduced the cosmological constant.

Since Einstein's time, we have learned to distrust this sort of aesthetic criterion. Our experience in elementary particle physics has taught us that any term in the field equations of physics that is allowed by fundamental principles is likely to be there in the equations. It is like the ant world in T. H. White's *The Once and Future King*: Everything that is not forbidden is compulsory. Indeed, as far as we have been able to do the calculations, quantum fluctuations by themselves would produce an infinite effective cosmological constant, so that to cancel the infinity there would have to be an infinite "bare" cosmological constant of the opposite sign in the field equations themselves. Occam's razor is a fine tool, but it should be applied to principles, not equations.

It may be that Einstein was influenced by the example of Maxwell's theory, which he had taught himself while a student at the Zürich Polytechnic Institute. James Clerk Maxwell of course invented his equations to account for the known phenomena of electricity and magnetism while preserving the principle of electric-charge conservation, and in Maxwell's formulation the field equations contain terms with only a minimum number of spacetime derivatives. Today we know that

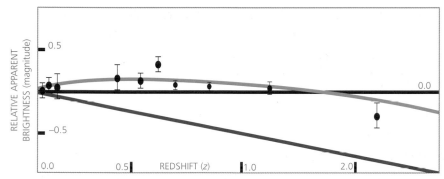

Measurements on distant supernovae show that the universe contains a preponderance of dark energy that behaves like a cosmological constant. Apparent brightness indicates distance; redshift is a measure of recessional velocity. For the blue curve that best fits the data, 70% of the cosmic energy density is attributed to a cosmological constant —which is absent for the red line.

the equations governing electrodynamics contain terms with any number of space-time derivatives, but these terms, like the higher-derivative terms in general relativity, have no observable consequences at macroscopic scales.

Astronomers in the decades following 1917 occasionally sought signs of a cosmological constant, but they only succeeded in setting an upper bound on the constant. That upper bound was vastly smaller than what would be expected from the contribution of quantum fluctuations, and many physicists and astronomers concluded from this that the constant must be zero. But despite our best efforts, no one could find a satisfactory physical principle that would require a vanishing cosmological constant.

Then in 1998, measurements of redshifts and distances of supernovae by the Supernova Cosmology Team and High-z Supernova Search Team showed that the expansion of the universe is accelerating, as de Sitter had found in his model. As discussed in the figure above, it seems that about 70% of the energy density of the universe is a sort of "dark energy," filling all space. This was subsequently confirmed by observations of the angular size of anisotropies in the cosmic microwave background. The density of the dark energy is not varying rapidly as the universe expands, and if it is truly time-independent then it is just the effect that would be expected from a cosmological constant. However this works out, it is still puzzling why the cosmological constant is not as large as would be expected from calculations of quantum fluctuations. In recent years the question has become a major preoccupation of theoretical physicists. Regarding his introduction of the cosmological constant in 1917, Einstein's real mistake was that he thought it was a mistake.

A historian, reading the foregoing in a first draft of this article, commented that I might be accused of perpetrating Whig history. The term "Whig history" was coined in a 1931 lecture by the historian Herbert Butterfield. According to Butterfield, Whig historians believe that there is an unfolding logic in history, and they judge the past

Hubble, Curtis and the Red Shift

The great debate in the United States in 1920 turned out to be not whether the universe was expanding, but whether the universe spanned more than our galaxy. Relativity as a topic for the William Ellery Hale Lecture at the 1920 April meeting of the National Academy of Sciences was considered—but turned down. C. J. Abbot, the NAS secretary, felt that few people understood relativity and responded in frustration to the suggestion, "…I pray to God that the progress of science will send relativity to some region of space beyond the fourth dimension, from whence it may never return to plague us."

A debate on "The Distance Scale of the Universe," with Harlow Shapley of Mt. Wilson Solar Observatory and Heber Doust Curtis of Lick Observatory was the final topic chosen by Edwin Hubble. Einstein, still in Germany, missed the debate. Shapley, like most astronomers at that time, thought the Milky Way covered the entire universe, but his estimate of its size at 300,000 light-years across, was the largest yet. Curtis favored a smaller universe divided among separate "island universes," the spiral nebulae, but thought the sun was the center of our own "island."

Shapley correctly placed the sun far from the center of the Milky Way. It was Hubble, looking through the 100 inch Hooker telescope at Mount Wilson in 1923, who proved Shapley's proprietary view wrong. Hubble identified Cepheid variable stars in the Andromeda Nebula and, using the method pioneered by Henrietta Leavitt, calculated their distance to be 100 million light years, far beyond Shapley's Milky Way. On receiving Hubble's letter, the flamboyant Shapley reportedly declared, "Here is the letter that destroyed my universe."

Hubble continued to examine the light, or line spectra, from different galaxies. He deduced his law from the observation that the more distant a galaxy, the greater its red shift and therefore the greater its velocity relative to the Milky Way. The "red shift" is comparable to the everyday "Doppler shift" that makes a train whistle's pitch lower as it moves away from you.

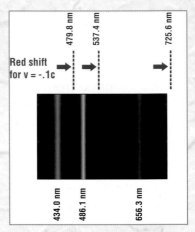

The spectrum lines show the chemical make-up of stars, but the lines are shifted due to the motion of the star away from Earth.

Similarly, light from a galaxy speeding away from Earth shifts to longer, more red, wavelengths. Hubble's work led to his 1929 announcement of an expanding universe.

Einstein, who in 1917 rejected the idea of an expanding universe, visited Hubble at Mount Wilson in 1931 to thank the astronomer and regret his "greatest blunder."   — PH

---

by the standards of the present. But it seems to me that Whiggery is to be avoided in political and social history (which is what concerned Butterfield), it has a certain value in the history of science. Our work in science is cumulative. We really do know more than our predecessors, and we can learn about the things that were not understood in their times by looking at the mistakes they made.

Contra quantum mechanics

The other mistake that is widely attributed to Einstein is that he was on the wrong side in his famous debate with Niels Bohr over quantum mechanics, starting at the Solvay Congress of 1927 and continuing into the 1930s. In brief, Bohr had presided over the formulation of a "Copenhagen interpretation" of quantum mechanics, in which it is only possible to calculate the probabilities of the various possible outcomes of experiments. Einstein rejected the notion that the laws of physics could deal with probabilities, famously decreeing that God does not play dice with the cosmos. But history gave its verdict against Einstein—quantum mechanics went on from success to success, leaving Einstein on the sidelines.

All this familiar story is true, but it leaves out an irony. Bohr's version of quantum mechanics was deeply flawed, but not for the reason Einstein thought. The Copenhagen interpretation describes what happens when an observer makes a measurement, but the observer and the act of measurement are themselves treated classically. This is surely wrong: Physicists and their apparatus must be governed by the same quantum mechanical rules that govern everything else in the universe. But these rules are expressed in terms of a wave function (or, more precisely, a state vector) that evolves in a perfectly deterministic way. So where do the probabilistic rules of the Copenhagen interpretation come from?

Considerable progress has been made in recent years toward the resolution of the problem, which I cannot go into here. It is enough to say that neither Bohr nor Einstein had focused on the real problem with quantum mechanics. The Copenhagen rules clearly work, so they have to be accepted. But this leaves the task of explaining them by applying the deterministic equation for the evolution of the wave function, the Schrödinger equation, to observers and their apparatus. The difficulty is not that quantum mechanics is probabilistic—that is something we apparently just have to live with. The real difficulty is that it is also deterministic, or more precisely, that it combines a probabilistic interpretation with deterministic dynamics.

Attempts at unification

Einstein's rejection of quantum mechanics contributed, in the years from the 1930s to his death in 1955, to his isolation from other research in physics, but there was another factor. Perhaps Einstein's greatest mistake was that he became the prisoner of his own successes. It is the most natural thing in the world, when one has scored great victories in the past, to try to go on to further victories by repeating the tactics that previously worked so well. Think of the advice given to Egypt's President Gamal Abd al-Nasser by an apocryphal Soviet military attaché at the time of the 1956 Suez crisis: "Withdraw your troops to the center of the country, and wait for winter."

And what physicist had scored greater victories than Einstein? After his tremendous success in finding an explanation of gravitation in the geometry of space and time, it was natural that he should try to bring other forces along with gravitation into a "unified field theory" based on geometrical principles. About other things going on in physics, he commented in 1950 that "all attempts to obtain a deeper knowledge of the foundations of physics seem doomed to me unless the basic concepts are in accordance with general relativity from the beginning." Since electromagnetism was the only other force that in its macroscopic effects seemed to bear any resemblance to gravitation, it was the hope of a unification of gravitation and electromagnetism that drove Einstein in his later years.

I will mention only two of the many approaches taken by Einstein in this work. One was based on the idea of a fifth dimension, proposed in 1921 by Theodore Kaluza. Suppose you write the equations of general relativity in five rather than four spacetime dimensions, and arbitrarily assume that the 5D metric tensor does not depend on the fifth coordinate. Then it turns out that the part of the metric tensor that links the usual four spacetime dimensions with the fifth dimension satisfies the same field equation as the vector potential in the Maxwell theory of electromagnetism, and the part of the metric tensor that only links the usual four spacetime dimensions to each other satisfies the field equations of 4D general relativity.

The idea of an additional dimension became even more attractive in 1926, when Oskar Klein relaxed the condition that the fields are independent of the fifth coordinate, and assumed instead that the fifth dimension is rolled up in a tiny circle so that the fields are periodic in that coordinate. Klein found that in this theory the part of the metric tensor that links the fifth dimension to itself behaves like the wave function of an electrically charged particle, so for a moment it seemed to Einstein that there was a chance that not only gravitation and electromagnetism but also matter would be governed by a unified geometrical theory. Alas, it turned out that if the electric charge of the particle is identified with the charge of the electron, then the particle's mass comes out too large by a factor of about 10^{18}.

It is a pity that Einstein gave up on the Kaluza-Klein idea. If he had extended it from five to six or more spacetime dimensions, he might have discovered the field theory constructed in 1954 by C. N. Yang and Robert Mills, and its generalizations, some of which later appeared as parts of our modern theories of strong, weak, and electromagnetic interactions. Einstein apparently gave no thought to strong or weak nuclear forces, I suppose because they seem so different from gravitation and electromagnetism. Today we realize that the equations underlying all known forces aside from gravitation are actually quite similar, the difference in the phenomena arising from color trapping for strong interactions and spontaneous symmetry breaking for weak interactions. Even so, Einstein would still probably be unhappy with today's theories, because they are not unified with gravitation and because matter—electrons, quarks, and so on—still has to be put in by hand.

Even before Klein's work, Einstein had started on a different approach, based on a simple bit of counting. If you give up the condition that the 4 x 4 metric tensor should be symmetric, then it will have 16 rather than 10 independent components, and the extra 6 components will have the right properties to be identified with the electric and magnetic fields. Equivalently, one can assume that the metric is complex, but Hermitian. The trouble with this idea, as Einstein became painfully aware, is that there really is nothing in it that ties the 6 components of the electric and magnetic fields to the 10 components of the ordinary metric tensor that describes gravitation, other than that one is using the same letter of the alphabet for all these fields. A Lorentz transformation or any other coordinate transformation will convert electric or magnetic fields into mixtures of electric and magnetic fields, but no transformation mixes them with the gravitational field. This purely formal approach, unlike the Kaluza-Klein idea, has left no significant trace in current research. The faith in mathematics as a source of physical inspiration, which had served Einstein so well in his development of general relativity, was now betraying him.

Even though it was a mistake for Einstein to turn away from the exciting progress being made in the 1930s and 1940s by younger physicists, it revealed one admirable feature of his personality. Einstein never wanted to be a mandarin. He never tried to induce physicists in general to give up their work on nuclear and particle physics and follow his ideas. He never tried to fill professorships at the Institute for Advanced Studies with his collaborators or acolytes. Einstein was not only a great man, but a good one. His moral sense guided him in other matters: He opposed militarism during World War I; he refused to support the Soviet Union in the Stalin years; he became an enthusiastic Zionist; he gave up his earlier pacifism when Europe was threatened by Nazi Germany, for instance urging the Belgians to rearm; and he publicly opposed McCarthyism. About these great public issues, Einstein made no mistakes.

References

1. G. Gamow, *My World Line—An Informal Autobiography*, Viking Press, New York, (1970), p. 44. I thank Lawrence Krauss for this reference.
2. A. Einstein, *Sci. Am.*, April 1950, p.13.
3. E. Hubble, *Proc. Natl. Acad. Sci. (USA)* 15, 168, (1929).
4. A.G. Riess et al., *Astrophys. J.* 607, 665 (2004).

The "Black Eye" Galaxy.

Beyond the Big Bang

———○———

by Marcia Bartusiak

On January 29, 1931, the world's premier physicist, Albert Einstein, and its foremost astronomer, Edwin Hubble, settled into the plush leather seats of a sleek Pierce-Arrow touring car for a visit to Mount Wilson in southern California. They were chauffeured up the long, zigzagging dirt road to the observatory complex on the summit, nearly a mile above Pasadena. Home to the largest telescope of its day, Mount Wilson was the site of Hubble's astronomical triumphs. In 1924 he had used the telescope's then colossal 100-inch mirror to confirm that our galaxy is just one of countless "island universes" inhabiting the vastness of space. Five years later, after tracking the movements of these spiraling disks, Hubble and his assistant, Milton Humason, had revealed something even more astounding: The universe is swiftly expanding, carrying the galaxies outward.

On the peak that bright day in January, the 51-year-old Einstein delighted in the telescope's instruments. Like a child at play, he scrambled about the framework, to the consternation of his hosts. Nearby was Einstein's wife, Elsa. Told that the giant reflector was used to determine the universe's shape, she reportedly replied, "Well, my husband does that on the back of an old envelope."

That wasn't just wifely pride. Years before Hubble detected cosmic expansion, Einstein had fashioned a theory, general relativity, that could explain it. In studies of the cosmos, it all goes back to Einstein.

Just about anywhere astronomers' observations take them—from the nearby sun to the black holes in distant galaxies—they enter Einstein's realm, where time is relative, mass and energy are interchangeable, and space can stretch and warp. His footprints are deepest in cosmology, the study of the universe's history and fate. General relativity "describes how our universe was born, how it expands, and what its future will be," says Alan Dressler of the Carnegie Observatories. Beginning, middle, and end "all are connected to this grand idea."

At the turn of the 20th century, 30 years before Einstein and Hubble's rendezvous at Mount Wilson, physics was in turmoil. X-rays, electrons, and radioactivity were just being discovered, and physicists were realizing that their trusted laws of motion, dating back more than 200 years to Isaac Newton, could not explain how these strange new particles flit through space. It took a rebel, a cocky kid who spurned rote learning and had an unshakable faith in his own abilities, to blaze a trail through this baffling new territory. This was not the iconic Einstein—the sockless, rumpled character with baggy sweater and fright-wig coiffure—but a younger, more romantic figure with alluring brown eyes and wavy hair. He was at the height of his prowess.

Among his gifts was a powerful physical instinct, almost a sixth sense for knowing how nature should work. Einstein thought in images, such as one that began haunting him as a teenager: If a man could keep pace with a beam of light, what would he see? Would he see the electromagnetic wave frozen in place like some glacial swell? "It does not seem that something like that can exist!" Einstein later recalled thinking.

He came to realize that since all the laws of physics remain the same whether you're at rest or in steady motion, the speed of light has to be constant as well. No one can catch up with a light beam. But if the speed of light is identical for all observers, something else has to give: absolute time and space. Einstein concluded that the cosmos has no universal clock or common reference frame. Space and time are "relative," flowing differently for each of us depending on our motion.

Einstein's special theory of relativity, published a hundred years ago, also revealed that energy and mass are two sides of the same coin, forever linked in his famed equation $E = mc^2$. (E stands for energy, m for mass, and c for the speed of light.) "The idea is amusing and enticing," wrote Einstein, "but whether the Almighty is ... leading me up the garden path—that I cannot know." He was too modest. The idea that mass could be transformed into pure energy later helped astronomers understand the enduring power of the sun. It also gave birth to nuclear weapons.

But Einstein was not satisfied. Special relativity was just that—special. It could not describe all types of motion, such as objects in the grip of gravity, the large-scale force that shapes the universe. Ten years later, in 1915, Einstein made up for the omission with his general theory of relativity, which amended Newton's laws by redefining gravity.

General relativity revealed that space and time are linked in a flexible four-dimensional fabric that is bent and indented by matter. In this picture, Earth orbits the sun because it is caught in the space-time hollow carved by the sun's mass, much as a rolling marble would circle around a bowling ball sitting in a trampoline. The pull of gravity is just matter sliding along the curvatures of space-time.

Einstein shot to the pinnacle of celebrity in 1919, when British astronomers actually measured this warping. Monitoring a solar eclipse, they saw streams of starlight bending around the darkened sun. "Lights All Askew in the Heavens. Stars Not Where They Seemed or Were Calculated to be, but Nobody Need Worry," proclaimed the headline in the *New York Times*.

With this new insight into gravity, physicists at last were able to make actual predictions about the universe's behavior, turning cosmology into a science. Einstein was the first to try. Yet as events showed, even Einstein was a fallible genius. A misconception about the nature of the universe led him to propose a mysterious new gravitational effect—a notion he soon rejected. But he may have been right for the wrong reasons, and his "mistake" may yet turn out to be one of his deepest insights.

For Newton, space was eternally at rest, merely an inert stage on which objects moved. But with general relativity, the stage itself became an active player. The amount of matter within the universe sculpts its overall curvature. And space-time itself can be either expanding or contracting.

When Einstein announced general relativity in 1915, he could have taken the next step and declared that the universe was in motion, more than a decade before Hubble directly measured cosmic expansion. But at the time, astronomers conceived of the universe as a large collection of stars fixed forever in the void. Einstein accepted this immutable cosmos. Truth be told, he liked it. Einstein was often leery of the most radical consequences of his ideas.

But because even a static universe would eventually collapse under its own gravity, he had to slip a fudge factor into the equations of general relativity—a cosmological constant. While gravity pulled celestial objects inward, this extra gravitational effect—a kind of antigravity—pushed them apart. It was just what was needed to keep the universe immobile, "as required by the fact of the small velocities of the stars," Einstein wrote in 1917.

Twelve years later, Hubble's discovery of other galaxies racing away from ours, their light waves stretched and reddened by the expansion of space-time, vanquished the static universe. It also eliminated any need for a cosmological constant to hold the galaxies steady. During his 1931 California visit, Einstein acknowledged as much. "The red shift of distant nebulae has smashed my old construction like a hammer blow," he declared. He reputedly told a colleague that the cosmological constant was his biggest blunder.

With or without that extra ingredient, the basic recipe for the expanding universe was Einstein's. But it was left to others to identify one revolutionary implication: a moment of cosmic creation. In 1931 the Belgian priest and astrophysicist Georges Lemaitre put the fleeing galaxies into reverse and imagined them eons ago merged in a fireball of dazzling brilliance—a "primeval atom," as he put it. "The evolution of the world can be compared to a display of fireworks that has just ended: some few red wisps, ashes and smoke," wrote Lemaitre. From this poetic scenario arose today's big bang.

Many were appalled by this concept. "The notion of a beginning ... is repugnant to me," said British astrophysicist Arthur Eddington in 1931. But evidence in its favor slowly gathered, climaxing in 1964, when scientists at Bell Telephone Laboratories discovered that the cosmos is awash in a sea of microwave radiation, the remnant glow of the universe's thunderous launch. Ever since then the image of the big bang has shaped and directed the work of cosmologists as strongly as Ptolemy's celestial spheres influenced astronomers in the Middle Ages.

In 1980 Alan Guth, now at the Massachusetts Institute of Technology, gave the big bang a boost, adding new particle physics to Einstein's flexible space-time. He realized that for its first trillionth of a trillionth of a trillionth of a second, the infant cosmos could have undergone a supercharged expansion—an instant of "inflation"—before settling into more measured growth.

Inflation would have helped smooth out the matter and energy in the universe and flattened its overall space-time curvature, just as satellites have found by making precise measurements of the cosmic microwaves. And these days some theorists believe inflation wasn't a flash in the pan. In an ongoing process of creation, space-time could be inflating into new universes everywhere and all the time—an infinity of big bangs.

Within our own universe, the high priests of astronomy have continued the cosmological quest initiated by Einstein and Hubble, first at Mount Wilson, then at the 200-inch telescope on California's Palomar Mountain, 90 miles to the south. How fast is the universe ballooning outward? they asked. How old is it? "Answering those questions," says Wendy Freedman, director of the Carnegie Observatories, "turned out to be more difficult than anyone anticipated."

Only at the turn of this century, with the help of a space telescope aptly named Hubble, did Freedman and others confidently peg the universe's current rate of expansion, as well as its age. A birthday cake for the universe would require some 14 billion candles.

Astronomers have found some strange objects in this expanding universe—and these too are Einstein's children. In the 1930s a young Indian physicist, Subrahmanyan Chandrasekhar, applied special relativity and the new theory of quantum mechanics to a star. He warned that if it surpassed a certain mass, it would not settle down as a white dwarf at the end of its life (as our sun will). Instead gravity would squeeze it down much further, perhaps even to a singular point. Horrified, Eddington declared that "there should be a law of Nature to prevent a star from behaving in this absurd way!"

There was no such law. Chandrasekhar had opened the door for others to contemplate the existence of the most bizarre stars imaginable. First there was a naked sphere of neutrons just a dozen miles wide born in the throes of a supernova, the explosion of a massive star. A neutron star's density would be equivalent to packing all the cars in the world into a thimble. Then there was the peculiar object formed from the collapse of an even bigger star or a cluster of stars— enough mass to dig a pit in space-time so deep nothing can ever climb out.

Einstein himself tried to prove that such an object—a black hole, it was later christened—could not exist. Like Eddington, he loathed what would be found at a black hole's center: a point of zero volume and infinite density, where the laws of physics break down. The discoveries that might have forced him to acknowledge his theory's strange offspring came after his death in 1955.

Astronomers identified the first quasar, a remote young galaxy disgorging the energy of a trillion suns from its center, in 1963. Four years later, much closer to home, observers stumbled on the first pulsar, a rapidly spinning beacon emitting staccato radio beeps. Meanwhile spaceborne sensors spotted powerful x-rays and gamma rays streaming from points around the sky.

All these new, bewildering signals are believed to pinpoint collapsed objects— neutron stars and black holes—whose crushing gravity and dizzying spin turn them into dynamos. With their discovery, the once sedate universe took on an edge; it metamorphosed into an Einsteinian cosmos, filled with sources of titanic energies that can be understood only in the light of relativity.

Even Einstein's less celebrated ideas have had remarkable staying power. As early as 1912 he realized that a faraway star can act like a giant spyglass, its gravity deflecting passing light rays and magnifying objects behind it. He eventually concluded that this tiny effect defied "the resolving power of our instruments" and had "little value."

With today's telescopes, astronomers are seeing galaxies and galaxy clusters act as powerful gravitational lenses, offering a peek at galaxies farther out. Since the light-bending depends on the mass of the lens, the effect also lets observers weigh the lensing galaxies. They turn out to have far more mass than can be seen. It's part of the universe's mysterious dark matter, the roughly 90 percent of its mass that can't be found in stars, gas, planets, or any other known form of matter.

A cosmic web of dark matter is now thought to have governed where galaxies formed. Dark matter is the universe's hidden architecture, and gravitational lensing is one of the few practical ways to "see" it. An effect Einstein thought insignificant has become a key astronomical tool.

Theorists have also dusted off his discarded cosmological constant to explain a startling new discovery, and now Einstein's "biggest blunder" is starting to look like one of his greatest successes. Astronomers had assumed that gravity is gradually slowing the expansion of the universe. But in the late 1990s two teams, measuring the distances to faraway exploding stars, found just the opposite. Like buoy markers spreading apart on ocean currents, these supernovae revealed that space-time is ballooning outward at an accelerating pace.

For Einstein, the cosmological constant was a way to steady the universe. But if its repulsive effect—now called dark energy—is big enough, it could also drive the acceleration. "The need came back, and the cosmological constant was waiting," says Adam Riess of the Space Telescope Science Institute, one of the discoverers of the acceleration. "It's totally an Einsteinian concept."

So is a prediction of general relativity that, if confirmed, could open new insights into the cosmos: ripples in space-time called gravity waves. To detect them, physicists have built three giant sensors, in south-central Washington State, Louisiana, and south of Pisa, Italy. In each one, laser beams run up and down miles-long pipes to measure the slight stretching and squeezing of space-time expected if a gravity wave passes by.

By triangulating these measurements, scientists might trace gravity waves back to their sources. Only stunningly violent events could cause space-time to shudder—a supernova, for example, or the titanic collision of two neutron stars or black holes. "If two black holes collided, gravity waves would be the only signals to come out," says Adalberto Giazotto, a scientist with the Pisa project.

The mighty jolt of cosmic birth probably also generated gravity waves, which would still be resonating through the cosmos. These remnant ripples could hold direct evidence of the fleeting moment when physicists believe all of nature's forces were united. If so, Einstein's gravity waves could at last offer clues to something he tried and failed to develop: a "theory of everything." Physicists are still seeking such a theory—a single explanation for both the large-scale force of gravity and the short-range forces inside the atom.

Catching these faint echoes of the big bang is a major goal of NASA's next generation of space astronomy missions, a plan the agency has tagged "Beyond Einstein."

Beyond Einstein? Not by a long shot. Einstein might be startled by the universe as we understand it today. But it is unmistakably his.

Einstein walking in Princeton on a winter's day in the early 1950s. The
solitary figure to the left further down the path is thought to be Gödel.

Einstein Mysterium: What We Still Don't Know about Einstein

A CONVERSATION WITH DENNIS OVERBYE

On May 3, 2006, we met with Dennis Overbye, science writer and editor of the New York Times *and Biographer of Einstein (*Einstein in Love*), in a restaurant in midtown Manhattan to discuss the myth and mystery of Albert Einstein. The Q&A that follows is based on his writings and on that conversation.*

Q It is now more than fifty years since Einstein's passing, and books continue to appear on virtually every aspect of his life—from the serious, such as your own work about his early life and career; to the trivial, such as works that have appeared on the search for Lieserl and on Einstein's FBI file; to the lurid, such as tell-alls on his romances and the peregrinations of his brain. How would you characterize what about Einstein remains still unknown, still a mystery?

A Well, we are talking about one of the most well-documented lives of modern times—a man who was in a spotlight for half his life. Einstein was also a prodigious letter writer, so by now we have a lot of material to go on (sometimes in spite of the zealousness of the guardians of the archives). But there are mysteries at the heart of every life and there are certainly things about Einstein that are mysterious—that were probably mysterious to Einstein himself; things about his life that he himself would have admitted were a source of puzzlement. For example, how was Einstein able to maintain that same innocent, naive sense of wonder and almost childish mode of thought that he had as a youngster wondering what it would be like riding on a light beam? This quality was a key ingredient in Einstein's nature and one he cherished and nurtured in himself, often at the expense of women he was involved with. Eric Erikson wrote about it in his essay "Psychoanalytic Reflections on Einstein's Centenary." The most common comment I get from friends, especially women, when they read some of Einstein's letters, is that he comes across as a spoiled child.

Some of the other so-called mysteries regarding Einstein are less profound and more gossipy. What happened to his and Mileva's illegitimate daughter, Lieserl? I believe that Einstein did think of Leiserl at various times over his life and wondered what had happened to her; he probably suffered much guilt about that episode in his life. On at least one occasion a woman claimed that she was his illegitimate daughter, and while Einstein made fun of her claim to his friend, he quietly set the sleuths on her to makes sure she wasn't Lieserl. To the extent that you can love somebody you've never seen, there's no question in my mind that he loved her — or thought he did—and thought often about her right to the end. In fact, I doubt he would have married Mileva in the first place had it not been for Lieserl.

Q Speaking of Mileva, speculation has also been voiced about the role Mileva had in the development of special relativity. How would you assess her contribution?

A Mileva was certainly adequately trained in mathematics to serve as a valuable sounding board to Einstein during this period, especially since the famous relativity paper requires no math beyond algrebra. She could easily and probably did go over the math before he sent it in. It would be a way to include her. The hard part of the paper is in the ideas, and those are his alone. Mileva, he once said, was not comfortable with abstract thinking. But she was important, I think, in convincing Einstein that he could do it—that he was really onto something and that his scientific work was important, revolutionary and path-breaking. I don't think this encouragement would have been as credible had Mileva not been more than just an adoring wife. She was a serious practitioner and trained professional, and her affirmation probably went a long way in getting Einstein through the dark days when he couldn't get a job. After all, she wanted to do this kind of work herself.

Q All the more surprising, then, that the marriage fell apart. How do you view what has come to light about Einstein's love life and his relations with women?

A What we've learned in the last 20 years, particularly from the love letters to Mileva, has completely transformed him. We now have the image of the man who invented relativity as a handsome swain, a teenager in love almost, immersed in a social, sexual and political nexus, where before we only knew his descendant, the frizzy-haired eccentric in Princeton. In some cases the pendulum has swung too far and now people only want to see Einstein as a cad towards women, a seducer and adulterer. But I don't think he was obsessed with notching his bedpost or anything; he was not what I would call a party animal. I think the problem people have in this area is with squaring the picture of Einstein as the befuddled, absent-minded professor and the worldly, sophisticate we usually associate with a man with an active social calendar.

The fact is, people are mistaken on both counts: Einstein was a very aware individual who looked after his own business affairs with care and competence—he had, after all, studied economics at university; and he cared deeply for a number of people throughout his life—men who were friends and confidants, and women toward whom he felt deep and lasting affection. It's always difficult to imagine our icons engaging in normal human behavior—no less true for a man idolized as the greatest scientific genius in history.

Q You've written that Einstein "became a symbol of both cosmic and moral mystery." The cosmic mystery is understandable enough; his science is famously challenging to the ordinary mortal. But in what way is Einstein a symbol of moral mystery?

A Einstein was a man who could embrace different ideas that on the surface seem incompatible—even contradictory. That was clear in the paper he produced in 1905: one aspect of those papers that is truly miraculous is that some advance a corpuscular view of light, and others advance a wave image. The same was true in other realms of life. We think of Einstein as a pacifist, but he was not the kind of pacifist who would turn the other cheek. If you knocked Einstein down, he'd get up and hit you back. He thought religion was, in his words, "a pack of lies," yet, he always supported Jewish causes and identified with the Jewish people. The invitation for him to become President of Israel, was, I believe, sincere on Ben Gurion's part, and considered seriously by him.

Then you come to the atom bomb. On the one hand, Einstein promoted America's development of the A-bomb; yet he was appalled when he learned of Hiroshima and worked tirelessly for disarmament. I think Einstein understood very early on that the modern era was, above all else, a period of paradox, contradiction and moral uncertainty. You almost get the sense that Einstein used that frizzy-haired sockless befuddled image as a shield to mask the contradictions and the ambiguities. I look at that wizened, weather-beaten face and that knowing gaze and see a twinkle in the eye of someone who is succeeding in putting one over on us—but for a good purpose: to allow us to accept the moral mysteries of life.

Q Which brings us to the famous letter to FDR and his relationship with Szillard. One can't help but wonder what was going through Einstein's mind when Leo Szillard presented him with the letter; how did he manage to reconcile in his mind the horror of atomic weaponry with his desire to promote peace and his humanitarianism?

A There are several aspects of that letter that one must bear in mind. First of all, Einstein had a real animus toward Nazi Germany. They had thrown him out after years of reviling him and his work—they had even put a price on his head. Second, the idea that Hitler would get the bomb first was simply too horrible a thought to consider and I'm sure Szillard drove that point home as only a firebrand

like him could. Although he was subsequently blackballed by the FBI from working on the bomb project, Einstein was pretty sure it was going on. Finally, recall that there was a second letter. This was written in 1945 to introduce Szillard, who wanted to argue Roosevelt out of dropping the bomb on Japan without a public demonstration of its terrible power first. It is said to have been on FDR's desk when the president died. But FDR had not read it, and, in fact, the America was well into the program already. I doubt whether the letter would have had much impact. Historians have argued that Japan was determined to fight to the bitter end. (Parenthetically, don't forget that Einstein's pacifism did not prevent him from consulting with the military during the war on explosives.)

Q Let's turn our attention to Einstein's later years. So often biographers have painted those years in, well, pathetic terms—a man alienated from his family, removed from the mainstream of contemporary science and obsessed with the pursuit of a chimera: the unification of all the forces and fields of physics. Is this a fair depiction?

A Nothing could be further from the truth. First of all, Einstein produced a land-mark work that lies at the core of modern discussion of quantum theory—the Einstein-Podolsky-Rosen (or EPR) Paradox paper, which he published in 1935, or when he was nearly 60. Not very many physicists produce earth-shaking material like that so late. Einstein wasn't dawdling during those years—he wasn't resting on his laurels and soaking in the money and accolades that were his for the asking. Fact is, Einstein wasn't interested in any of that—he wasn't even interested in getting a better table in a restaurant! He was interested in physics, pure and simple, and he saw it as a key to understanding life, the world... all right, perhaps also creation and God. This way of thinking, now one of the build-

ing blocks of the edifice of modern physics, has its origins in Einstein and gushes forth from the scientific work of his later years.

In addition, Einstein spent two hours reading the *New York Times* every day and spent hours talking about politics with his cronies, grousing about McCarthy.... He knew his name had enormous power and he used it judiciously to support human rights and civil liberties.

Q What do you think the future of Einstein's image will be? How will he be perceived decades or even generations from now?

A In my darkest days I fear that he will be worshipped. My favorite Einstein story is the one where he is playing the violin with a group of friends in a chamber group one afternoon and it's not going well. Finally the conductor turns to him and yells angrily, "Einstein! Can't you count!?" Now the reason I like this story is that I have heard it so many times from so many people who claim to have been there when it happened, that one can only conclude that Einstein must have been bawled out by conductors once a month and that he played his violin in stadium-sized living rooms. Clearly, the myth has taken on a life of its own and will overtake and obscure the reality—if one were in an ironic mood, one would say it will eclipse the reality—until only the mythic image remains.

Einstein was the fountainhead of several strains of thought that are still at the core of physics today. It may not be clear what he would have thought of string theory, for example, but some of the intellectual values that he championed—elegance of formulation; coherency of thought; simplicity of design; accessibility to human intellect—saturate all of modern science, even as million-dollar experiments are performed to test one or another of his theories. That indicates real staying power—the kind of influence and inspiration that are likely to be with us for a long, long time.

An Einstein Chronology

1879   Albert Einstein is born on March 14, at 11:30 a.m. in the German town of Ulm, the first child of the Jewish couple Hermann and Pauline (Koch) Einstein. James Clerk Maxwell, the Scottish physicist who developed modern electromagnetic theory and postulated the existence of ether, dies that same year.

1880   The Einstein family moves to Munich in June. Hermann and his brother found an electro-technical company, Einstein & Cie.

1881   Albert's sister Maria (called Maja) is born in Munich on November 18.

1884   While Albert is recuperating in bed, his father gives him a compass; Albert is fascinated by the realization that unseen forces move the needle. Einstein will later refer to this as the "first wonder" of his youth.

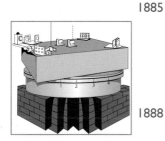

1885   Albert attends Petersschule, a Catholic elementary school in Munich. He is a good pupil (and the only Jewish student in his class). At home, he receives lessons in Jewish religion (as required by law) and in the violin. He is fascinated by religion, but only slowly comes to enjoy playing the violin. Niels Bohr, the Danish physicist with whom Einstein would debate quantum theory, is born.

1888   Einstein attends Luitpold Gymnasium (grammar school) in Munich from October 1. Here he is prepared for his bar mitzvah by the principal, Heinrich Friedmann. (One year earlier, the Michelson-Morley experiment is performed, for the first of many times, in Cleveland, Ohio.)

1889   Max Talmud (later changes name to Talmey), a Jewish medical student and family friend, begins tutoring young Albert and exposes him to popular science journals and books. He will be a frequent visitor to the Einstein home for the next 5 years. In April, Adolf Hitler is born in Austria.

1890   Albert devises his own proof of the Pythagorean Theorem and also becomes interested in Jewish ritual and religion.

1891   Albert's lifelong interest in mathematics begins; he is impressed with the "holy geometry book," which he calls the "second wonder" of his youth. He prepares earnestly for his bar mitzvah ceremony.

1892   Albert refuses to attend his bar mitzvah, declaring himself a "free thinker."

1894   The Einsteins move to Italy, settling in Milan, but Albert is left behind in Munich to complete his schooling. He is lonely and miserable and becomes a problem student; his teachers are relieved when he abruptly leaves Munich and joins the family in Milan.

1895   Albert does not pass the non-mathematical entrance examination for the "Poly"—the Federal Polytechnical School (later called Swiss Technical College, ETH) in Zurich at the beginning of October. He attends the trade department of the school in Aargau, living in the home of the headmaster, Jost Winteler. He writes his first scientific work—on the effect of magnetism on the ether—and sends it to his uncle, Caesar Koch.

1896   At the age of 17, Albert gives up his German citizenship with his father's permission and is stateless for the next 5 years. In October, he passes his Matura (examination required for university entrance) and enters the Polytechnic in Zurich that same month. His aim is to achieve the certificate of a subject teacher for mathematics and physics. His fellow students include Mileva Maric and Marcel Grossmann.

1897   Albert meets Michele Angelo Besso, beginning a lifelong friendship. (For more on Einstein's friends see "Albert Einstein in Leiden" by Dirk van Delft.)

1898   In October, Einstein successfully passes his intermediate test for his ETH diploma.

1899   In Zurich, Einstein files an application for Swiss citizenship.

1900   Albert Einstein finishes his studies at the ETH and receives a teacher's diploma for mathematics and physics. He applies for a job as an assistant at the Polytechnic and at other universities but is not successful in securing a position. In December, he submits his first scientific work to the *Annalen der Physik* (Annals of Physics).

1901   Einstein is granted Swiss citizenship in February. His first scientific work is published in the *Annalen der Physik* in March. He continues to fail at securing a position at a university. He finds work as a teacher in two private schools while he works on his dissertation. He hands in his dissertation in November. In December, his friend, Marcel Grossmann, helps him secure a job at the Swiss Patent Office in Bern. (For more on Einstein's friends see "Albert Einstein in Leiden" by Dirk van Delft)

1902   In January, Lieserl, the illegitimate daughter of Einstein and Mileva Maric, his former classmate, is born in Hungary while Einstein is in Bern. Einstein withdraws his dissertation. To support himself, he advertises in newspapers that he gives private lessons in physics and mathematics. On June 23, he assumes the post of third-class technical expert at the Patent Office in Bern. That October, Hermann Einstein dies in Milan.

1903   Einstein marries Mileva Maric on January 6 against the wishes of both their families. In spring, he founds the "Olympia Academy" in Bern, together with Maurice Solovine and Conrad Habicht. Lieserl is believed to be put up for adoption in Hungary that autumn. (For an in-depth account of Mileva's relationship with Einstein, see "The Forgotten Wife" by Andrea Gabor. For more on Einstein's philosophical influences, see

"Albert Einstein as a Philosopher of Science" by Don Howard)

1904   Einstein's first son, Hans Albert, is born in Bern on May 14. Einstein's job at the Patent Office is made permanent.

1905   Einstein's "annus mirabilis." Five groundbreaking papers by Einstein are published in *Annalen der Physik*—they revolutionize the foundations of physics around 1900. One of his works, *Zur Elektrodynamik bewegter Körper ("On the Electrodynamics of Moving Bodies")*, contains the special theory of relativity. In another work he deduces the famous formula $E = mc^2$. In April, Einstein submits his dissertation, *Eine neue Bestimmung der Moleküldimensionen ("A New Determination of Molecular Dimensions")* at the University of Zurich. It is accepted at the end of July (For an extensive discussion of the ideas in Einstein's 1905 papers see the essays in Part Two: Einstein's Early Science. For a science oriented discussion, see " 'What Songs the Syrens Sang': How Einstein discovered Special Relativity" by John Stachel; and for a down-to-earth explanation of the consequences of special relativity, see "Einstein's Bovine Dreams" by João Magueijo. For a chronological history of Einstein's discoveries including a brief history of his teenage years, see "Albert Einstein: A Laboratory in the Mind" by Owen Gingerich; and for an explanantion of the famous twin paradox, see "When Time Slows Down" by Donald Goldsmith.)

1906   In the middle of January Einstein is awarded a doctorate by the University of Zurich and in April he is promoted to the position of second-class technical expert at the Patent Office. (For a discussion of how a journal receiving five revolutionary papers in 1905 from an unknown patent clerk could decide on their validity, see "How Can we Be Sure that Albert Einstein Was Not a Crank?" by Jeremy Bernstein.)

1907   Einstein starts to think about the general theory of relativity and discovers the principle of equivalence of mass and energy for continuously accelerated systems. His application for a doctorate is rejected by the University of Bern; his dissertation is deemed insufficient. (For a discussion of general relativity's principle of equivalence, see "An Old Man's Toy" by A. Zee.)

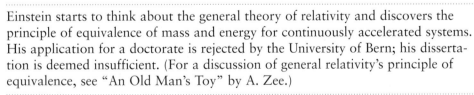

1908   Einstein submits a new dissertation to the University of Bern and is awarded a doctorate; he becomes a private lecturer at the university and delivers his first lectures.

1909   In July, Einstein is awarded his first (of many) honorary doctorates from the University of Geneva. He leaves the Patent Office and becomes an associate professor for theoretical physics at the University of Zurich in October.

1910   Einstein's second son Eduard ("Tete") is born in Zurich on July 28.

1911   Einstein becomes a full professor at the German University in Prague. At the end of October he takes part in the first Solvay Congress in Brussels. He calculates the deflection of the light in the gravitational field of the sun and recognizes the experimental possibility of observing this deflection during a total solar eclipse.

1912   He begins a love affair with his divorced cousin, Elsa Löwenthal, in Berlin. He accepts an invitation to become a full professor of theoretical physics at the ETH and returns to Zurich in August. He begins his collaboration with Marcel Grossmann, then a professor of mathematics at ETH, about the principles of general relativity.

1913   Einstein is nominated to the Prussian Academy of Sciences (requiring approval of Wilhelm II, German Emperor and King of Prussia). Max Planck and Walther Nernst visit Einstein in Zurich and recruit him for Berlin. He is offered a senior position in the newly formed Kaiser Wilhelm Institute for Physics. Einstein accepts the offer December 7.

1914   Einstein arrives in Berlin in April; Mileva and their sons Hans Albert and Eduard arrive one month later. Albert Einstein and Mileva soon separate, and in June she returns to Zurich with the two boys. Erwin Freundlich organizes an expedition to the Crimea to photograph starlight deflection by the sun during a solar eclipse, testing Einstein's early predictions in his first version of general relativity. World War I begins on August 1. The Freundlich expedition is arrested and placed in a POW camp as spies. Einstein becomes interested in politics for the first time. He becomes member of a pacifist organization and signs the pacifist "Manifesto to the Europeans." (For an account of the exchanges between Einstein and Freundlich, see "The Crimean Expedition" by Amir Aczel.)

1915   Einstein begins experiments on the Einstein-de Haas effect in gyromagnetics. In November, he completes work on the general theory of relativity and presents it in a 4-part speech at the Prussian Academy of Sciences.

1916   "The Formal Foundation of the General Theory of Relativity" is published in the *Annalen der Physik* on March 20. Einstein becomes head of the German Physical Society on May 5, succeeding Max Planck. In December, he writes *On the Special and General Theory of Relativity, A Popular Account*. (For closer look at general relativity and its consequences, see "Einstein's Dream" by Stephen Hawking and "A Farewell Look at Gravity" by John Archibald Wheeler.)

1917   Einstein's health deteriorates as he suffers from, among other things, liver disease and a stomach ulcer. His cousin Elsa cares for him, but it will be several years before he recovers. He completes a work on cosmology that includes the cosmologic constant, a term which implies a static universe. He will later refer to this as his "biggest blunder." He assumes control of the Kaiser Wilhelm Institute for Physics in October. (For a discussion of the cosmological constant, see "Einstein's Mistakes" by Steven Wienberg and "Beyond the Big Bang" by Marcia Bartusiak.)

1918   In August, Einstein rejects generous offers from the Swiss Polytechnic and the University of Leiden. Germany surrenders on November 9 and the Republic is proclaimed. Einstein is pleased with this turn of events.

335

1919   Albert Einstein and Mileva Maric are divorced on February 14. Einstein has his first discussion on Zionism with Kurt Blumenfeld. Arthur Stanley Eddington's expedition observing a solar eclipse on May 29 confirms Einstein's prediction about the deflection of light in the gravitational field of the sun. News reports circulate the globe, making Einstein famous overnight. Einstein marries his cousin Elsa Löwenthal on June 2; he becomes the stepfather to her two daughters from her first marriage, Ilse and Margot. (For more on Eddington's expedition, see "The Crimean Expedition" by Amir Aczel, and on Einstein's fame and influence, see "Einstein and the Shaping of Our Imagination" by Gerald Holton.)

1920   In February, Einstein's mother Pauline dies after a severe illness in Berlin. The Danish physician Niels Bohr visits Einstein in Berlin for their first meeting. Anti-semitic comments against Einstein and the theory of relativity increase in the European press, causing Einstein to consider leaving Germany. At a scientific conference in Bad Nauheim on September 23, a heated discussion with Nobel physicist Philipp Lenard, the most vocal German critic of relativity, occurs. Einstein assumes his position as visiting professor at the University of Leiden on October 27. (For an account of Einstein's years in Germany and his years in Leiden, see "Einstein's Germany" by Fritz Stern and "Albert Einstein in Leiden" by Dirk van Delft.)

1921   Einstein and Chaim Weizmann visit the United States from April 2 to May 30; it is Einstein's first visit to the U.S. The ostensible reason for the visit is to raise money for the newly formed Hebrew University in Jerusalem. Einstein delivers four lectures at Princeton University about the theory of relativity which are later published in book form, and he is awarded an honorary doctorate. He is welcomed by President Warren G. Harding in the White House in Washington.

1922   In January, Einstein completes his first work on unified field theory. He visits France that spring and works to normalize relations between France and Germany. Einstein travels and lectures extensively in Europe—until the June 24 assassination of German Foreign Secretary Walther Rathenau, after which he cancels all public speeches and presentations. (Einstein writes an emotional obituary for Rathenau.) Einstein and his wife Elsa visit Japan on October 8 for an extensive tour. On November 9, during his Japan tour, Einstein learns that he has been awarded the Nobel Prize for Physics for the year 1921. He receives the prize not for his theory of relativity but for his explanation of the photoelectric effect. Einstein is not present for the award ceremony on December 10; he is represented by the German ambassador to Sweden.

1923   On his return from Japan in February, Einstein visits Palestine and becomes the first honorary citizen of Tel Aviv. In June he supports Friends of New Russia and becomes a member of its board. In July, he travels to Sweden and delivers his Nobel Address in Gothenburg in ther presence of the king on July 11. (For more on Einstein's opinions and writings pertaining to Judaism and religion, see "Einstein and the Jews" by Abraham Pais.)

1924   With the Indian physicist Satyendra Nath Bose, Einstein discovers Bose-Einstein-condensation. In December, the Einstein Tower, designed by Erich Mendelsohn, is completed in Potsdam. Einstein is appointed Chairman for Life of the Board of Trustees of the Einstein Institute.

1925   Einstein and Bose create Bose-Einstein statistics forcertain types of particles and publish their landmark work on Bose-Einstein condensation at the end of the year. From April to June, Einstein travels through South America, visiting Argentina, Brazil and Uruguay. He joins Gandhi and other luminaries in signing a manifesto against compulsory military service.

1926   On February 12, Einstein is awarded the gold medal of the Royal Astronomical Society in London. Quantum mechanics is formulated by, among others, Werner Heisenberg, Max Born and Erwin Schrödinger. Einstein expresses his doubts about the theory. (To understand more about Einstein's objections to quantum mechanics, see "Einstein's Dilemma" by Shimon Malin.)

1927   At the fifth Solvay Congress in Brussels in October, intensive discussions take place between Einstein and Niels Bohr about the foundations of quantum mechanics. (For a discussion of Einstein's thoughts on quantum theory and its implications, see "Quantum Trickery" by Dennis Overbye.)

1928   Einstein suffers from a heart illness due to exhaustion. He is confined to bed for several months, and does not recover until the next year. Helen Dukas begins to work for Albert Einstein as a secretary-assistant on April 13.

1929   In March, Einstein celebrates his fiftieth birthday. He arranges for a summer home to be built in Caputh near Potsdam; he will summer there until his emigration in December 1932. During his stay at the Solvay Congress in Brussels, he visits the Royal Family. A friendship begins, resulting in a lifelong correspondence with his "Beloved Queen," Queen Elizabeth of Belgium. He is awarded the Max Planck medal by his supporter Max Planck on June 28. (For more on Einstein's friendship with the Belgian queen see "Albert Einstein in Leiden" by Dirk van Delft.)

1930   Einstein's first grandson, Bernhard Caesar, the son of Hans Albert and Frieda Einstein is born. In May, Einstein signs a petition demanding world disarmament and develops an intense interest in pacifism. Einstein visits the U.S. for the second time—this time the visit lasts a year and he visits many places, including New York and Cuba. The reason for his visit is to lecture and do research at the California Institute of Technology (Caltech) in Pasadena.

1931   Einstein returns from his visit to the U.S. in March. Observation causes him to reject the cosmological term (of 1917). In May the University of Oxford awards him an honorary doctorate (Dr. h. c.) in science. He spends the summer months in his house in Caputh. In December, he once again travels to the U.S., spending most of his time at Caltech in Pasadena. It is his third visit to the U.S. (For more on Einstein's visit with Edwin Hubble in California in January 1931, see "Beyond the Big Bang" by Marcia Bartusiak.)

1932   At the beginning of March, Einstein returns to Berlin. He accepts Flexner's invitation to join the newly-founded Institute for Advanced Study in Princeton, New Jersey. He plans to spend half the year in Berlin and half the year in Princeton. Inspired by the Commission for Intellectual Cooperation, he begins to correspond with Sigmund Freud on the subject, *Why war?* This conversation is published in 1933. In December he travels to Caltech in Pasadena, California, again. He plans to go back to Germany in March 1933, but the rise of Nazism in Germany causes him to resolve never to set foot in Germany again.

1933   On January 30, Adolf Hitler becomes chancellor in Germany. Einstein declares on March 10 that he will not return to Germany; on March 28, he resigns from the Prussian Academy of Sciences and severs all contact with German institutions with which he had worked. He stays in Belgium, Switzerland and England, then emigrates to the U.S., stopping in New York City, and then settling into his new home in Princeton, New Jersey. He wife Elsa, his secretary, Helen Dukas, and his assistant, Walther Mayer accompany him. On October 17, he begins his tenure at the Institute for Advanced Study. (For an account of events in Germany at this time, see "Einstein's Germany" by Fritz Stern.)

1934   Einstein's step-daughter Ilse dies in Paris; his other step-daughter Margot moves to Princeton. A collection of his popular essays is published under the title, *The World As I See It.*

1935   The Einstein-Podolsky-Rosen Paradox paper is published in May. Einstein receives the Franklin Medal in Philadelphia on May 15. He and his wife Elsa move into a modest new house in Princeton at 112 Mercer Street in September. His step-daughter Margot and his secretary, Helen Dukas live with them. (For more on the E.P.R. paradox, see "Quantum Trickery" by Dennis Overbye.)

1936   Einstein's friend Marcel Grossmann dies on September 7 and Elsa Einstein dies after a long illness in their house on December 20.

1938   Einstein and his old friend, Leopold Infeld, publish a book (largely written by Infeld), *The Evolution of Physics.*

1939   Einstein signs a letter to president Franklin D. Roosevelt on August 2, alerting him to the possibility of an atomic bomb. Germany attacks on Poland—thus beginning World War II—on September 1. Einstein's sister Maja moves in with Einstein in Princeton. (To read more about how Einstein came to write the letter to Roosevelt, see "That Famous Equation and You" by Brian Greene.)

1940   Einstein is sworn in as an American citizen on October 1, 1940. He maintains, however, his Swiss citizenship.

1941   The "Manhattan Project" begins in November, aimed at developing an atomic bomb. Einstein is regarded as a security risk and is not allowed to take part in the project. Japan attacks Pearl Harbor on December 7 and the U.S. enters the war.

1943   The U.S. Navy appoints Einstein "adviser for highly explosive materials."

1944   Einstein is officially given emeritus status at the Institute for Advanced Study, but he maintains an office at the Institute until his death.

1945   Einstein is shocked by the news of the two atomic bombs dropped on Hiroshima (on August 6) and Nagasaki (on August 9). The second World War ends soon afterwards. At a dinner on December 10 in New York he delivers a widely publicized speech: "The war is won, but the peace is not."

1946   In an open letter to the United Nations, Einstein promotes the formation of a world government. He becomes head of the Emergency Committee of Atomic Scientists, organized to promote the control of nuclear weapons and the peaceful use of nuclear energy. Einstein's sister Maja suffers a stroke.

1947   Einstein intensifies his activities for nuclear arms control and on behalf of world government.

1948   Einstein's first wife, Mileva Maric, dies in Zurich on August 4. In December, Einstein is diagnosed as having an aortic aneurysm and receives surgery immediately.

1949   Einstein leaves the hospital in January. His *Autobiographical Notes*, written in 1946 for the Library of Living Philosophers, is published. It contains little about his private life, focusing on his scientific career. On Einstein's 70th birthday, mathematician Kurt Gödel presents him with a solution to the general relativity equations that hints at a possibility for time travel. (For more on Einstein and Gödel, see "Time Bandits: What were Einstein and Gödel Talking About?" by Jim Holt)

1950   Einstein finalizes and signs his will on March 18. Dr. Otto Nathan and Einstein's secretary Helen Dukas are made his administrators. His papers and scientific notebooks are bequeathed to the Hebrew University in Jerusalem. He publishes *Out of My Later Years*, a collection of popular essays and speeches of the last 20 years.

1951   Maja, Einstein's sister, dies in Princeton on June 25.

1952   After the death of Chaim Weizmann, Einstein is offered the presidency of Israel; he declines.

1954   Einstein publicly supports J. Robert Oppenheimer against accusations of the U.S. government concerning his loyalty. Einstein is diagnosed with haemolytic anaemia.

        Einstein's friend Michele Angelo Besso dies in Geneva on March 15. In April, Einstein signs a letter to Bertrand Russell in which he declares himself ready to join Russell in imploring all nations to abandon nuclear weapons—the document becomes known as the Einstein-Russell Manifesto. This manifesto gives birth to the international Pugwash movement.

1955   On April 15, Einstein is brought to the hospital in Princeton; it is determined that his aneurysm had burst. Albert Einstein dies in the hospital on April 18 at about 1:15 a.m., at age 76. His body is cremated the same day and the ashes are scattered at an unknown location (reportedly, a nearby river) after a private and modest funeral service.

—INDEX—

— CONTRIBUTORS —

Amir Aczel is an associate professor of Mathematical Sciences at Bentley College in Massachusetts. He is the author of many popular science and math books including *Entanglement*.

Marcia Bartusiak is a visiting professor with the Graduate Program in Science Writing at the Massachusetts Institute of Technology. She has written a number of popular science books including *Einstein's Unfinished Symphony*.

Jeremy Bernstein is a professor of physics at the Stevens Institute of Technology and was a staff writer for the *New Yorker* until 1993. He has written many books on popular science and mountain travel, including *Secrets of the Old One: Einstein, 1905*

Dirk van Delt is an associate professor of history of science at the Leiden University Observatory in the Netherlands. He is also a senior science editor of the NRC *Handelsblad*, a daily newspaper published in Rotterdam

Andrea Gabor is an associate professor in the Master's Program in Business Journalism at the Weissman School of Arts & Sciences, Baruch College, New York. She is the author of three books, including *Einstein's Wife: Work and Marriage in the Lives of Five Great Twentieth Century Women*

Owen Gingerich is Professor Emeritus of Astronomy and History of Science at the Harvard-Smithsonian Center for Astrophysics and is the author of *The Book Nobody Read: Chasing the Revolutions of Nicolaus Copernicus*

Donald Goldsmith received his Ph.D. in astronomy from the University of California at Berkeley in 1969. The author or co-author of more than a dozen books on astronomy and astrophysics, he has received lifetime achievement awards for astronomy popularization from the American Astronomical Society and the Astronomical Society of the Pacific

Brian Greene is a professor of physics and mathematics at Columbia University, New York. He is widely regarded for a number of groundbreaking discoveries in superstring theory and wrote the best selling book *The Elegant Universe,* which explains string theory for a general audience.

Stephen Hawking is the Lucasian Professor of mathematics at the University of Cambridge and a fellow of Gonville and Caius College, Cambridge. He is best known for his research into black holes. He is the author of several books including the best-selling *A Brief History of Time.*

Jim Holt writes for *The New Yorker*, the *New York Times*, and other publications. He was a former visiting journalist at U.C. Berkeley and an American correspondent for BBC Wales.

Gerald Holton is Mallinckrodt Research Professor of Physics and Research Professor of History of Science at Harvard University. For several years he supervised the conversion of the collection of Einstein's largely unpublished correspondence and manuscripts into an archive suitable for scholarly study. He is the author and editor of many books on science including *Einstein, History and Other Passions*.

Don Howard is a professor of philosophy and director of the graduate program in history and philosophy of science at the University of Notre Dame in South Bend, Indiana.

Joao Magueijo is a cosmologist and lecturer in theoretical physics at Imperial College London. He is the author of *FasterThan the Speed of Light*.

Shimon Malin is a professor of physics at Colgate University. He is a leading authority on quantum mechanics, general relativity and cosmology and philosophy. He is author or co-author of three books including *Nature Loves To Hide*.

Dennis Overbye is Deputy Science Editor of *The New York Times* and author of the recently published *Einstein in Love* and the best-selling *Lonely Hearts of the Cosmos*

Abraham Pais is Detlev W. Bronk Emeritus Professor at Rockefeller University, New York. A theoretical physicist and founding father of particle physics he now devotes himself to the history of science.

John Stachel is Professor of Physics Emeritus and Director of the Center of Einstein Studies at Boston University. He has edited or co-edited a number of books dealing with Einstein and relativity.

Fritz Stern is Professor Emeritus at Columbia University and is an American historian of German history and Jewish history. His recent books include *Einstein's German World*.

Steven Weinberg was co-recipient of the 1979 Nobel Prize in Physics. He holds the Josey Chair in Science at the University of Texas at Austin, where he is a member of the physics and astronomy departments and heads the physics department's Theory Group.

John Archibald Wheeler is a professor at the University of Texas at Austin. He is one of the leading figures in the development of general relativity and quantum gravity and coined the term "black hole." He is the author of a number of books on gravity and black holes, including *Journey into Gravity and Spacetime*.

A. Zee is a permanent member of the Kavli Institute for Theoretical Physics at the University of California, Santa Barbara. Besides as *Old Man's Toy* he is the author of a number of other books including *Fearful Symmetry*, *Swallowing Clouds*, *Quantum Field Theory in a Nutshell* and a forthcoming book on *Dark Energy*.

Photo Credits

p. 8   Courtesy of the Archives, California Institute of Technology

p. 11   The Pittsburgh Post-Gazette

p. 13   SCIENCE PHOTO LIBRARY / Photo Researchers, Inc.

p. 14   Collections of the Historical Society of Princeton

p. 26   Courtesy of AIP Emilio Segrè Visual Archives. Reprinted with permission from Physics Today, April 2006, © 2006, American Institute of Physics

p. 28   Reprinted with permission from Physics Today, April 2006, © 2006, American Institute of Physics

p. 31   Reprinted with permission from Physics Today, April 2006, © 2006, American Institute of Physics

p. 34   (top left) Besso Family, courtesy AIP Emilio Segrè Visual Archives, (top right) Image Archive ETH-Bibliothek Zurich, (bottom right) AIP Emilio Segrè Visual Archives

p. 38   The Albert Einstein Archives, The Hebrew University of Jerusalem, Israel

p. 42   The Albert Einstein Archives, The Hebrew University of Jerusalem, Israel

p. 52   Swiss National Library/SNL, Berne

p. 60   AIP Emilio Segrè Visual Archives

p. 66   Image Archive ETH-Bibliothek Zurich

p. 70   Courtesy of Dr. George M.H. van de Velde

p. 87   The Albert Einstein Archives, The Hebrew University of Jerusalem, Israel

p. 88   The Illustrated London News

p. 96   (top) U.S. National Archives & Records Administration, (right) bpk-Bildarchiv Preußischer Kulturbesitz, Berlin, (bottom) Photograph by Charles Holdt, Courtesy of the Leo Baeck Institute, New York

p. 112   (left) AIP Emilio Segrè Visual Archives, (right) Cartoon by Josef Plank, Courtesy Library of Congress, Washington, D.C., (bottom) bpk-Bildarchiv Preußischer Kulturbesitz, Berlin

p. 119   Courtesy of the Archives, California Institute of Technology

p. 120   Wiley-VCH Verlag GmbH & Co.

p. 127   Courtesy of the Archives, California Institute of Technology

p. 128   (top right) AIP Emilio Segrè Visual Archives

p. 136   Cartoon by Low/courtesy Caltech Archives

p. 180   Drawn by U. Schleicher-Benz, Lindauer Bilderbogen,reprinted in Alan Reinberg and Michael H. Smolensky, et al., *Biological Rhythms and Medicine: Cellular, Metabolic, Physiopathologic, and Pharmacologic Aspects*, New York, NY, Springer-Verlag, 1983. With kind permission of Springer Science and Business Media

p. 198   NASA

p. 206   NASA

p. 211   Courtesy of the Archives, California Institute of Technology

p. 212   NASA

p. 216   Courtesy of Bob Kahn, Gravity Probe B, Stanford University

p. 231   AIP Emilio Segrè Visual Archives

p. 235   (bottom) Reconstruction of the old man's toy by Tim Westbury, Zero G art + design (2005)

p. 236   *Quantum Entanglement, Spooky Action at a Distance, Teleportation and You* © 2003 Jim Ottaviani and Roger Langridge, courtesy of G.T. Labs

p. 240   AIP Emilio Segrè Visual Archives

p. 246   AIP Emilio Segrè Visual Archives

p. 261   (left) Photograph by Hanson Robotics Inc., Courtesy of Wired, Condé Nast Publications Inc., (right) Timecapsule Toys Inc. and Toypresidents Inc.

p. 262   Courtesy of the Archives, California Institute of Technology

p. 264   (top) U.S. National Archives & Records Administration, (bottom) Photo by Harry Burnett/courtesy Caltech Archives

p. 266   © The New Yorker Collection 1929 Rea Irvin from cartoonbank.com. All Rights Reserved.

p. 270   (bottom) Lion Brand Yarn Company (www.lionbrand.com)

p. 273   Photographer Arthur Sasse, © Bettmann/CORBIS

p. 275   From *Herblock's Here and Now*, (Simon & Schuster, 1955)/Courtesy The Herb Block Foundation

p. 276   Courtesy of the Archives, California Institute of Technology

p. 291   (top) Time & Life Pictures/March Of Time/Getty Images, (bottom) Letter from Albert Einstein to President Franklin D. Roosevelt, August 2, 1939, courtesy of the Franklin D. Roosevelt Library Digital Archives

p. 294   AIP Emilio Segrè Visual Archives

p. 300   bpk-Bildarchiv Preußischer Kulturbesitz, Berlin

p. 308   Courtesy of the Archives, California Institute of Technology

p. 310   AIP Emilio Segrè Visual Archives

p. 318   NASA

p. 326   Alan Richards, photographer. Courtesy of the Archives of the Institute for Advanced Study, Princeton, New Jersey, U.S.A.

p. 332   bpk-Bildarchiv Preußischer Kulturbesitz, Berlin